Optical Methods in
Engineering Metrology

ENGINEERING ASPECTS OF LASERS SERIES

Series Editor
Dr T. A. Hall
Reader in Physics
University of Essex

SERIES EDITOR'S PREFACE

In the late 1960s and early 1970s the laser was still something of a scientific curiosity with only a limited practical use. The extent of the four volumes in this series shows the enormous change that has happened since that time. The laser is now an indispensable addition to the toolbox of the engineer and scientist. The progress from the time when the laser was often dubbed 'a solution in search of a problem' to today, when engineers of all disciplines frequently use lasers as a matter of course, is a remarkable transformation. Even so, the use of lasers in engineering and other walks of life is still in its infancy and has been held back partly by their relatively high cost and in some cases by their inconvenience in use. As these problems are overcome lasers will find wider and wider application and there is an ever increasing need for engineers and scientists, who perhaps have little interest in lasers themselves, to have access to an authoritative source which not only acts as an introduction but also takes the reader up to the latest developments in laser applications.

The four books of the series 'Engineering Aspects of Lasers' arose from a series of laser workshop courses which have been held annually at the University of Essex since 1979. These courses have evolved very considerably since their inception but aspects of their organization have not changed – the contents of the courses have always been coordinated by the recognised international authority in each subject area and the lectures given by experts in the particular field from industry, government laboratories or universities. When the idea of publishing a series of books based upon the contents of these courses was first suggested the course coordinators at that time became the editors of each volume and the lecturers were asked to contribute.

The workshop courses are self-supporting courses which also form part of the MSc degree in Lasers and their Applications. There are many people who have contributed much to these courses over the years and have made them the success that they have been. I would like to express my gratitude to them all. The courses and the

MSc were the brainchild of T. P. Hughes who was then Reader in Physics at Essex University. Without his foresight, hard work and determination in setting up the courses, this series of books would not have been written.

Other titles in series

Laser Processing in Manufacturing
Edited by R. C. Crafer and P. J. Oakley

Nonlinear Optics in Signal Processing
Edited by R. W. Eason and A. Miller

Advances in Optical Communications
Edited by N. Doran and I. Garrett

Contents

List of contributors		xv
Preface		xvii
1	**Introduction**	**1**
	J. M. Burch	
	1.1 The optical metrologist	1
	1.2 Optical measurement	3
	1.3 A wider view	7
	References	9
2	**Laser beam geometry and its applications**	**11**
	D. C. Williams	
	2.1 Analysis of laser beam size	11
	2.1.1 Characteristics of a laser beam	11
	2.1.2 Analysis using skew rays	16
	2.1.3 Some applications of the theory	21
	2.2 Increasing the precision	25
	2.2.1 Three-point alignment system	25
	2.2.2 Fresnel zone plates	27
	2.2.3 Variable-focus zone plate	31
	2.2.4 Variable-focus lens systems	36
	2.2.5 Afocal lens systems	38
	2.2.6 Bessel beam	40
	2.3 The effect of the atmosphere	43
	2.3.1 Bending of a light beam	44
	2.3.2 Effect of temperature and pressure gradients	45
	2.3.3 Two-colour methods	47
	2.4 Detection of beam position	49
	2.4.1 Photoelectric detection methods	49
	2.4.2 Phase-sensitive systems	50
	2.4.3 Bore straightness measurements	52
	2.4.4 Quadrant detectors	57

	2.4.5	Axial position detection	58
	2.4.6	Photopotentiometers	60
	2.4.7	Intensity-modulation methods	61
	2.4.8	Measurement of large movements	63
2.5	Problem solving with optical devices		65
	2.5.1	Scanning and triangulation techniques	65
	2.5.2	Two-mirror and related devices	67
	2.5.3	Surface profilometer	74
	2.5.4	Profilometry in two dimensions	76
	2.5.5	Profilometry refinements	78
	2.5.6	Roller parallelism measurements	79
	2.5.7	Intensity-based measurements	82
	References		85

3 Alignment metrology 87
B. S. Pearn

3.1	Instrumentation for alignment		88
	3.1.1	Micro-alignment telescope	88
	3.1.2	Lasers	93
3.2	Telescope applications		94
	3.2.1	Crankshaft bearings	95
	3.2.2	Transport and process rollers	98
	3.2.3	Coaxial spindles	99
3.3	Flatness or level		99
	3.3.1	Flatness on-board ship	100
	3.3.2	Alignment on an offshore platform	101
3.4	Autocollimation		103
	3.4.1	Straightness by autocollimation	103
	3.4.2	Electronic autocollimators	105
	3.4.3	Flatness of surface tables	106
	3.4.4	Angle comparison	107
3.5	Practical problems		108
	3.5.1	Environmental problems	108
	3.5.2	Obstructed sight line	109
	3.5.3	Parallax	110
	3.5.4	Conclusion	112
	Reference		112

4 Photogrammetry in industrial measurement 113
R. A. Hunt

4.1	The technique of photogrammetry		113
4.2	How to make a measurement		116
	4.2.1	Choice of method	117

	4.2.2	Mathematical modelling	117
	4.2.3	Experimental design	118
	4.2.4	Prediction of uncertainties	118
	4.2.5	Assessment of the method	118
	4.2.6	The experiment	118
	4.2.7	Estimation of experimental uncertainties	119
4.3		Mathematical formulation	119
	4.3.1	The bundle solution	123
	4.3.2	Least squares revisited	123
	4.3.3	The Newton-Raphson method	124
	4.3.4	Treatment of distortions	125
	4.3.5	Error ellipsoids: fact or fiction?	126
	4.3.6	Singularities in the solution	127
	4.3.7	Design of experiments	128
	4.3.8	Calculation of initial values	129
4.4		Imaging devices	129
	4.4.1	Metric and semi-metric cameras	130
	4.4.2	The NPL monocentric axicon camera	132
	4.4.3	The role of the measuring machine	134
	4.4.4	Finding the image centres	134
4.5		Towards real time	135
	4.5.1	The coming of computer vision	136
	4.5.2	Video theodolites	138
	4.5.3	Location of images	139
	4.5.4	True video photogrammetry	141
4.6		Applications for photogrammetry	144
	4.6.1	Survey of an infrared telescope	145
	4.6.2	The destruction of a bridge	148
	4.6.3	Measurement of a fuel tank	149
		References	152

5 Laser interferometry for precision engineering metrology   **153**
P. Gill

5.1		General principles	153
	5.1.1	Optical interference by amplitude division	153
	5.1.2	Limitations to fringe formation	154
	5.1.3	Refractive index compensation	156
	5.1.4	Single-wavelength fringe-counting interferometry	157
5.2		Stabilized laser wavelength sources	158
	5.2.1	International definition of the metre	158
	5.2.2	He-Ne stabilization techniques	159
	5.2.3	Laser wavelength calibration and traceability	162
5.3		Heterodyne interferometry	163

5.4	Machine tool characterization		164
	5.4.1	Angle measurement	164
	5.4.2	Straightness measurement	165
	5.4.3	Plane mirror interferometers	165
5.5	Laser interferometer measurement errors		166
	5.5.1	System specification	166
	5.5.2	Laser wavelength errors	166
	5.5.3	Compensation errors	166
	5.5.4	Interferometer verification	167
	5.5.5	Operator errors	167
5.6	Multiple wavelength interferometry		168
	5.6.1	Method of excess fractions	168
	5.6.2	The NPL gauge-block interferometer	169
5.7	Multiple- and swept-wavelength interferometry using laser diodes		171
	5.7.1	Laser diode properties and stabilization	171
	5.7.2	EDM equivalence	174
	5.7.3	Multiple-wavelength diode theory and application	175
	5.7.4	Swept-wavelength diode theory and application	176
	5.7.5	Future trends	176
	References		177

6 Laser vibrometry — 179
N. A. Halliwell

6.1	Laser Doppler velocimetry		179
	6.1.1	Principles of operation	180
	6.1.2	Frequency-shifting devices	183
	6.1.3	Doppler signal processing	186
6.2	Solid surface vibration measurement		187
	6.2.1	Laser speckle effects and theory of operation	188
	6.2.2	Measurements on rotating targets	190
	6.2.3	Choice of vibrometer: practical considerations	191
6.3	Torsional vibration measurement		194
	6.3.1	Cross-beam torsional vibrometer	195
	6.3.2	The laser torsional vibrometer	197
	6.3.3	Practical considerations	201
6.4	Fibre optic vibration sensors		203
	6.4.1	Intrinsic vibration sensors	204
	6.4.2	Extrinsic vibration sensors	206
	6.4.3	A practical all-fibre laser vibrometer	207
	6.4.4	The fibre optic vibrometer: practical considerations	208
	References		210
		Laser Doppler velocimetry texts	210
		Other references	210

x *Contents*

7	**Industrial application of holographic interferometry**	**213**
	R. J. Parker	
7.1	Holography	213
	7.1.1 Recording the whole picture	214
	7.1.2 Two-beam interference and holography	214
	7.1.3 Recording transmission holograms	216
	7.1.4 Viewing in white light	217
7.2	Equipment for holography	218
	7.2.1 Lasers	218
	7.2.2 Laser safety	219
	7.2.3 Recording materials	220
	7.2.4 Other holographic equipment	226
7.3	Measurement with holograms	228
	7.3.1 Holographic interferometry	228
	7.3.2 Fringe pattern interpretation and analysis	231
	7.3.3 Resolving components of displacement	232
	7.3.4 Automatic analysis	233
7.4	Applications of holographic interferometry	235
	7.4.1 Non-destructive testing	236
7.5	Vibration analysis by holography	242
	7.5.1 Time-averaged holograms	242
	7.5.2 Large-object or large-amplitude analysis	247
	7.5.3 Holography of rotating objects	250
7.6	Flow visualization	253
	7.6.1 Fringe interpretation	254
	7.6.2 Two-dimensional flows	255
	7.6.3 Three-dimensional flows	259
	7.6.4 Real-time flow visualization	259
	7.6.5 Rotating flows	261
7.7	Holographic contouring	262
	7.7.1 Two-angled illumination	264
	7.7.2 Two-refractive-index techniques	266
	7.7.3 Two-wavelength techniques	268
	7.7.4 Holographic contouring - new potential	270
7.8	Conclusion	270
	References and bibliography	270
8	**Television holography and its applications**	**275**
	J. C. Davies and C. H. Buckberry	
8.1	Limitations of conventional holography	275
	8.1.1 The need for holography	276
	8.1.2 The benefits and limitations of holographic analysis	277
8.2	Principles of television holography	278

		8.2.1	Resolution requirements	278
		8.2.2	Forming TV holograms	279
		8.2.3	The effect of the aperture stop	281
		8.2.4	Operating modes	284
		8.2.5	Fringe formation	284
	8.3	Systems and their operation		286
		8.3.1	Optical arrangement	286
		8.3.2	Operating parameters	289
		8.3.3	Fibre optic systems	291
		8.3.4	Setting up a system	292
		8.3.5	Alternative operating geometries	293
	8.4	Modulation techniques		295
		8.4.1	Limitations of a basic system	296
		8.4.2	Intensity modulation	296
		8.4.3	Phase modulation	298
		8.4.4	Combined modulation	300
	8.5	Automatic fringe analysis		301
		8.5.1	Digital processing	301
		8.5.2	The engineering need	302
		8.5.3	Phase-measurement algorithms	302
		8.5.4	Application of phase stepping to TV holography	306
		8.5.5	Practical considerations	308
		8.5.6	Practical examples	311
		8.5.7	Conclusions	316
	8.6	Examples of applications		316
		8.6.1	Automobile applications	316
		8.6.2	*In-situ* measurement of stone degradation	325
		8.6.3	Structural integrity testing	328
		8.6.4	Pulsed TV holography	330
		8.6.5	Surface contouring	332
		8.6.6	Future developments	335
		Acknowledgements		336
		References		337
9	**Moiré methods in strain measurement**			**339**
	C. Forno			
	9.1	Conventional moiré measurements		339
		9.1.1	The moiré phenomenon	339
		9.1.2	Out-of-plane moiré techniques	340
		9.1.3	Imaged moiré	342
		9.1.4	In-plane measurement analysis	344
		9.1.5	In-plane measurement techniques	347
	9.2	High-resolution moiré photography		348

xii *Contents*

	9.2.1	Tuning the lens response	349
	9.2.2	Photographic recording	352
9.3	Application of modified camera		354
	9.3.1	Spatial filtering of moiré fringe patterns	355
	9.3.2	Fringe-pattern analysis	356
	9.3.3	Examples of engineering structures	358
	9.3.4	Deformation measurement in a timber structure	358
9.4	Types of surface pattern		362
	9.4.1	Printed pattern	362
	9.4.2	Stencilled patterns	366
	9.4.3	Random patterns	369
9.5	Additional features of the technique		372
	9.5.1	Optical differentiation	372
	9.5.2	Three-dimensional movement	374
	9.5.3	Pattern frequency multiplication	374
	9.5.4	Optical regeneration of defective gratings	374
	9.5.5	Displacement sensitivity direction	375
9.6	Moiré interferometry		375
	9.6.1	Two-beam interference	375
	9.6.2	Surface grating	379
	9.6.3	Moiré fringe analysis	381
	Acknowledgement		382
	References		383

10 Automatic analysis of interference fringes **385**
G. T. Reid
10.1	Intensity-based methods		386
	10.1.1	Dedicated systems	386
	10.1.2	Fringe tracking	390
10.2	Phase-based methods		395
	10.2.1	Electronic heterodyning	395
	10.2.2	Phase stepping	397
	10.2.3	Spatial phase measurements	405
	References		410

11 Monomode fibre optic sensors **415**
J. D. C. Jones
11.1	Measurement using optical fibres		415
11.2	Transduction mechanisms		417
	11.2.1	Sensor transfer function	418
	11.2.2	Phase modulated sensors	418
	11.2.3	Polarization-modulated sensors	419

11.3 Optical processing 420
    11.3.1 Two-beam interferometers 421
    11.3.2 Multiple-beam interferometers 425
    11.3.3 Polarimetric techniques 427
    11.3.4 Spectral techniques 429
11.4 Modulators and components 430
    11.4.1 Phase modulation 431
    11.4.2 Polarization modulation 432
    11.4.3 Frequency modulation 434
    11.4.4 Directional couplers 436
11.5 Electronic processing 439
    11.5.1 Active homodyne processing 440
    11.5.2 Passive homodyne processing 442
    11.5.3 Heterodyne processing 445
    11.5.4 Noise considerations 445
11.6 Applications 450
    11.6.1 Intrinsic sensors 450
    11.6.2 Extrinsic sensors 457
    References 459

Index   **465**

Contributors

C. H. Buckberry
Rover Group, Gaydon, Warwickshire, UK

J. M. Burch
Cranfield Institute of Technology, UK

J. C. Davies
Rover Group, Gaydon, Warwickshire, UK

C. Forno
National Physical Laboratory, Teddington, UK

P. Gill
National Physical Laboratory, Teddington, UK

N. A. Halliwell
Loughborough University of Technology, UK

R. A. Hunt
National Physical Laboratory, Teddington, UK

J. D. C. Jones
Heriot-Watt University, Edinburgh, UK

R. J. Parker
Rolls-Royce plc, Derby, UK

B. S. Pearn
Formerly with Rank Taylor Hobson, Leicester, UK

G. T. Reid
Formerly with National Engineering Laboratory, East Kilbride, UK

D. C. Williams
National Physical Laboratory, Teddington, UK

Preface

'Now I'm an engineer, and I come along and say "now look...".' This reaction came from an engineering graduate with many years of experience in tackling real-world industrial problems. His comment was prompted during the development of one of the optical devices described in this book, when he saw it being carefully aligned by a physicist who had surrounded it with a bewildering array of ancillary optical components.

The incident highlights a common dilemma. The physicist believes that his new creation has the potential to revolutionize an inspection process and when met with a cool reception he complains bitterly about the reluctance of industry to adapt to new technology. The engineer already has an armoury of highly developed practical processes which have stood the test of time and which are well suited to the requirements of production. He does not wish to invest in alternatives unless the benefits are clear.

For the calibration and performance verification of three-axis measuring machines, considerable effort was expended on an optical device which defined points in three-dimensional space by reflection from a flat plate placed on the table of the machine. The machine owners showed little interest, because the defined points were not tangible with well-defined locations; such calibrations are normally carried out using traditional mechanical artefacts such as ball plates and step gauges. However, optics has been able to make a genuinely useful contribution in this field by replacing the mechanical touch probe with an optical probe. Removing the need for physical contact between the probe and the artefact allows the machine to make settings without halting its motion, which gives considerable increases in both speed and reliability when measuring surface profiles.

Thus there are areas where optics is quite inappropriate, and others where it has much to offer. The most useful devices are often those which are ingenious and yet basically simple. It is my hope that this book may help to bridge the gap between optics and engineering by indicating how to approach problems which may be amenable to optical solution in a genuinely constructive way. The authors all have considerable experience in successfully applying optical methods to metrological problems of various kinds. As well as presenting a range of optical techniques and principles, ways to approach the solution of new problems are indicated. It is assumed that the reader already has a basic knowledge of optics; it is better for the engineer without this background to refer to one of the many textbooks that are available rather than being given an inferior and inadequate summary.

The programme of the Workshop Course on which the book is based was organized by the National Physical Laboratory and several of the authors are members of its staff. The content is therefore inevitably biased towards those areas of activity with which the NPL has been particularly associated, but the coverage is nevertheless quite broad. An attempt has been made at least to mention topics that are not treated in detail. The NPL has always been a centre for direction of enquiries about measurement problems which cannot be solved by established methods. In the parlance of today, such requests would be described as 'challenges', but for those of us who are fortunate enough to have memories of a less self-orientated society, they tend to be regarded as just pleasurable. Successful solutions depend not on management documents, objectives and targets, but on an enthusiasm for problem-solving coupled with a sense of dedication to meeting the needs of the customer.

A measure of cheating has taken place, in that while the majority of the text is concerned with laser methods, there is some discussion of optical techniques which use traditional light sources. However, this may serve to emphasize that the optimum technique is not always the one which uses the most up-to-date technology. In many areas, the fundamental principles have changed little in recent years; the latest developments have been in the application of new technology. The chapters differ markedly in their approaches to their subjects, but this serves to illustrate that individualism is the basis of creativity. I am grateful to all of the authors for their contributions; many of them were preoccupied with new responsibilities at the time of writing. Thanks are also due to various individuals and organizations who have provided information and illustrations.

It would be wrong to conclude this preface without special mention of Professor Jim Burch. He was the original organizer of the Workshop Course, and it is a measure of his knowledge of the subject that in the earlier years he delivered almost all of the lectures himself. Most organizations depend for their success on a few outstanding individuals, and Jim was one such at the NPL. His comments on the efforts of the rest of us were always stimulating and frequently astonishing. Many of the topics covered throughout the book owe their origin or success to his contributions. The breadth and depth of his knowledge of optics and his enthusiasm for it were complemented by warmth, humour and an unfailing courtesy shown to everyone with whom he came into contact. I am delighted that he has been able to contribute an introductory chapter, for it is he and not I who should be the editor. He will find many errors and omissions, but I hope that he will not be too displeased.

David C. Williams

1
Introduction

J. M. BURCH

One of the advantages of using optical methods for measurement is the parallel manner in which several separate channels provide information. But when one is asked to introduce or contribute to a multi-author volume, this parallel operation begins to look more like a hazard. It is reassuring, perhaps, that metrology is concerned, not so much with opinions or theories, as with verifiable methods of coming as close as possible to an objective truth. So even if readers find themselves offered more than one path to follow, they should have confidence that each of these paths can lead to a valid result. Indeed, this kind of duplication was considered an instructive feature of the earlier Laser Workshop Courses on which this book is based.

Experience of those courses suggests to me that only a few of our readers will be practising metrologists. More of them may be mature engineers for whom a particular area of measurement has acquired overriding importance, while others may be at the outset of a career, still wondering where to direct their efforts. With both these groups in mind, I propose to begin with some discussion of the role of a metrologist. We shall then consider, from the engineer's point of view, what optics has to offer, and some of the choices that may need to be made. Then we shall take a wider look at other methods and other fields of application, and conclude with a look towards the future.

1.1 THE OPTICAL METROLOGIST

What personal traits are needed for a career as a metrologist? Judging from a splendid diversity of colleagues observed at NPL and elsewhere, I would say that, while energy and enthusiasm may not come amiss, optimism is suspect, and it may be better to replace it with pessimism, or at least with a prudent caution. Some people are attracted into the career almost by accident. (I myself began as an amateur telescope maker, a hobby which rapidly becomes an obsession but seems to be a good training since it combines principles with practice and discourages one from dropping things!)

It may be helpful if metrologists have some training in physics and mathematics, but mainly they rely upon common sense. Before they undertake measurements, they

Optical Methods in Engineering Metrology. Edited by D.C. Williams. Published in 1993 by Chapman & Hall, London. ISBN 0 412 39640 8

ask themselves, 'What can go wrong? Is the equipment safe and free from vibration and from thermal or atmospheric effects? Are all the auxiliary controls and fasteners which are not going to be used set as they should be? Is the illumination and viewing system properly focused, and are we looking at the right image? Are the electric circuits and software all in order? etc.' The metrologist who ignores such basic precautions can be compared with the rugby full-back who unnecessarily allows an elliptical ball to bounce, or the bridge declarer who does not cater for a 4–0 trump break, thus forfeiting the respect of partner or team-mates.

Equally insidious, sometimes, is the fact that measurements can suffer from systematic as well as random sources of uncertainty. The latter can perhaps be reduced tenfold if one has the patience to repeat the same measurement 100 times. The effort may be far better spent, however, in trying to foresee any sources of systematic error, and then following a schedule in which, for example, one of the components is reversed or displaced or rotated between measurements. Taking the difference as well as the mean of such measurements (section 2.2.1) makes it possible not only to eliminate an error but also to determine its size and obtain better understanding of the equipment. Some of these procedures are as old as the pyramids, but their diagnostic power is now enhanced enormously by the spectacular ease with which a mass of redundant data can be analysed by computer.

Even if a given career seems to provide an interesting challenge, one needs to be reassured that training is available and that there is a market where the jobs to be done are adequately paid. Readers should be warned that in the case of metrology the situation is improving but still leaves much to be desired!

For the older generation of metrologists, much of their working wisdom has been acquired either by feats of reinvention or via informal contact with colleagues, a process which works best in large establishments. More recently, efforts have been made to give beginners a flying start, mainly by means of short courses, and this book derives partly from one such effort. Some other books have been produced; among these are Frankowski *et al.* (1991), Gasvik (1987), Luxmoore (1983) and Soares (1987). But degree courses in metrology are still not widely available. At the time of writing, Brunel University offer a one-year MSc course, and for five years Cranfield Institute of Technology have been offering a two-year MSc course in Metrology and Quality Assurance. There has been no lack of well-qualified applicants for both of these courses, most of them coming from overseas.

It may be that here we have a 'chicken and egg' situation in which it is difficult to assign blame. So far as traditional universities are concerned metrology is an interdisciplinary subject which is unlikely to be championed by a single department. So far as UK industry is concerned there are areas where quality-assurance staff may be just as valuable as financial watchdogs, but they tend to lack formal qualifications or boardroom status. Even if asked, they may question the idea that by recruiting graduates they could achieve a general upgrading of their own profession, advantages which they themselves have been denied. In other areas, fortunately, a more enlightened and altruistic attitude prevails; students are sponsored, and each success breeds another.

When work is to be done under contract, a resourceful metrologist should be able to consider several possibilities. Which method is finally chosen will then depend on the requirements of the customer. Are measurements to be made repeatedly on a production line, or on only one or a few items? Where will the measurements be done, and will there be adverse effects from a disturbing environment? What accuracy and what time-scale is being sought, and how much money is available? Can the work be done with existing equipment, or must new devices be designed, and if so who will own any resulting patents? Will the results be made available to others, or must they be kept confidential? Should the work be undertaken elsewhere? Clearly the success of a project depends not just on technical factors but also on establishing a sound business relationship between customer and contractor.

1.2 OPTICAL MEASUREMENT

Having discussed the role and responsibilities of the metrologist, let us now consider some of the ways in which optics can be used for measurement. In general light is emitted from a source, passes through an intervening optical system, and reaches a detector whose output provides useful information. The light may come from the sun or sky, from a lamp or quite often from a laser.

Let us assume first that we have a simple optical system which generates a picture whose changing details need to be followed by the detector. If this is the human eye, the information is passed directly to the brain and analysed with remarkable efficiency, for example in reading this book, but persistence of vision and no doubt other factors must limit the speed of response. Alternatively, data can be acquired quickly and with high resolution on a photographic plate or hologram (Chapter 7), even on cinefilm, but there will be a processing delay of minutes or hours before such data can be analysed, and this may be a serious disadvantage. More immediate operation can be secured by using vidicons or charge-coupled detector arrays whose output can be handled in many cases at television frame rates (Chapters 4, 8, 10). If very fast observations are needed, individual detectors can cover frequencies up to about 1 GHz, for example in fringe counting (section 5.1.4) or in Doppler measurement (section 6.1). Where ultra-fast phenomena need to be studied, flashlamps and strobes have now been supplemented by lasers whose pulse duration may range from nanoseconds down to a small fraction of a picosecond, with peak powers reaching gigawatts or even terawatts.

Let us now consider, for more static situations, the types of information that can be provided by a complete optical measuring system:

SOURCE >> OPTICAL SYSTEM >> DETECTOR

There are of course several scientific applications, for example in spectroscopy or astronomy, where the aim may be to measure the spectral or spatial distribution of the energy coming from the source, or even the variations in quantum efficiency of a new photodetector. In such cases the intervening optical system may be a spectrograph or a telescope (coupled usually with the atmosphere). For nearly all engineering applications, however, the intervening optical system is itself the object of study. In some cases we wish to test the shape of one of the components, and in others we need to measure their relative position in order to deduce misalignment or displacement or rotation. Moreover, as we shall see in this book, there are important cases where the 'optical system' has to include not only lenses and polished mirrors but also engineering structures such as a large radio antenna (section 4.6.1), a rough brick wall (section 9.4.1), or the side of a vibrating engine block (section 8.6.1). Vibration and deformation of such 'optical components' (and sometimes their absolute geometry) can be measured with excellent accuracy.

We should perhaps sound here a note of metrological caution. Efforts should always be made to design a system so that it measures only what is relevant. It should respond only to variations in the quantity being measured and should ignore changes in other factors. If this cannot be done then those other factors should preferably be fixed, or at any rate monitored so that corrections can be made. In measuring displacement, for example, with modern lasers almost any system will readily produce exquisite interference fringes which display the movement of a mirror with great sensitivity (Chapter 5). But unless one is quite careful in designing the system, its output will be influenced also by several other effects (Abbe errors, slide and laser misalignment, atmospheric changes, vibration and third mirror feedback). The latter especially may be so unpredictable as to make the system useless except as an intruder alarm.

Let us now list briefly five types of optical method that may be useful in dimensional metrology.

1. First (and still going from strength to strength) is the direct use of an imaging system. For parallel operation it should be possible to resolve at least 1000 × 1000 separate channels. Scanned systems using laser sources can also be very attractive (section 2.5.1).
2. An important group of methods uses some kind of phase measurement, as in interferometry, polarimetry or moiré techniques (section 10.2). Most of these rely heavily on the spectral purity or longitudinal coherence of the light coming from a laser, but others may function quite well even with an extended white light source.
3. Diffraction methods used for alignment (Chapter 2) or spatial filtering require light that is directional or spatially coherent, and this is obtained usually from a laser in order to secure adequate brightness.
4. Holography (Chapter 7) and, with a few exceptions, speckle techniques (Chapter 8) need illumination which is both spatially and spectrally coherent, so a laser is essential.
5. Starting perhaps with optical fibres (Chapter 11), gradient index (GRIN) lenses and integrated optics, a new group of 'confined field' techniques has emerged

which exploit the phenomena of frustrated total internal reflection, evanescent waves and photon tunnelling.

If engineers or students wish merely to make use of some of these techniques, they may not need to become experts in geometrical aberration theory or in laser physics. But if they really wish to understand them and take part in their development, if they aspire to be metrologists, they should at least learn to calculate the propagation of a laser Gaussian beam, using the geometrical method presented in the next chapter, or ray-transfer matrices and the easily remembered ABCD-rule (Gerrard and Burch, 1975). Second, they should study some of the applications of Fourier methods in optics (section 2.1.1), with especial emphasis on the convolution theorem for Fourier Transforms.

Given that they have equipped themselves in this way, what are the tasks in dimensional metrology that may need to be tackled? Perhaps the simplest are those in which only a single quantity, such as a length, a diameter or an angle needs to be measured. Alternatively, it may be required to measure, for example, the height y for a whole series of positions along the length of a long straight-edge pointing in the z-direction, in other words to sample $y(z)$. If a carriage is sliding in the z-direction, both vertical-displacement errors $y(z)$ and horizontal errors $x(z)$ will need to be measured. The same carriage may also suffer from the three angular errors of pitch, yaw and roll and, indeed, its z-transducer may also be reporting wrong z-values because of an imperfectly divided scale. If three such slideways are combined in a coordinate measuring machine, their axes may not be perfectly orthogonal to each other, so altogether there may be 21 different sources of error to be determined! Even more important, there may be little point in attempting such a calibration unless each of the slideways gives repeatable performance unaffected by friction or working forces or by uncompensated changes in weight distribution. It is hardly surprising that such machines are expensive.

Rather more tractable is the problem of measuring the shape of a single polished surface, for example a large optical flat or a concave spherical mirror. If this is arranged to face in the z-direction then we need to measure $z(x, y)$ and it is usually possible to obtain an interference contour map of the whole surface, supplemented if necessary with phase-stepping methods. For testing large spherical mirrors a common-path design can be used for the interferometer (Burch, 1953, 1969), and there is no need to fabricate a 'perfect' comparison surface. Small departures from sphericity can then be measured to within better than 10 nm, even with an incoherent source, but only in nearly perfect air conditions. Wherever an optical beam has to travel through several metres of air, one usually finds that it is atmospheric refraction, whether regular or irregular, which determines the accuracy with which the phase or the slope of the wavefronts can be measured (sections 2.3 and 5.1.3).

In outer space, of course, as the designers of the Hubble Space Telescope were aware, it is no longer true that 'The sky is the limit', and the performance of a good optical system can be superb. The serious error in shape of the Hubble primary mirror, discovered only after launch, is a tragic example of how even the best metrology can be vitiated by a single mistake (in this case the wrong reflected image had been selected during preliminary set-up of the two-element corrector used for null-testing).

Engineers are of course concerned with more complicated shapes than just flats and spheres. Interferometric null-tests can in fact be designed also for (polished) cylinders, cones and involute helicoids (such as screws and helical gear teeth). In practice, however, it is much better to rely on mechanical systems in which accurately measured translation and rotation are combined with a suitable method for probing the surface. If all of these measurements, from moiré fringe transducers or laser interferometers as well as from the probes, are sent into a computer, such a device can be generalized into a versatile coordinate measuring machine (CMM) capable of probing the external shape of a complicated component or an assembled structure.

Let us list some of the limitations to this approach to 3 D measurement.

1. For an accurate CMM, located in its own temperature-controlled enclosure, measurements can be made only on components which are compact and portable enough to be brought in and positioned on the measuring table.
2. Even if a component is small enough, a sequential point-by-point method of inspection is bound to take time, so it will be difficult to measure anything but a static geometry.
3. A typical CMM will give excellent results when measuring along a single axis, but for slanted directions its 'volumetric accuracy' can be much less impressive. Discrepancies of this kind can be revealed by comparing several measurements made with the same rigid spaceframe (e.g. a tetrahedral arrangement of four balls) mounted in various orientations.
4. CMM measurements acquire sequentially and rather slowly the absolute geometry of selected areas of the surface. In many applications, the engineer is interested not so much in the absolute shape of a component as in the *changes* that occur when it is loaded or pressurized or vibrated.

One of the main purposes of this book will be to show how most of these limitations can be removed or at least alleviated by using optical methods.

For absolute measurement of shape, the techniques of photogrammetry (Chapter 4) have no upper limits of size, and with flash illumination all the measurements are acquired simultaneously. Processing and analysis of the data takes time, but for some targeted structures a volumetric accuracy close to one part in a million has been achieved.

For differential measurements, there are now several whole-field optical methods based on holography, laser speckle and moiré techniques. Apart from offering excellent accuracy, they have been applied outside the laboratory and with great success in a wide variety of engineering projects. In the chapters that follow, moreover, they are described by authors who have been at the forefront in achieving this progress.

So far as dimensional metrology is concerned, we have now identified seven types of measurement task (length, angle, straightness, flatness, sphericity, 3D shape and 3D deformation), and we have also listed five types of optical method that could be used. It follows that any known technique, described in this book or elsewhere, could be sorted into the appropriate box in a 5×7 rectangular array. According to one recipe for invention, any box that remains empty at the end of this process can be regarded as a promising area where, sooner or later, and thanks to market pull and technology push, a new and useful development will occur.

Before enthusiastic inventors or investors rush off to construct such a table, we should point out that the situation is not that simple. There are several other technical choices or options whose pros and cons need to be considered before one can home in on a useful candidate project. I have listed thirteen of these below, and perhaps this is a number which will prompt the reader to 'replace optimism with prudent caution'.

1. Does the measuring system operate in 'real time' or in separate stages?
2. Is accuracy achieved by **averaging** (using many lines of a moiré grating or an extended white source) or by a **pointwise approach** (using a laser with a sharply defined position and wavelength)? (Compare similar choices in mechanics; do we use **overconstrained** lapping processes and hydrostatic bearings, or do we rely on single-point diamond machining and **kinematic location**?)
3. Will the optical system use conventional mirrors and lenses, or specially fabricated holographic elements, tuned cameras and axicons?
4. Will it function inside a laboratory, or on a large scale (e.g. outdoors)?
5. Will it cover a whole field, or only selected small areas?
6. Will the field be observed simultaneously, or by scanning?
7. Will measurements be made directly, or in the Fourier domain?
8. Will the system operate single-passed or double-passed?
9. In optical testing, are even and odd aberrations to be separated?
10. Will optical path differences be measured incrementally, or absolutely?
11. Will a null-sensing system use d.c. methods, or synchronous detection?
12. Will height errors be measured directly, or by integrating slope values?
13. Will a test surface be illuminated at normal or at oblique incidence?

While this purports to be a binary check-list, there will clearly be some optical measuring projects where the answer to a few of the above questions may be 'Neither' or 'Both' or merely 'N/A'. Nevertheless there may be others which will repay attention. For example, it would seem that **slope sensing** at **oblique incidence** (section 2.5.5) could be used, at high train speeds, to monitor corrugations on a railway track. *Verbum sapienti sufficit!*

1.3 A WIDER VIEW

To conclude this introduction, I should like to encourage readers briefly to take a wider look outside the topics of this book. Engineers who are concerned with small components will already be aware of the exciting progress that has been made in the field of nanotechnology. For some time we have had scanning electron microscopes whose depth of field and resolution beats the performance of any optical microscope. But now we have three new types of system where a piezoelectric 3D scanner moves a small specimen surface past an extremely sharp probe. The very small spacing between probe and surface is servo-controlled in such a way as to maintain constant either a minute atomic force, an electron tunnelling current or a photon tunnelling flux. After a raster scan has been completed, images of superb quality are displayed. Although in their infancy, these new microscopes have demonstrated startling resolution, displaying individual atoms with little or no surface preparation, and operating in some cases on biological molecules immersed in water. Commercial versions of this exciting new probe technology are already available.

If we turn now from the atomic scale to the galactic, here too there is much to interest the metrologist. There are, for example, distant radio 'point sources' whose direction can be measured by means of a large antenna array with nanoradian precision. This makes it possible to detect changes in latitude and hence to monitor N–S continental drift to within about 6 mm – of interest to the seismologist at any rate. In the field of frequency measurement, there is a neutron star, the 'millisecond pulsar' which rotates about 600 times each second (Ashworth *et al.*, 1983). The radio pulses from this are very weak, and they arrive early or late as the earth rotates daily on its axis, and around the sun. But after allowance is made for these effects their frequency is found to keep step with our best caesium frequency standards to within about one part in 10^{13}. If two more such pulsars can be found we shall have the basis for an interstellar navigation system with a 3D resolution of about 300 m!

This agreement between a man-made frequency standard and a rotating neutron star is reassuring and slightly humbling. But in justice to our colleagues in scientific metrology we should point out that current work, using a few atoms trapped and cooled in the laboratory, may well result in optical frequency standards whose linewidth is sharper still by perhaps two orders of magnitude. It is difficult to relate achievement of this level of accuracy to practical requirements, but the same was probably said a century ago about the Michelson–Morley experiment. It is very hard to predict all the consequences of such advances. Certainly a career in scientific metrology does not lack excitement.

Let us consider one more example of celestial metrology, in this case using modest geometrical optics. In February 1987 the observed outburst of a supernova attracted much public interest. Over four years later the site of this explosion is seen to be surrounded by a nearly circular halo where outlying interstellar gas has been excited to emit light by the expanding shell of X-rays, etc. On comparing the angular diameter of this halo with the time elapsed since the explosion was observed we can calculate approximately the absolute distance to the supernova. This calculation may be accurate only to about one per cent since it depends on some assumptions about the gas distribution, but it is unaffected by attenuation caused by dust absorption. (A simplistic approach suggests that, if the diameter of the halo is measured to be 10 milliradians, then we must be observing a sphere of radius 4 light years from a distance of 800 light years. But an aficionado of holography will realize that, for any given time of observing, the apparent locus of the expanding shell is not a sphere but an elongated ellipsoid having the earth and the supernova as its two foci, as in one of Nils Abramson's holo-diagrams (Abramson, 1981). Far from being outside such a shell, we are inside it, but fortunately the X-rays at our end of the ellipsoid are highly attenuated by the inverse square law!)

Some of the examples chosen for this wider look illustrate the unexpected repercussions which work in one field like nanotechnology or radio-astronomy may have on another such as biology or seismology. In general almost any invention or discovery can provide the metrologist either with new working tools or with new requirements for measurement.

So far as the future is concerned, optical metrologists can expect that as industries such as precision engineering and optical communication grow in importance they will continue to be kept busy with new areas of work. Equally, they can be confident of acquiring better lasers and detectors and faster methods for processing information. Finally, if sensible attitudes prevail, they should find the career rewarding as well as interesting.

REFERENCES

Abramson, N. (1981) *The Making and Evaluation of Holograms*, Academic Press.

Ashworth, M., Lyme, A. G. and Smith, F. G. (1983) The 1.5 millisecond pulsar PSR1937+21. *Nature*, **301**, 313–14.

Burch, J. M. (1953) Scatter fringes of equal thickness. *Nature*, **171**, 889.

Burch, J. M. (1969) Interferometry with scattered light. In *Optical Instruments and Techniques* (ed. J. H. Dickson) Oriel Press, pp. 213–29.

Frankowski, G., Abramson, N. and Füzessy, Z. (ed.) (1991) *Application of Metrological Laser Methods in Machines and Systems*, Akademie-Verlag.

Gasvik, K. J. (1987) *Optical Metrology*, Wiley.

Gerrard, A. and Burch, J. M. (1975) *Introduction to Matrix Methods in Optics*, Wiley.

Luxmoore, A. R. (1983) *Optical Transducers and Techniques in Engineering Measurement*, Elsevier (Applied Science).

Soares, O. D. D. (ed.) (1987) *Optical Metrology*, Martinus Nijhoff.

2
Laser beam geometry and its applications

D. C. WILLIAMS

Crown copyright

2.1 ANALYSIS OF LASER BEAM SIZE

In popular conception, a laser beam is often thought to be equivalent to a straight weightless thread of unlimited extent. Indeed, for the alignment of the large particle accelerators at CERN in Geneva, tensioned nylon threads and laser beams are both used in different circumstances. In a magazine for computer users, an article outlining the functioning of laser printers has stated that 'if the ray of light leaving the laser was 1 mm wide, it would still be 1 mm wide thousands of miles away'. In fact, nothing could be further from the truth. A laser beam has a finite size which changes as it propagates, and a proper understanding of these changes is needed in order to use such a beam successfully in alignment applications (Williams 1983).

The first part of this chapter is devoted to a pseudo-geometrical method for designing the optics necessary to optimize the beam characteristics for a particular application. The method is a straightforward extension of the more familiar process of tracing ordinary light rays through systems of lenses. Some related technologies are then discussed, with examples of their use in solving various measurement problems. Some optical principles which can be particularly valuable are introduced as they arise. The applications are mainly specialized ones with which the author has been directly concerned; they are intended as indications of how a little ingenuity enables the basic principles to be adapted to suit demanding requirements. A recurring theme is the importance of choosing the optimum technique, which may not be the one which first comes to mind, and a number of examples of not so good and better approaches are given.

2.1.1 Characteristics of a laser beam

For most practical users, an understanding of the internal operation of a laser and the underlying physics is not necessary. All that is needed is some knowledge of the properties of the beam which emerges. The theory of beam propagation which will now be described can usually be applied directly to gas lasers, and particularly to the

Optical Methods in Engineering Metrology. Edited by D.C. Williams. Published in 1993 by Chapman & Hall, London. ISBN 0 412 39640 8

12 Laser beam geometry and its applications

helium-neon laser which is widely used for metrological applications. For these lasers it predicts the behaviour of the beam with a good degree of accuracy. Diode lasers offer great advantages in compactness and convenience, but at the present stage of development the beam quality can be inadequate for applications where a high degree of precision is required.

The diameter of a gas laser plasma tube and the curvatures of the cavity end mirrors are normally chosen so that the laser is constrained to oscillate only in the fundamental transverse mode classified as TEM_∞ (Bloom, 1968). Ideally, the distribution of irradiance (optical power per unit area) across any diameter of the beam is then Gaussian, having the same form as the normal distribution curve of random errors in a repeated measurement. In practice the distribution is perturbed by the finite width of the plasma tube, but it is normally closely Gaussian except for the detailed structure in the outer regions where the irradiance is low. It can also be contaminated by secondary reflections from optical surfaces, and by wavelets scattered from blemishes and particles of dirt. These disturbances usually have wavefront curvatures which differ from that of the main beam, and therefore produce spurious patterns of interference rings.

The radial distribution of amplitude is described by

$$A = A_c \exp - (r^2/w^2), \qquad (2.1)$$

where A_c is the amplitude on the central axis, A is the amplitude at a distance r from the axis, and w is a parameter usually termed the **spot size**. The irradiance I in terms of the axial value I_c is obtained by squaring the amplitude, so that equation (2.1) becomes

$$I = I_c \exp - (2r^2/w^2). \qquad (2.2)$$

At a radial distance r from the axis equal to the spot size w, the amplitude has fallen to $1/e$ and the irradiance to $1/e^2 = 0.135$ of the central value. If w is to be regarded as a measure of the beam diameter, then for the effective edge of the beam $2r = w$ and the irradiance reduction is $1/e^{0.5} = 0.444$.

As the beam propagates in a direction z, the spot size varies according to the expression

$$w = w_0 \sqrt{[1 + (z/z_0)^2]} \qquad (2.3)$$

in which the origin of z has been chosen so that the beam has a minimum spot size w_0 at the origin position. The distance z_0 is also a constant parameter for a particular beam. As shown in Fig. 2.1, the beam converges as it approaches the origin and then diverges symmetrically, the region of minimum size being known as the **waist**. To maintain constancy of the total power in the beam as it propagates, the axial irradiance reduces as the beam size increases according to

$$I_c = (w_0/w)^2 I_0, \qquad (2.4)$$

where I_0 is the value at the waist.

For large values of z such that $|z| \gg z_0$, equation (2.3) gives approximately

$$w = sz, \tag{2.5}$$

where the parameter s is defined by the equation

$$w_0 = sz_0. \tag{2.6}$$

This region is referred to as the **far field**, in which the beam tends asymptotically to a condition where it has a constant angular divergence s from an origin point at the waist. For the **near field** region, with $|z| \ll z_0$, the spot size w is scarcely greater than w_0. This form for the beam wave can be shown to be a solution of the scalar wave equation provided that the divergence is small (Kogelnik and Li, 1966).

The value of z_0 is related to the value of w_0 by

$$z_0 = \pi w_0^2/\lambda, \tag{2.7}$$

where λ is the wavelength of the laser light, so that the axial extent of the near-field region is proportional to the square of the waist size. There is a family of possible laser beam profiles corresponding to different pairs of values of w_0 and z_0; Fig. 2.1(a) and (b) show two typical members of such a family. The significance of the slanting lines will be explained later. Eliminating z_0 between equations (2.6) and (2.7), we find that

$$s = \lambda/\pi w_0 \tag{2.8}$$

so that the far-field divergence is inversely proportional to the spot size at the waist.

In any plane containing the axis, equations (2.2) and (2.3) show that the loci of points for which I/I_c is constant are hyperbolic in form, the lines in Fig. 2.1 for $r = w$ being typical ones. These curved loci can be regarded as 'effective rays', and they are normal to the wavefronts in the propagating beam, which are weakly spherical. At radius w, the slope of the effective ray and therefore of the normal to the wavefront is dw/dz. Extending the wave normal until it meets the axis, one sees that the radius of curvature ρ of the wavefront is given by

$$w/\rho = dw/dz, \tag{2.9}$$

which by differentiation of equation (2.3) yields

$$\rho = z + (z_0^2/z). \tag{2.10}$$

In the far field, $\rho \approx z$ and the centres of curvature of the wavefronts coincide with the waist. For wavefronts which are closer to the waist, the centres of curvature lie beyond it, and at the waist they tend to infinity. The radius of curvature ρ has a minimum value of $2z$ when $z = z_0$, the wavefront position then being intermediate between the near and far fields. For this position, the centre of curvature of the wavefront on one side of the waist coincides with the corresponding wavefront on the other side, this being known as the **confocal condition**.

14 Laser beam geometry and its applications

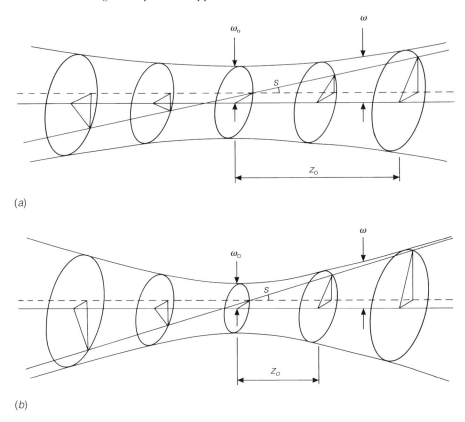

Fig. 2.1 Laser beam shapes: (*a*) larger waist, (*b*) smaller waist.

It is instructive to see how the relationship between the waist and far-field distributions accords with the general theory of optical diffraction. The reader for whom this, and other optical principles which will be mentioned later, are unfamiliar is encouraged to consult a basic textbook such as Hecht (1987). When one wishes to calculate the distribution in a plane further on which results when a light wave passes through a pinhole or other obstruction, each elementary area across the aperture can be regarded as a secondary point source emitting a spherical wave with the same phase as the original one. At any point in the new plane, one sums the contributing amplitudes due to all the spherical waves, taking account of their relative phases as they arrive.

When the new plane is very distant, the contributing disturbances will be practically in phase on the axis, giving maximum resultant amplitude. At points away from the axis, the contributing phases will differ because the distances from the edges of the aperture become different. When they differ by about a wavelength, the resultant amplitude falls to a low value. For a smaller aperture, one has to go further from the axis before this happens, so that a smaller aperture produces a larger spread. It can also be seen that the new far-field wavefront will be centred on the centre of the aperture, since the mean distance travelled by the secondary wavelets to different places on the new wavefront will then be nearly constant.

If the wavefront across the aperture is plane and the amplitude is symmetrical about the centre, the secondary sources can be grouped in opposite pairs. In the far field, one obtains pairs of wavefronts with a slight angle between them which interfere to give systems of straight, parallel, equispaced interference fringes whose spacing is inversely proportional to the distance off axis of the sources. The far-field distribution can then be regarded as a Fourier summation of these sinusoidal distributions, whose amplitudes are proportional to those at the corresponding points in the aperture.

It can be shown quite generally that a Fourier transform relationship exists between the amplitude distributions in the aperture and in the far field; in the laser beam case, the Fourier transform of a Gaussian distribution is another Gaussian distribution of inversely proportional width. For light travelling from the far field towards the waist the argument can be inverted, secondary wavelets from the converging spherical wave combining to produce the distribution at the waist.

The phase of the beam along the axis basically varies according to

$$\phi = 2\pi z/\lambda, \qquad (2.11)$$

as in the case of a uniform plane wave. However, there is a correction term $\Delta\phi$ to be subtracted from ϕ which is given by

$$\tan \Delta\phi = z/z_0. \qquad (2.12)$$

As the beam propagates from the far field on one side of the waist to the waist position there is a phase shift of $\pi/2$, and a further phase shift of $\pi/2$ when it reaches the far field on the other side. This accords with a general prediction of scalar diffraction theory.

When a laser beam is employed in a two-beam interferometer using division of amplitude, as described in Chapter 5, the interference takes place between two similar copies of the beam between which there is some path difference. The difference in the wavefront curvatures of the two beams produces a system of concentric circular interference fringes which varies along the axis. However, if the whole of the optical power is received by a photodetector, it follows from conservation of energy that the signal from the detector must be independent of its position along the axis of the composite beam. As the path difference l is increased, it can be shown that the periodic modulation M of the signal due to the interference is reduced according to

$$M = M_0/\sqrt{[1 + (l/z_0)^2]}, \qquad (2.13)$$

where M_0 is the modulation at zero path difference. The phase of the interference signal is also affected by the beam geometry, so that the apparent path difference l is reduced by an amount Δl which is given by

16 *Laser beam geometry and its applications*

$$\tan(2\pi \Delta l/\lambda) = l/z_0. \tag{2.14}$$

For $l \gg z_0$, Δl tends to a maximum value of $\lambda/4$.

2.1.2 Analysis using skew rays

For many metrological applications, it is necessary to tailor the laser beam so that the waist has a particular size and lies at a particular position along the axis. In general, this can be achieved by means of two separated thin lenses, given the parameters of the original beam as it emerges from the laser. One therefore needs a method for calculating the effect on a laser beam of passage through a lens of given power.

In traditional optics, this calculation in the paraxial approximation is very straightforward (Fig. 2.2). A point source on the optical axis of the system at C_1 generates spherical waves, so that a wavefront falling on a thin lens has its centre of curvature at the source position. The lens changes the curvature of the wavefront by an amount proportional to its power (the reciprocal of its focal length). Thus if the lens is sufficiently strong the new wavefront converges towards a focal point C_2 further along the axis.

A ray from the source is everywhere perpendicular to the spherical wavefronts. If it strikes the lens at P, a distance y_L from the axis, it is bent through an angle $\alpha = y_L/f$ by the prismatic action of the lens, where f is the focal length. If the source and image points are distant z_1 and z_2 from the lens, one has

$$y_L/z_1 + y_L/z_2 = y_L/f \tag{2.15}$$

leading to the familiar lens formula

$$1/z_1 + 1/z_2 = 1/f. \tag{2.16}$$

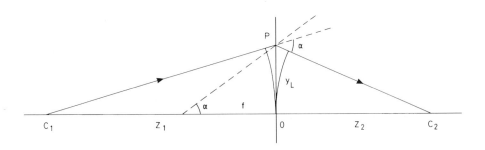

Fig. 2.2 Imaging of a point source.

If the source point lies off the axis, in any direction, the ray to the same point on the lens is still bent by the same amount. The above theory can therefore still be applied relative to an auxiliary axis passing through the source point and the optical centre of the lens (where its surfaces are parallel). The refracted ray lies in the plane containing the incident ray and the auxiliary axis. Also, if the ray paths are projected on to another plane still containing the axis but at an azimuthal angle θ relative to the plane containing the rays themselves, the same analysis is still applicable, all angles and transverse distances such as y_L being reduced by a factor $\cos\theta$.

In the case of a Gaussian laser beam, the curvature of a spherical wavefront is changed in just the same way as it passes through a lens. However, because the effective rays are not straight, the simple lens formula is not directly applicable, and algebraic formulae relating the waist positions and sizes for the two sides of the lens are found to be cumbersome. The simple theory only approximates to the truth if the waists are very small so that the near-field regions are very short.

Various calculation methods have therefore been devised, as mentioned in section 1.2, including a 2×2 transfer matrix with complex coefficients (Gerrard and Burch, 1975) and a graphical approach using a chart which is analogous to the Smith chart developed for waveguide applications (Collins, 1964). However, the physical interpretation of these can be difficult to visualize, and we shall now show that the ray tracing method of traditional optics can still be applied in modified form (Arnaud, 1985).

We have seen that in general the effective rays are hyperbolic in shape, so that for all azimuths they collectively form a hyperboloid of revolution about the axis. We now show that although the effective rays are curved, there nevertheless exist skew lines which lie entirely in the surface of the hyperboloid and which are straight.

The $1/e^2$ points of the Gaussian beam lie at the radius $r = w$. If we take transverse horizontal and vertical coordinates x and y, measured from the axis of the beam, we have

$$r^2 = x^2 + y^2. \tag{2.17}$$

It then follows from equations (2.3) and (2.6) that points which lie on the $1/e^2$ hyperboloid of revolution will satisfy the equation

$$x^2 + y^2 = w_{01}^2 + s_1^2 z^2, \tag{2.18}$$

where the suffix 1 is used for beam parameters on the input side of the lens. In particular, this equation is satisfied by the skew straight line formed by points such that

$$x = w_{01}, \; y = s_1 z. \tag{2.19}$$

The line lies in a vertical plane which is at a horizontal distance w_{01} from the axis. The line is level with the axis at the waist, and its slope angle in the plane is s_1. Figures 2.1(a) and (b) show two such skew lines; in the second case w_0 is smaller and s is larger.

18 Laser beam geometry and its applications

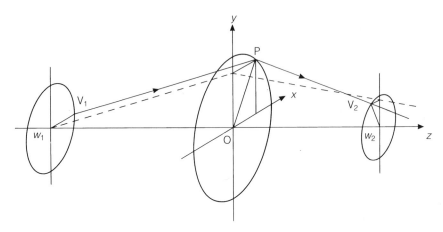

Fig. 2.3 Passage of a skew ray through a lens.

If the skew line is regarded as a light ray, this ray will be bent by the lens through the same angle as the real effective ray which passes through the lens at the same point. The new skew ray then lies in the hyperboloidal surface associated with the modified Gaussian beam, enabling its parameters to be determined. These rays are shown in Figure 2.3, where the dashed line is their projection on to the vertical yOz plane. Projections on to the vertical and horizontal planes are also shown in Figs. 2.4 (a) and (b).

When the skew ray V_1P reaches the lens with its centre at O in Fig. 2.3, we have

$$x_L = w_{01}, \quad y_L = s_1 z_1, \tag{2.20}$$

where the suffix L is used for quantities measured at the lens, and z_1 is the distance W_1O from the waist to the lens. Considering now the projections of the skew ray in Figs. 2.4, the angles of bending caused by the prismatic action of the lens are x_L/f and y_L/f. Therefore the slope angles associated with the projections of the new skew ray beyond the lens are

$$s_{x2} = x_L/f = w_{01}/f, \tag{2.21}$$

$$s_{y2} = y_L/f - s_1 = s_1(z_1/f - 1), \tag{2.22}$$

where the suffix 2 applies to quantities on the output side of the lens.

There is, as for the input ray, a plane containing the new skew ray, PV_2 in Fig. 2.3, which is at a constant distance from the axis as before, but no longer vertical. The distance of the new plane from the axis is w_{02}, the new waist size. Since the horizontal and vertical projections of the new ray have slopes s_{x2} and s_{y2}, it follows as shown in Fig. 2.5 that the resultant overall slope in the new plane is s_2 given by

$$s_2^2 = s_{x2}^2 + s_{y2}^2, \tag{2.23}$$

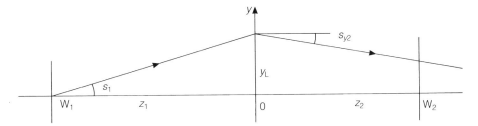

(a)

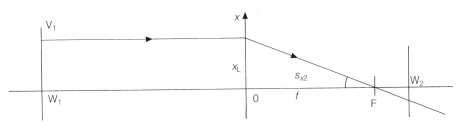

(b)

Fig. 2.4 Projections of skew ray: (*a*) vertical plane, (*b*) horizontal plane.

the plane being inclined at an angle θ to the vertical such that

$$\sin\theta = s_{x2}/s_2, \cos\theta = s_{y2}/s_2. \tag{2.24}$$

Figures 2.5 and 2.6 both show views in the direction of the axis, looking towards the lens from the output side.

Taking new coordinates x' and y' which are perpendicular and parallel to the new plane, one has at the lens position

$$x_L' = w_{02}, \ y_L' = s_2 z_2, \tag{2.25}$$

where z_2 is the distance from the lens of the new waist at W_2. The new coordinates are then related to the old ones by

$$x_L' = y_L \sin\theta - x_L \cos\theta, \tag{2.26}$$
$$y_L' = x_L \sin\theta + y_L \cos\theta, \tag{2.27}$$

as can be seen in Fig. 2.6.

Substituting in equation (2.26) from equations (2.20) to (2.25), we find that

$$w_{02} = w_{01} s_1/s_2, \tag{2.28}$$

20 *Laser beam geometry and its applications*

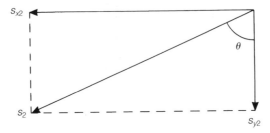

Fig. 2.5 Slope of output skew ray, viewed along the optical axis.

a simple expression for the new waist size which confirms that it has to be inversely proportional to the far-field divergence. Substituting in equation (2.27) we find that the distance to the new waist is given by

$$z_2 = (w_{01}s_{x2} + s_1z_1s_{y2})/s_2^2. \tag{2.29}$$

Once s_2, w_{02} and z_2 have been determined, we can consider the appropriate skew ray in a vertical plane as before, and apply the same analysis to the next lens in the system.

As in ordinary lens calculations, care is required to observe a self-consistent sign convention; in the above analysis, quantities have been taken to be positive when the beams on both sides of the lens have real waists. An alternative way to derive z_2,

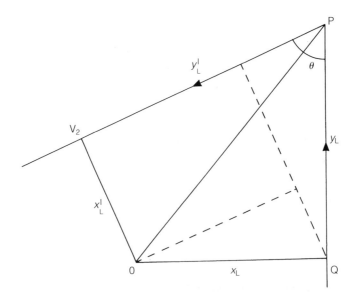

Fig. 2.6 Input and output transverse coordinate systems.

which requires a square root to be taken but avoids some of the problems with sign convention, is to equate two values for the square of the radius OP. Thus

$$w_{02}^2 + s_2^2 z_2^2 = w_{01}^2 + s_1^2 z_1^2, \tag{2.30}$$

in which all quantities except z_2 are known. This result can be obtained by squaring and adding equations (2.26) and (2.27). It can also be used as a check if z_2 has been obtained from equation (2.29).

The reader may find it helpful to have a numerical example; the accompanying diagrams are drawn according to the example, but with unequal longitudinal and transverse scales. Numbers of equations used to obtain the derived quantities are given in parentheses.

Input quantities

Laser wavelength	$\lambda = 633$ nm
Lens focal length	$f = 0.82$ m
Input beam waist size	$w_{01} = 0.47$ mm
Waist to lens distance	$z_1 = 1.30$ m

Derived quantities

Input beam far-field slope	$s_1 = 0.429$ mm/m (2.8)
Components of output slope	$s_{x2} = 0.573$ mm/m (2.21)
	$s_{y2} = 0.251$ mm/m (2.22)
Output beam far-field slope	$s_2 = 0.626$ mm/m (2.23)
Output beam waist size	$w_{02} = 0.322$ mm (2.28)
Lens to waist distance	$z_2 = 1.045$ m (2.29)

2.1.3 Some applications of the theory

This geometrical approach may at first sight seem no simpler than any other, but it does enable some results to be derived rather easily. For example, if the waist sizes are both very small, the input and output skew rays are everywhere close to the axis in the horizontal direction. Thus the rays behave almost as if they were in the vertical plane through the axis, and it can be seen that the variation of beam size nearly accords with geometrical optics. Two further illustrations are now given.

Figure 2.7 shows that if the waist of the input beam is at a distance of one focal length f from the lens, centred at O, then so also is the waist of the output beam. The input skew ray $V_1 P$, in its vertical plane, is level with the axis at V_1, distant f from the lens; $V_1 W_1$ is the waist radius. The output ray PV_2 is therefore bent downwards at P into the horizontal direction and lies in a horizontal plane. Because the vertical plane containing the input ray is parallel to the axis, the ray PV_2 is also bent sideways

22 Laser beam geometry and its applications

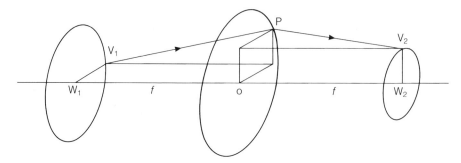

Fig. 2.7 Beam waists in Fourier planes of lens.

towards the axis, and lies above it at the point V_2 distant f from the lens. Thus the new waist is at this position, its radius being V_2W_2.

It can be seen from the diagram that

$$w_{01} = s_2 f, \quad w_{02} = s_1 f \tag{2.31}$$

so that, recalling equation (2.8),

$$w_{01}w_{02} = f\lambda/\pi. \tag{2.32}$$

This result again accords with Fourier theory. The amplitude distributions in the two focal planes of a lens can be shown to be Fourier transforms of each other, and the transform of a plane wavefront with a Gaussian distribution is another plane wavefront with a Gaussian distribution of inversely proportional width. If, while maintaining the symmetrical condition, the waist sizes are made equal but very small, then their distance from the lens becomes $2f$ rather than f, in accordance with the corresponding condition of geometrical optics (see section 2.2.3).

A second result which can readily be demonstrated is that the size of the output beam at one focal length from the lens depends only on the size of the waist of the input beam, and not on its position along the axis. Figure 2.8 shows vertical and horizontal projections of the corresponding skew ray for three positions of the waist W_1. In Fig. 2.8(b), the ray always reaches the axis at the focal point F. In Fig. 2.8(a), the input rays appear oblique but parallel to each other, and therefore the output rays always pass through the same focal point. Thus the beam radius at F always has the same value. Although the size of the beam at F does not depend on the axial position of W_1, the entity which does change at F is the curvature of the wavefront.

A likely requirement for alignment applications is that a laser beam should be as narrow as possible over a given range. Symmetry considerations suggest that the waist should be located at the centre of the range, with the largest spot size at the ends. For a total distance $l = 2z$, this spot size r is given by equation (2.17). With x

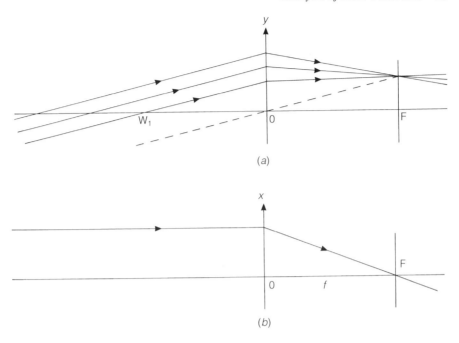

Fig. 2.8 Constancy of focal spot size: (*a*) vertical plane, (*b*) horizontal plane.

equal to w_0 and y inversely proportional to w_0, as shown by equations (2.19) and (2.8), r has its smallest value r_{min} when x and y are equal. Then

$$r_{min}^2 = 2w_0^2 = \lambda l/\pi \qquad (2.33)$$
$$\approx 0.2l \text{ mm}^2$$

for helium-neon red light (wavelength $\lambda = 633$ nm) when l is in metres. From equation (2.3), we see that this corresponds to $z = z_0$, the confocal condition. Thus the smallest spot size achievable at the ends is $\sqrt{2}$ times the central waist size, increasing in proportion to the square root of the distance. Over 5 m it is about 1 mm, increasing to 10 mm for a range of 500 m.

The light emerging from a laser will usually have a divergence s of about 1 mm per metre, with a virtual waist inside the laser whose size w_0 is about 0.4 mm. For laboratory applications the beam may well be used unmodified, or with its divergence reduced by a simple collimating lens. For outdoor applications over longer distances, it may be necessary to fit a beam expander with a power appropriate to the range. This is essentially a telescope used in reverse, with optics specially designed for the laser wavelength.

In the Galilean type of expander (Fig. 2.9(*a*)), a strong negative lens to increase the divergence of the beam is followed by a weaker positive lens to collimate it. In the astronomical telescope equivalent (Fig. 2.9(b)), the strong lens is positive and focuses the beam to a small waist before it diverges. This gives a longer tube length, but has the advantage that the beam can be 'spatially filtered' at the waist. Secondary beams of laser light reflected from lens surfaces or scattered by dirt particles often produce a pronounced structure by interference with the main beam. Because the distribution at the waist is the Fourier transform of that in the far field, the energy corresponding to the detail of the structure falls outside the main beam waist and can be blocked off by a small circular aperture slightly larger than the waist size.

If the main beam does not have a perfectly Gaussian irradiance distribution, then the distribution will vary as it propagates. However it can be shown, by resolving the beam into a Fourier summation of infinite plane wavefronts propagating in different directions, that the centroid or 'centre of gravity' of the beam still travels in a straight line. A photoelectric device which detects the centroid will therefore always give a correct result in straightness measurement if the beam is unperturbed by the atmosphere.

A practical limitation of a real laser beam is that it may change direction due to thermal effects in the laser head. For a typical 2 mW helium-neon laser, the beam has been observed to drift exponentially towards a final position during initial warm-up, with a time constant of 15 min. As a general rule, therefore, the laser should be switched on at least one hour before any accurate measurements are undertaken. One way of countering directional drift is to monitor it by means of a position-sensing detector (section 2.4) at the far end of the alignment path; appropriately scaled corrections can then be applied to measurements taken at other places. However, there are also optical solutions which make such a procedure unnecessary.

If directional stability is more important than positional stability, it may sometimes be possible to achieve this by invoking a simple optical principle. If a point source is placed in the focal plane of a lens, all rays of light from the source passing

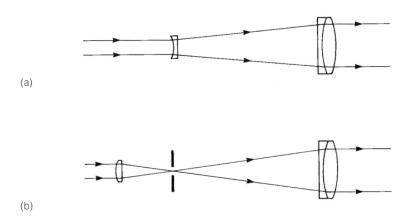

Fig. 2.9 Beam expanding telescopes: (a) Galilean, (b) terrestrial.

Fig. 2.10 Confocal etalon giving erect and inverted beams.

through the lens are bent to emerge parallel to each other, so that there is a complementarity between positions in the focal plane of the lens and directions on the other side. If the laser beam is changing its direction in a particular azimuth, there must be a position somewhere along the axis where there is no corresponding lateral movement. This position may well lie inside the laser head. A lens placed one focal length from the position will then give an output beam which does not change direction, at least in the same azimuth. Combinations of lenses may be used in the same way, provided that the stationary point is finally imaged to infinity.

Another possibility is to place in front of the laser an optical device which, by means of at least one partially reflecting surface, produces two versions of the laser beam which are mutually inverted. These will then drift in opposite directions, the centre line being defined by the optical device rather than by the laser. One such device has been constructed as a multi-element prism (Gates, 1975). If the beams are superimposed rather than separated, interference fringes will be formed between them, and these fringes rather than the whole composite beam can be used as the alignment reference. A system based on this principle (Gates and Bennett, 1968) uses a confocal etalon, consisting essentially of two partially reflecting spherical concave surfaces separated by their common radius of curvature (Fig. 2.10). However, the performance of such a system will be limited by imperfections of the optical components as well as by any instability of the mounting. A radically different optical solution is described in the next section.

2.2 INCREASING THE PRECISION

As the physics of the propagation of a Gaussian beam determines the minimum size that can be obtained over a given range, it might be supposed that this sets a fundamental limit to the alignment accuracy which can be achieved. However, the apparent limitation can in fact be overcome (Harrison, 1973; Chrzanowski *et al.*, 1976) in a quite simple way which is particularly appropriate for metrology.

2.2.1 Three-point alignment system

If the beam emerging from a laser is given a waist with a fairly small size, it will expand steadily as it propagates. If a weak lens of appropriate focal length is then placed in the relatively large beam, it will cause it to reconverge as it travels further, gradually contracting, eventually to form another small waist at the far end of the alignment path (Fig. 2.11(a)). The waist regions at the ends of the path, being small, have centres whose positions are well defined, and the position of the lens is also well defined because its outer circumference provides a mechanical reference.

This arrangement is sometimes described as a 'three-point' alignment system, since the three points defined by the centres of the two waists and the lens are known to lie on a straight line. If the far waist is allowed to fall as an image on a translucent screen or photoelectric detector, movements of the lens can be monitored by observing the corresponding movements of the image. These will be magnified in the ratio $(z_1 + z_2)/z_1$, z_1 and z_2 being defined as in Fig. 2.2. Strictly, z_1 should be measured from the centre of curvature of the wavefront entering the lens; this will be close to the waist if the beam has expanded considerably.

The optical centre of the lens lies where its two surfaces are exactly parallel, being on the line joining their centres of curvature. If the lens has not been accurately edged, the optical centre may not coincide precisely with its mechanical centre defined by the circumference. This potential source of error can readily be eliminated by rotating the lens one half turn about the mechanical centre, and taking the mean of two readings of the image position before and after the inversion. Asymmetry in the image caused by imperfections of figure of the lens surfaces is also reversed and therefore eliminated. This principle of inversion which enables systematic errors to be eliminated is widely applicable in optical metrology, and some further examples will be found in this chapter and the next one, particularly in section 3.1.1.

As well as improving the accuracy of setting, the three-point configuration also substantially overcomes another practical difficulty already mentioned, namely angular instability of the laser. With the lens well into the far field of the beam,

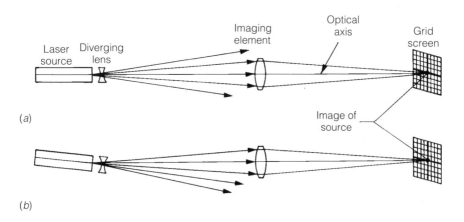

Fig. 2.11 Three-point alignment system: (a) laser aligned, (b) laser tilted.

angular drifts of the beam itself or of the laser mounting are likely to take place about a point not far removed from the centre of curvature of the wavefronts reaching the lens. Therefore, the wavefront tilt when the beam moves across the lens will be almost unaltered and the place to which the transmitted wavefront converges will scarcely be affected, as shown in Fig. 2.11(b). This remains true whether the laser beam is larger or smaller than the lens, the only limitation being that the beam must not drift so far that a substantial proportion falls outside the lens aperture. The size of the image will be approximately in inverse proportion to that of the laser beam or the lens aperture, whichever is the smaller, or with bad misalignment it may depend on the combination of both.

2.2.2 Fresnel zone plates

The three-point system is particularly suitable for monitoring movements in structures such as bridges and dams, because the source and detector can be mounted at the ends of the structure where it is known to be stable. For these large-scale applications, suitable lenses would have surfaces with very weak curvatures, and such lenses are relatively difficult and expensive to manufacture. The quality must be such that the emergent wavefront is spherical to a fraction of a wavelength to produce a diffraction-limited image which has maximum brightness and negligible distortion.

It is therefore advantageous in these situations to make use of the Fresnel zone plate (Fig. 2.12). This consists of a series of alternately transparent and opaque annuli, the radii of the zones increasing in proportion to the square roots of the natural numbers. It forms an image at a position along the optical axis such that the path travelled by the light to the image centre increases by one wavelength for successive zones, or one half wavelength for successive zone boundaries. The contributions from all zones then have the same phase, and combine constructively to give maximum brightness in the image.

Referring to Fig. 2.13, we see that the optical path $<C_1PC_2>$ via the mth boundary with radius r_m exceeds that along the axis $<C_1OC_2>$ by an amount Δ_m given by

$$\begin{aligned}\Delta_m &= <C_1PC_2> - <C_1OC_2> \\ &= (z_1 \sec s_1 + z_2 \sec s_2) - (z_1 + z_2) \\ &\approx \frac{1}{2}r_m^2(1/z_1 + 1/z_2) \\ &\approx \frac{1}{2}m\lambda.\end{aligned} \quad (2.34)$$

Comparing with equation (2.16), the boundary radii are related to the focal length f of the zone plate by

$$r_m^2 = mf\lambda. \quad (2.35)$$

With a 10 metre focal length and a helium–neon red laser, for example, $r_m = 6.3\sqrt{m}$ mm. Thus the zone plates, unlike lenses, become larger and therefore easier to

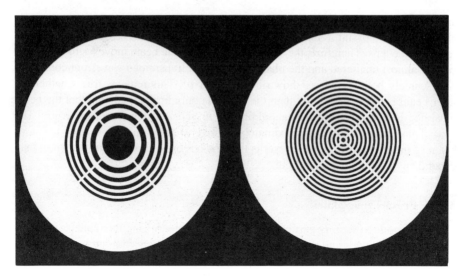

Fig. 2.12 Fresnel zone plate and circular grating.

manufacture as the distances increase. For distances above a few metres they can be made by photoetching of steel shim.

Since the area enclosed by the mth boundary of a circular zone plate is πr_m^2, proportional to m, all the transparent zones have equal areas. The zone plate can therefore be regarded as the equivalent with circular symmetry of a linear diffraction grating having alternate uniformly spaced bars and spaces. This analogy enables some characteristics of the image to be readily understood. Just as a grating produces a series of diffracted orders propagating in different directions, a zone plate gives an equivalent set of diffracted wavefronts differing in curvature and producing

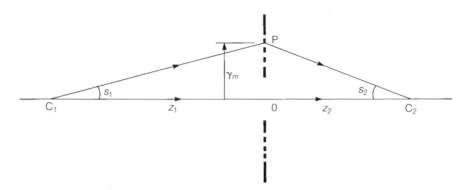

Fig. 2.13 Imaging by a zone plate.

secondary foci at various places along the axis. The prime focus corresponds to the first-order wavefront diffracted inwards. This wavefront on its own is indistinguishable from that produced by the equivalent lens with the same clear aperture as the zone plate, so the size and profile of the image are almost the same. The higher-order wavefronts with different curvatures interfere with each other in the image plane to produce a ring structure surrounding the main image (Fig. 2.14).

It is shown in standard optical texts that a linear grating diffracts a fraction $1/\pi^2$ of the incident light energy into each of the two first orders. It follows that the intensity of the zone plate image is about $1/\pi^2$ or 10% of that which would be given by the equivalent lens. As in the case of a line-and-space grating, the intensity can be increased by arranging for the opaque annuli to contribute to the image-forming process. To achieve this they are made transparent but with an optical thickness such that the paths through them differ from those through the other zones by a half wavelength. This can be realized by etching the zone plate pattern into a glass substrate. The amplitude in the image is thereby doubled, so that the intensity is quadrupled, becoming $4/\pi^2$ or 40% relative to the equivalent lens. However, the

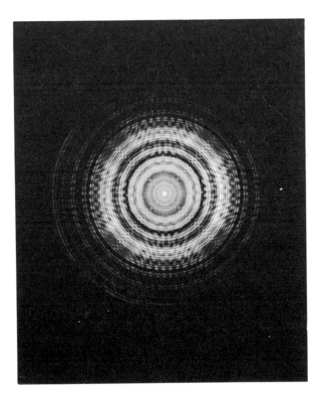

Fig. 2.14 Structure surrounding zone-plate image.

30 *Laser beam geometry and its applications*

advantage given by the zone plate that the light does not have to pass through a material medium is lost, and the cost is considerably increased.

Figure 2.15 shows the 300-metre long Wimbleball dam in the West Country and Figs 2.16 show the components of a zone-plate alignment system installed on the crest for monitoring any structural movement (Wallis, 1984). The laser beam passes through a 1 mm aperture to define its position reproducibly, and the zone plate is mounted on a staff with a plunger which closely fits a series of sockets grouted into the structure.

For alignment in one transverse direction, it is also possible to use a grating with a series of straight and parallel bars and spaces whose widths are described by equation (2.35); such a device forms a line focus rather than a point focus, being the equivalent of a cylindrical lens. Gratings of this type with two perpendicular sets of

Fig. 2.15 Wimbleball dam.

Increasing the precision 31

Fig. 2.16 Monitoring system on Wimbleball dam: (*a*) laser

bars superimposed have been used for particle accelerator alignment (Herrmansfeldt *et al.* 1968).

2.2.3 Variable-focus zone plate

Considering again equation (2.16), substitution of $z_2 = l - z_1$ yields

$$f = z_1(1 - z_1/l). \tag{2.36}$$

32 *Laser beam geometry and its applications*

Fig. 2.16 Monitoring system on Wimbleball dam: (*b*) zone plate

This shows that, for a path of length l, the focal length f required for the lens or zone plate is a quadratic function of its position along the path. At the centre, substitution of $z_1 = l/2$ shows that f has a maximum value which is one quarter of the overall length l; near the ends, f is nearly equal to the distance of the lens z_1 from the source or z_2 from the screen. Although there is a considerable range in the central region where a single lens or zone plate will suffice, it is generally necessary

Fig. 2.16 Monitoring system on Wimbleball dam: (*c*) screen.

to have a set of focal lengths available if measurements are to be made at a series of positions.

To meet the need for a choice of focal lengths in a flexible way, a zone plate of variable focal length has been devised (Burch and Williams, 1977). It is formed as a moiré pattern between a pair of identical barrel-shaped grids (the basic principles of moiré effects are discussed in Chapter 9). The boundaries of the grid lines are described by the cubic equation.

$$\phi(x, y) = ax[(x^2 + 3y^2) + b] = m_g, \qquad (2.37)$$

each boundary being the locus of the point (x, y) for a particular integer value of m_g. The constant a is a scaling factor, while b determines the number of lines in the grid without affecting the form of the moiré pattern. If the grids are superimposed and then displaced in the x direction by equal and opposite amounts e, the resultant moiré fringe loci are given by the quadratic equation

$$\begin{aligned}\Delta\phi &= \phi(x+e, y) - \phi(x-e, y) \\ &= 2ae[3r^2 + e^2 + b] = m_z\end{aligned} \qquad (2.38)$$

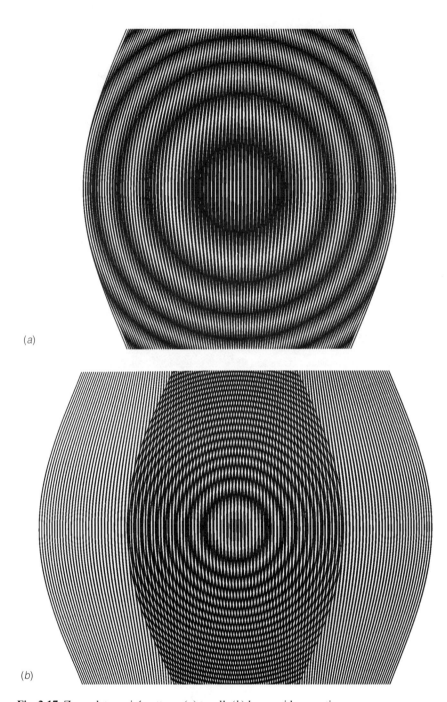

Fig. 2.17 Zone plate moiré pattern: (*a*) small, (*b*) large grid separation.

for integer values of m_z, where $r^2 = x^2 + y^2$. Comparison with equation (2.35) shows that the moiré pattern is a zone plate of focal length

$$f = 1/(6ae\lambda). \tag{2.39}$$

Figure 2.17 shows the patterns produced by two different grid separations.

In this case only one quarter of the combined area is transparent, the energy in the image being $1/\pi^4$ or 1% of that from the lens with equivalent aperture. However, if both grids are etched into glass blanks so that the whole area contributes to the image-forming process, the amplitude is quadrupled so that the relative energy becomes $16/\pi^4$ or 16%, which is a very worthwhile and necessary gain. Not only does the main image become sixteen times brighter, but the spurious energy in the surrounding ring structure is considerably reduced. Figure 2.18 shows a typical image with diffraction structure surrounding it.

Figure 2.19 shows a mount containing a pair of grids of this type; the knob on the side turns a lead screw which has left- and right-handed threads to obtain the opposing motions. Each grid is 20 mm square, with 100 lines as in Figure 2.17, giving a focal length which is variable from 2 to 20 metres. The mount has three feet at top and bottom so that it can be inverted and relocated kinematically to reverse systematic errors in the manner already described. In laboratory tests, an alignment accuracy of 0.1 mm over a range of 80 m has been demonstrated. It is also possible, but less practical, to generate grid pairs which give a similar variable zone plate effect when one of them is rotated about its centre (Bara et al., 1991).

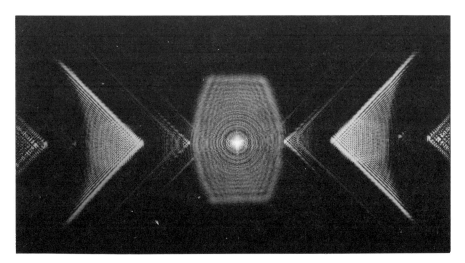

Fig. 2.18 Structure surrounding moiré zone plate image.

Fig. 2.19 Variable-focus zone plate unit.

2.2.4 Variable-focus lens systems

The variable zone plate is intriguing scientifically, and an equivalent lens of variable focal length has also been described; this consists of a pair of glass elements each figured on one side so that the surface height varies over the area in proportion to $\phi(x, y)$. However, a lens unit of variable focal length can be made in a much simpler way, giving a device which is more realistic for practical applications (Williams and Penfold, 1986). If two lenses with equal and opposite powers, one converging and the other diverging, are placed in contact with each other, the combination is virtually equivalent to a parallel-sided glass plate and has no focusing power. But if the two lenses have a finite separation, ξ, the focal length F of the combination is large and positive, being given by

$$F = f^2/\xi, \tag{2.40}$$

where f is the focal length of the individual lenses.

Increasing the precision 37

If, for example, f is 0.5 m and ξ is varied from 5 mm to 50 mm, a range of focal lengths F from 50 m to 5 m is obtained. This is very convenient, because lenses which are relatively easy to manufacture by standard methods can be combined to obtain the more demanding longer focal lengths. Any distortion of the transmitted wavefront must be considerably less than the wavelength of the light if the image is to be diffraction-limited, but the brightness of the image will be nearly equal to that which would be obtained from a single lens with the same aperture.

Across a diameter of a lens, the prismatic angle between the two surfaces varies linearly with position, producing a varying amount of bending of the light. It follows that if one of the lenses is moved laterally rather than longitudinally, the transmitted beam as a whole is given a proportionate angular deflection and the image therefore moves transversely. This effect can be useful for providing angular adjustment in an optical system; if the lenses are almost in contact they simply form the equivalent of a glass wedge with a small and variable apex angle. For a displacement η, the angular deflection is η/f. In an alignment application, any pitching or yawing of the mount containing the separated lenses will produce a relative shear between them,

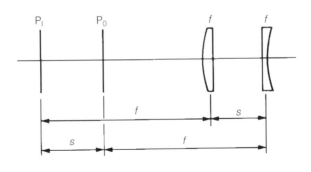

(a)

(b)

Fig. 2.20 Variable-focus lens systems: (*a*) two elements, (*b*) three elements.

giving a spurious transverse movement of the image. This is in contrast to the case of a single thin lens, for which small rotations about any axis through the centre scarcely affect the image.

Referring to Fig. 2.20(a), standard geometrical optical analysis shows that, for light incident on the positive lens, the input and output principal planes P_I and P_O lie at a distance f in front of the positive lens and the negative lens, respectively. The combination is thus nearly equivalent to a single weak lens distant f from it. If the movement of a part of a structure that is locally rigid is to be monitored, immunity to tilt can thus be achieved by mounting the lens unit a distance f from the place that is of interest. One can achieve a small separation with reduced tilt sensitivity by giving the two lenses slightly different focal lengths, but it is convenient to be able to figure the convex and concave surfaces using matching laps. Another possibility would be to use glasses with suitably differing refractive indices.

The tilt problem can also be overcome, at the expense of increased mechanical and optical complexity, by effectively having two of the twin lens units back-to-back. This leads to the arrangement in Fig. 2.20(b), two positive lenses of focal length f having a negative lens of focal length $f/2$ mid-way between them. If s is the separation of adjacent lenses, the focal length of the combination is

$$F = f^2/[2s(1 - s/f)]$$
$$\approx f^2/2s. \qquad (2.41)$$

The principal planes are now symmetrically disposed about and close to the central lens, their distance from it being

$$p = s^2/(f - s). \qquad (2.42)$$

The tilt effect is thus reduced by a factor of approximately $2(s/f)^2$.

2.2.5 Afocal lens systems

The schemes described thus far are primarily intended for making alignment measurements at fixed positions. But they could also be used to monitor the straightness of the motion of a continuously moving carriage if the focus adjustment was geared appropriately to the longitudinal position. However, it would be more straightforward to have an optical device which did not require adjustment, and suitable properties are possessed by afocal systems of unit magnification (Burch and Williams, 1974). An afocal arrangement of lenses has no focal length associated with it, because a parallel beam of light entering the system emerges still collimated. For unit magnification, the input and output beams have the same diameter, although the output beam may be inverted. It can be shown quite generally for such a system that the image formed of an extended object is the same size as the object and a constant distance from it, irrespective of the position of the lens system along the axis relative to the object.

Figure 2.21(a) shows a basic inverting system which is suitable for alignment applications. A pair of similar converging lenses are separated by twice their common focal length f, and a third diverging lens with focal length f_c is placed between them at the common central focal plane. A parallel input beam would come to a focus in this plane, diverge again and emerge from the system parallel but inverted, the central lens having no influence on the afocal condition. It does, however, enable the distance between object and image to be changed. This distance is given by

$$l = 4f + f^2/f_c \tag{2.43}$$

so that if f_c is small and negative, the lens being strong and diverging, the distance l can be made large compared with the length $2f$ of the system. For a laser source, the input and output cone angles of the beam and therefore the size of the diffraction-limited image are constant for all positions of the unit, being determined by the diameter of the central stop.

Since transverse movement of the source point produces an equal and opposite transverse movement of the image, it follows that lateral movement of the unit at any axial position produces a doubled movement of the image, the behaviour being similar to a three-point system with the lens at the centre of the range. It also follows that if the unit pitches or yaws at any axial position, there will be no image movement if the centre of rotation is at the centre of the range. Unlike the case of a single lens, therefore, tilts cause errors which become larger when the unit is nearer either end of the travel.

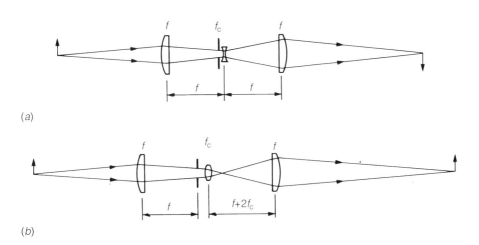

(a)

(b)

Fig. 2.21 Afocal lens systems: (a) displacement sensitive, (b) tilt sensitive.

40   *Laser beam geometry and its applications*

The tilts can be monitored conveniently, enabling corrections to be applied, by means of a second non-inverting afocal unit which can be mounted alongside the main one (Fig. 2.21(*b*)). In this case a parallel input beam is brought to a focus twice on either side of the central lens, which is now converging. The beam thus emerges collimated and not inverted, the separation of the outer lenses being $2f + 4f_c$. One now has

$$l = 4(f + f_c) + f^2/f_c. \qquad (2.44)$$

Lateral movement of the source now gives an identical lateral movement of the image, which is therefore unaffected by transverse displacements of the unit. Simple geometry shows also that pitch or yaw will always give an image movement that is the product of the distance *l* and the angle of tilt. Thus the device is an optical tilt meter with very high sensitivity, and could be used to measure the flatness of surfaces by stepping along in the same way that a mirror is used with an auto-collimator (section 3.4.1).

A pair of afocal units of this type has been used to measure the accuracy of movement of the main 4 m slide of a highly precise three-axis measuring machine. The outer lenses were achromatic doublets with $f \approx 200$ mm, which requires $f_c \approx 10$ mm for the inner lenses. To obtain adequate power, a pair of plano-convex singlets and a glass sphere were used in the inverting and non-inverting cases. Using photoelectric detection, and taking second sets of readings with the units inverted, an accuracy of ≈ 1 μm was achieved. Figure 2.22(*a*) shows straightness results, which revealed a slight hysteresis when the direction of travel was reversed, and Fig. 2.22(*b*) shows the pitch and yaw of the moving member. This order of accuracy, 1 part in 4 million, would scarcely be achieved by any other method.

2.2.6   Bessel beam

The previous sections have mainly been concerned with maximizing the energy in a small image spot. This is clearly necessary for most outdoor applications, where background daylight, atmospheric attenuation and the desirability of a low-power laser for battery operation and safety reasons are the main considerations. However, there are other applications, such as mining and tunnelling, where a weaker central spot may be acceptable if it can be made to propagate over a considerable range without increasing in size.

Another intriguing optical phenomenon in this context, which again overcomes the apparently fundamental size limitation of a Gaussian beam, is the so-called **Bessel beam** (Durnin, 1987; Turunen *et al.*, 1988). As in the earlier consideration of Fourier methods in section 2.1.1, we suppose that a pair of hypothetical collimated beams of infinite lateral extent are inclined at equal and opposite angles α to an optical axis. The corresponding wavefronts will be infinite planes also inclined at +α and −α to the axis. In any plane perpendicular to the axis, the resulting set of straight, parallel, equispaced interference fringes will have an amplitude distribution given by

Increasing the precision 41

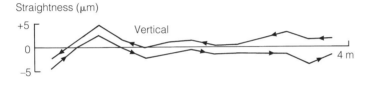

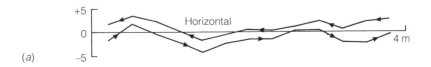

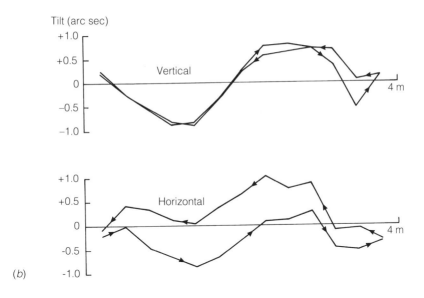

Fig. 2.22 Results obtained with afocal units: (*a*) straightness, (*b*) tilt.

$$A = A_c \cos([2\pi/\lambda]x \sin \alpha), \qquad (2.45)$$

the *x*-coordinate being in the direction perpendicular to the fringes. The spacing between successive fringes, that is between the maxima of A^2, is $\lambda/(2 \sin \alpha)$. It is assumed that the two beams have the same phase on the axis where *x* has its origin, so that the central fringe is bright. The phase over the plane, being the mean of the two component phases, is constant apart from apparently abrupt changes of π for alternate fringes, corresponding to the changes of sign of the amplitude. Such fringes, which can be generated by diffraction from a pair of separated slits, can themselves be used for alignment (van Heel, 1961).

We now consider the summation of an infinite set of these fringe patterns, orientated at all azimuths around the axis. On the axis, the bright fringes will combine to produce a bright spot of about the same width as themselves; elsewhere they will be smeared and the resultant amplitude reduced. In polar coordinates (r, ϕ), we have $x = r\cos\phi$. At any particular point, therefore, the radial distribution of amplitude is given by

$$\begin{aligned} A &= \int_0^{2\pi} [A_c/2\pi]\cos(\Omega\cos\phi)\,d\phi \\ &= A_c J_o(\Omega), \end{aligned} \qquad (2.46)$$

where J_o is the Bessel function of the first kind of order zero and the parameter Ω is given by

$$\Omega = [2\pi/\lambda]r\sin\alpha.$$

The irradiance is then

$$\begin{aligned} I &= I_c J_o^2(\Omega) \\ &\approx I_c \quad \text{if } r \text{ is small} \\ &\approx [2I_c/\pi\Omega]\cos(\Omega - \pi/4) \text{ if } r \text{ is not small.} \end{aligned} \qquad (2.47)$$

This summation is analogous to that occurring in time-averaged holographic interferometry for a sinusoidally vibrating subject, as discussed in section 7.5.1. The distribution consists of a series of nearly equispaced dark and bright rings whose spacing is almost the same as that of the original straight fringes. The irradiance falls off approximately as $1/r$, so that all rings contain nearly the same energy, their circumferences being proportional to r.

Because the component beams in the summation do not change their character along the axis, the Bessel beam itself propagates without any change in the irradiance distribution. An interesting if irrelevant property is that the obliquity of the component beams implies that the speed of propagation is less than the velocity of light, the reduction factor being $\cos\alpha$.

An infinitely wide beam is not realizable in reality, particularly as it would contain infinite energy, and a practical approximation is required. If the composite beam is restricted by a circular aperture, then the component beams will propagate in a cone of directions, eventually separating to form an annulus with a peaked radial irradiance distribution. The composite envelope wavefront formed by the component wavefronts is then conical in form, the apex angle being $\pi - 2\alpha$. If a lens is introduced, the component wavefronts will converge to form a narrow diffraction-limited ring of coherent light in the focal plane.

Conversely, an approximation to the Bessel beam can be produced by means of a ring lens or equivalent holographic element, the ring of light it produces being arranged to be in the focal plane of a collimating lens. Here the generation of the Bessel distribution is analogous to the formation of ring images by the spherical lens for close-range photogrammetry described in section 4.4.2. However, a simpler and more efficient alternative is to pass a collimated beam through a conical lens, sometimes called an **axicon** (McLeod, 1960), in which one of the surfaces is a shallow cone rather than spherical. With a radial distribution over the lens aperture which is fairly uniform rather than peaked, the distribution in the region where the converging conical beam interferes with itself will be a poor approximation to the J_o^2 function, but it will still show a sequence of approximately equispaced rings for a considerable range of positions along the axis.

The zone plate equivalent of the conical lens is a grating in which the zone radii vary linearly rather then quadratically, so that all zones have the same width (Fig. 2.12). This device has been exploited for alignment over a range of distances, and a reflecting version known as the **Rodolite** was produced commercially (Tillen, 1965).

2.3 THE EFFECT OF THE ATMOSPHERE

The accuracy which can be achieved in optical alignment is often limited by the presence of the earth's atmosphere. The air has a significant refractive index n of about 1.000 28, exceeding the vacuum value of unity by about 3 parts in 10 000. This reduces the speed of propagation of the light, and therefore the spacing between successive wavefronts, in proportion to $1/n$. Turbulence in the air produces a small random spatial and temporal variation of the index. There is a well-developed theory of the optical effects of turbulence (Strohbehn, 1978), but this is beyond the scope of this book. It is also applicable primarily to propagation in the outdoor atmosphere well away from the ground. In general terms, if the scale of the turbulence is larger than the beam diameter, the main effect is a wander of the beam direction, as successive elements of the path behave like weak prisms. If the turbulence structure contains scales smaller than the beam diameter, an initially smooth wavefront gradually becomes distorted as it propagates. When the amplitude of the corrugations becomes comparable with the wavelength of the light, further propagation produces random variations of intensity as well as of phase.

A laser image spot formed by an optical system will deviate randomly from its mean position, and may break up into a number of speckles. At any instant in time, the random variations of phase across the system aperture are analogous to those produced by scatter from a rough object and exploited in speckle interferometry (Chapter 8). The size of the image speckles is therefore comparable to the resolution limit imposed by the system aperture, and thus to the size of the undisturbed image itself. If the incoming light overfills the aperture so that it is not all collected, there will be fluctuations of the total energy in the image plane, which can be quite large.

44 Laser beam geometry and its applications

The frequency spectrum of the fluctuations typically extends to a few kilohertz, the limit increasing with the crosswind speed.

2.3.1 Bending of a light beam

Superimposed on the variation due to turbulence, and even in the absence of significant turbulence, there will be a mean value of the refractive index whose changes in space and time are large scale and gradual. While the turbulence produces random uncertainties in a measurement, the quasistatic variations produce a systematic error (Johnston, 1991). A clear distinction must be made here between the effects on distance measurement and on alignment measurement. In distance measurement, as in laser interferometry (Chapter 5), the relevant quantity is the number of waves in the length of the path. For a coordinate z measured along the physical path, this is proportional to the optical path length given by $\int n \, dz$, or to the average refractive index along the path. In alignment and geometrical measurements, it is the direction of travel of the beam which is relevant, and this is only affected if there are refractive index changes across – rather than along – the beam. Different parts of a wavefront then travel different distances, so that the wavefront changes its direction and the propagating beam is bent. The magnitude of this effect can be readily analysed (Harrison et al., 1972).

We consider a small segment of a ray path with length s, as shown in Fig. 2.23, the refractive index along it having a uniform value n. A neighbouring ray path separated by dy from the first one has length $s + ds$ between the same initial and final wavefront positions, the refractive index now being $n + dn$. The optical paths along these two rays must be the same, so that

$$(n + dn)(s + ds) = ns,$$

giving

$$n\,ds + s\,dn = 0. \tag{2.48}$$

Since the rays must be at right angles to the wavefronts, they are bent into an arc of a circle of radius Y. It is then clear from similar triangles that

$$\frac{1}{Y} = \frac{1}{s}\frac{ds}{dy}. \tag{2.49}$$

Because Y will be very large, it is more convenient to consider the beam curvature $1/Y$, which can be interpreted as the change in direction ds/dy per unit distance of travel s. Since n is not very different from unity, it follows to a sufficient approximation from equations (2.48) and (2.49), ignoring a negative sign, that

$$\frac{1}{Y} = \frac{dn}{dy}. \tag{2.50}$$

It should be noted that dn/dy is the index gradient transverse to the path, which may be at any azimuth. For a gradient in a general direction, the horizontal and vertical components transverse to the path of travel of the beam may be equated to the

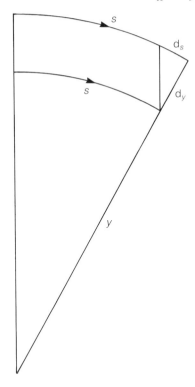

Fig. 2.23 Bending of a beam by the atmosphere.

horizontal and vertical components of curvature. For a beam which is inclined to the horizontal, the 'vertical' component will be equally inclined to the vertical.

2.3.2 Effect of temperature and pressure gradients

We now have to consider the dependence of the refractive index on the temperature, pressure and composition of the air. More detail is given in section 5.1.3 and in Edlén (1966), but with a sufficient accuracy for the present purpose it may be stated that the difference from unity $(n - 1)$, for a particular wavelength of light, is simply proportional to the density of the air. In other words this quantity, which is known as the **refractivity**, is proportional to the number of molecules in unit volume but is independent of their velocity.

If the atmospheric density is ρ, the pressure is p and the absolute temperature is T, the gas laws give, again to fully adequate accuracy,

$$n - 1 \propto \rho = \frac{p}{RT}, \tag{2.51}$$

where R is the specific gas constant for unit mass of gas. Differentiating logarithmically, one has

$$\frac{\mathrm{d}n}{n-1} = \frac{\mathrm{d}\rho}{\rho} = \frac{\mathrm{d}p}{p} - \frac{\mathrm{d}T}{T}$$

and therefore

$$\frac{\partial n}{\partial y} = (n-1)\left(\frac{1}{p}\frac{\partial p}{\partial y} - \frac{1}{T}\frac{\partial T}{\partial y}\right). \tag{2.52}$$

For the temperature term, a typical value of the absolute temperature might be 290 K (17 °C), so that

$$\frac{\partial n}{\partial y} \approx -\frac{\partial T}{\partial y} \times 10^{-6} \tag{2.53}$$

where T is measured in °C. Thus not only is the path curvature approximately equal to the index gradient, but also its value in microradians per metre is approximately equal to the transverse temperature gradient in °C per metre.

For the pressure term, which only applies in the vertical direction, we consider the equilibrium of a thin horizontal lamina of air of thickness dy. For this to be in hydrostatic equilibrium, the pressure difference dp between the upper and lower sides must be equal to the weight per unit area, so that

$$\mathrm{d}p = g\rho\partial y, \tag{2.54}$$

where g is the acceleration due to gravity. Therefore

$$\frac{\partial n}{\partial y} \approx \frac{g}{R} \times 10^{-6}. \tag{2.55}$$

The value of this quantity is 32 μrad/km, which is about one fifth of the earth's curvature, its mean radius being 6370 km. This result, and also equation (2.53) for temperature, apply at low altitudes.

In situations where the earth's curvature can be neglected, the pressure gradient effect can certainly be neglected. The other factor which can possibly be significant is a large gradient of the partial water vapour pressure, but usually only the temperature effect need be considered. This manifests itself in extreme cases as the mirage phenomenon, which shows that a sight line should be as far away from material surfaces as is practical, and that the effects of solar radiation should be minimized by appropriate choice of sight line, weather conditions and time of day.

To obtain the actual displacement of a beam due to bending, suppose that it is projected from the origin along the z-axis, and suffers small transverse displacements $y(z)$. The curvature can then be interpreted as the rate of change of slope $\partial y/\partial z$, so that

$$\frac{\partial}{\partial z}\left(\frac{\partial y}{\partial z}\right) = \frac{1}{Y} \approx \frac{\partial T}{\partial y} \times 10^{-6}. \tag{2.56}$$

This has to be integrated twice; for a uniform transverse temperature gradient along a path of length l, one has

$$y = \frac{l^2}{2} \frac{\partial T}{\partial y} \times 10^{-6}. \tag{2.57}$$

Over 100 m, for example, a gradient of $0.2\,°\text{C m}^{-1}$ produces a displacement of 1 mm. It is difficult to predict orders of magnitude, but at a distance of 1 metre from a surface, a gradient of $0.1\,°\text{C m}^{-1}$ might be encountered typically.

2.3.3 Two-colour methods

One can attempt to measure the temperature gradients by means of arrays of sensors such as thermistors, but such measurements have doubtful reliability and are not necessarily representative of the whole path. If the sight line can be contained within a tube, there are several ways in which the gradients can be reduced. The ideal but expensive solution is to evacuate the tube, while another possibility is to fill it with helium, which has about one-tenth of the refractivity of air for the same temperature and pressure. To reduce the gradients in the air itself, the tube can be slowly rotated or the air can be gently stirred or replenished through a series of distributed jets.

It would be more satisfactory if a direct optical measurement of the bending could be made. This is in principle possible (Williams and Kahmen, 1984) by exploiting the fact that air has dispersive properties in the visible region of the spectrum which are rather similar to those of glass, blue light being bent a little more than red light. Elaborating equation (2.51) slightly, one has

$$n - 1 \propto \phi(\lambda)\,\rho, \tag{2.58}$$

the refractivity being the product of density ρ and a separate factor ϕ which is dependent only on wavelength. Its values for red and blue light have a difference $\Delta\phi$ which is a little less than 2% of ϕ.

Figure 2.24 shows typical paths taken by initially coincident red and blue rays from a two-colour source. Since, from equations (2.50) and (2.52), the curvature is proportional to $(n - 1)$, it follows that to a good approximation the paths have the same basic shape, differing only in transverse scale. The ratio of their separation to their departure from the unperturbed line is always $\Delta\phi/\phi$, so that measurement of the separation enables the departure to be determined. The relationship applies to all beam displacements and directions, and equally to the images formed at the focus of a telescope.

The smallness of the dispersion implies that, for red and blue beams, the separation of their centres must be measured with an uncertainty nearly 100 times smaller than that required for the unperturbed position. This makes extreme demands on the performance of the equipment, which must be carefully designed to minimize systematic errors. In attempting to obtain high accuracy, early methods for geodetic

48 *Laser beam geometry and its applications*

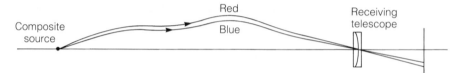

Fig. 2.24 Paths travelled by beams of two different colours.

purposes used interference fringes formed by pairs of widely separated apertures to determine the difference in direction of two large beams, but such instrumentation was crude and cumbersome.

A recent investigation (Huiser and Gächter, 1989) has used a single position-sensing detector to receive both beams, which are modulated temporally in a way which allows their signals to be separated. This is much better than using two discrete detectors, since it minimizes any possibility of systematic error in the measured separation. The second beam is obtained by second-harmonic generation from the first, which ensures that both beams are coincident initially. The authors have shown both theoretically and experimentally that the noise induced by turbulence in the difference between the two beam positions decreases more rapidly than the noise on an actual individual position as the integrating time is increased.

An elegant possibility (Williams and Kahmen, 1984) is to construct a dual-beam sensor so that coincidence between the two colours indicates automatically the position that the composite beam would have occupied in the absence of refraction. Figure 2.25 shows the principle. Red and blue beams which are slightly separated and not quite parallel due to refraction pass through a dispersive lens which bends them by different amounts before they reach the detector. Settings are made by moving the lens transversely until the beams coincide on the detector; in the absence of atmospheric refraction it would then be centred on the composite beam. If refraction is now introduced, the lens can be designed so that the separated beams are still brought to coincidence, as shown in Fig. 2.25, making the setting independent

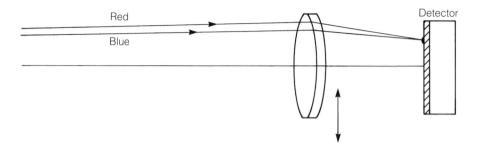

Fig. 2.25 Lens acting as refraction compensator.

of the refraction. Taken in combination with the work mentioned above, this is an attractive possibility for the future.

2.4 DETECTION OF BEAM POSITION

There is a considerable choice of techniques for determining the position of a beam or image, and the method adopted will depend on a number of factors. These include the size and nature of the image and the structure surrounding it, the influence of turbulence, the range of movement and the accuracy required, and the minimum and maximum time scales over which results are needed. Most of the possibilities are rather obvious ones, but some might be overlooked if they were not explicitly catalogued.

In some situations, visual observation on a translucent screen bearing a rectilinear or circular pattern of grid lines is quite adequate, and careful repeated observations can yield surprisingly high accuracy even in the presence of turbulence. For the highest accuracy it is preferable to employ a null method, in which the screen is moved transversely by a graduated control until a particular grid line is centred on the image. An adjustment of this type can be seen in Fig. 2.16(c). Movements in two orthogonal directions can of course be provided if required. Background daylight can be reduced by coloured filter material placed close to the screen so that it cannot distort the image significantly. However, this reduces the colour difference which can be helpful in distinguishing the image light. In large-scale applications, the size of the image is such that auxiliary magnifying optics are unnecessary, so measurements tend to be more objective than when sighting through a telescope.

2.4.1 Photoelectric detection methods

Even in circumstances where visual observations would be possible, the cost of making suitable staff available may dictate that photoelectric detection and automatic recording must be used. Many of the photoelectric methods, particularly the simpler ones, only give varying signals for movements of the image which are small compared with its size. To accommodate larger movements it is again necessary to adopt a null technique, the detector being moved by a mechanical stage or stages. The stages can be driven by servo motors receiving amplified signals from the detector. This introduces additional complexity, but errors due to variations in electronic gain or source intensity become much less significant, being divided by the loop gain of the servos.

Perhaps the simplest photoelectric method is to allow the image spot to fall on a circular aperture of comparable size and pass through to a detector. The detector signal then has a maximum value when the spot is centred. This arrangement is virtually insensitive to small departures of the spot from the centred condition and gives a very coarse accuracy. It can, however, be useful as an alarm device to trigger when movement of a structure exceeds a millimetre or more, particularly as several apertures can be positioned along a single beam. If only the transverse horizontal direction is significant, vertical slits rather than holes can be used.

Much better sensitivity to the null condition can be obtained if it is taken to be on the flank of the image profile, in the region where the rate of change of transmitted energy with aperture position is greatest. The null will then correspond to a defined signal level, which will also be dependent on the brightness of the beam. Furthermore, the null indicated by a hole will not be unique, but could correspond to any position on a circle surrounding the image centre. A much better arrangement is to have two separated apertures, each with its own detector, operating on opposite flanks of the image. Subtraction of the two signals gives a resultant which passes through zero when the apertures are symmetrically positioned about the image centre. The setting is also independent of the source intensity and of the spot position in the orthogonal direction. It requires the two detectors to give similar responses, but this can be checked by interchanging them.

Instead of an aperture, it is possible to use a simple straight edge separating opaque and transmitting regions. This will give best sensitivity if it lies centrally across the image for the null condition. To obtain a balanced pair of single edges, the whole of the light in the image spot can be used by allowing it to fall on the apex edge of a reflecting Porro prism (see section 2.5.2), whose faces direct the light to the two detectors. One can of course arrange for two pairs of detectors to operate in orthogonal directions. Although the null is independent of beam intensity for a balanced pair of detectors, the error signal is still proportional to it; this can be overcome by monitoring the total intensity separately and performing an electronic division. Instead of a stationary edge, one can obtain one or more moving edges by introducing a rotating chopper disc with one or more blades. The beam position can then be obtained from timing electronics referenced to the chopper position. If the chopper has many blades, so that the spaces between them become comparable in width with the beam, one obtains an a.c. signal whose phase is the measure of beam position.

2.4.2 Phase-sensitive systems

Two separated apertures can be used with a single detector if the signals which they produce are suitably multiplexed. A common equivalent procedure is to cause the hole or slit to execute a periodic sinusoidal motion between two separated positions. The aperture can be formed in a thin shim which is mounted on a pair of parallel flexible strips and driven at the resonant frequency by a small electromagnetic coil; this type of arrangement has been used in commercial autocollimators, as described in section 3.4.2. Alternatively, the light spot rather than the aperture can be moved by reflecting the beam from a small mirror executing a rotational oscillatory movement. Commercial units similar in principle to galvanometers are available for this purpose.

If the aperture is off-centre, it produces a periodic alternating signal whose amplitude is proportional to the intensity gradient in the region being scanned. On the opposite side of the centre the sign of the gradient changes, so that the peaks of the signal become troughs and the troughs become peaks. The first row in Fig. 2.26 shows typical waveforms. With the aperture centred, the a.c. signal at the scan frequency vanishes. However, because the aperture passes over the peak twice during one cycle of movement, there is a residual signal at twice the scan frequency.

If the a.c. component of the signal were simply rectified and smoothed to obtain the amplitude, it would have a small minimum value at the null position. This would be similar to but sharper than the maximum obtained with a stationary aperture, and it would be much more satisfactory to generate a signal which passed through zero, as in the case of two apertures and detectors. This can be achieved by the technique of **phase-sensitive rectification**. Its popularity is indicated by the many other names given to it, such as **synchronous demodulation** and **homodyne detection**.

In phase-sensitive rectification, a square-wave reference derived from the sinusoidal wave driving the aperture is used to reverse the polarity of the detector signal twice in each cycle. This means that on one side of the null positive-negative becomes positive-positive, while on going to the other side of the null negative-positive becomes negative-negative, as shown in the second row of Fig. 2.26. The full-wave rectified version of the resultant signal then has to be smoothed using a time constant considerably larger than the periodic time of the oscillation.

The a.c. signal will be accompanied by a d.c. level corresponding to the mean power of the light. To show the effect of this, the mean level of the original signal in Fig. 2.26 has been given a deliberate offset from the line representing the mean d.c.

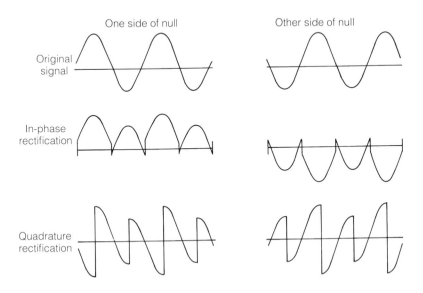

Fig. 2.26 Phase-sensitive detector waveforms.

level. It can be seen that this d.c. bias would be converted to an a.c. signal of large amplitude at the switching frequency, and it should therefore be rejected by high-pass prefiltering.

Elaborate commercial units are available for phase-sensitive rectification, generally designed to extract weak signals which are coherent with a reference in the presence of large amounts of noise. In the present case, however, a circuit giving adequate performance can be constructed in a very simple way. The signal is fed to both inverting and non-inverting inputs of an instrumentation amplifier chip, via field-effect transistors with antiphase reference signals applied to their gates. The switching process is equivalent to multiplication of the signal by the reference, and can also be performed by an analogue multiplier chip, using a square wave or sine wave as reference.

The phase of the reference signal must be correct relative to that of the detector signal, so that switching occurs as the signal passes through its mean value. If the switching occurs when the signal has its extreme values, then it swings from fully positive to fully negative or *vice versa* for successive samples, so that the smoothed result is zero, as shown in the third row of Fig. 2.26. Correct phasing can be assured by deriving the reference from a sensor directly attached to the scanner.

The dependence on phase can be exploited in a convenient way to obtain two-dimensional sensing. The aperture or the light spot is made to perform a circular motion. The aperture can be mounted eccentrically in a rotating component; the light beam can be reflected from two mirrors oscillating in phase quadrature about orthogonal axes, or from a simple mirror mounted not quite perpendicularly on a motor shaft. Decentring then produces a signal whose phase depends on the azimuth of the eccentricity. Error signals for two perpendicular directions can therefore be derived by using two phase-sensitive detectors with their references in phase quadrature. Each of these will extract the appropriate in-phase component and reject the quadrature component.

Dyson and Noble (1964) have described a related scheme in which both source and detector have cruciform masks with vertical and horizontal slits. The image of the source cross is made to gyrate in a circle relative to the detector one, so that four pulses are produced for each revolution when the horizontal or vertical slits coincide. The spacings in time of the pulses are then a measure of the image position.

2.4.3 Bore straightness measurements

The circular motion principle has been applied in a prototype system for measuring the straightness of rifle barrels. For moving the light beam, the system also exploited the properties of a parallel-sided glass plate, whose function is complementary to that of the glass wedge mentioned in section 2.2.4. Whereas the wedge produces bending of a beam without lateral displacement, the plate gives displacement without bending.

Detection of beam position

Suppose that a plate in a light beam is tilted by an angle i, so that the angle of incidence on the first face is also i, as shown in Figure 2.27(a). Then the angle of refraction r between the beam direction inside the glass and the surface normal is given according to Snell's law by

$$n \sin r = \sin i, \tag{2.59}$$

where n is the refractive index of the glass. If the angles are small, we have

$$nr \approx i.$$

At the second face, the angle of incidence becomes r, the angle of refraction on emergence becomes i, and the beam travels parallel to itself but laterally displaced by an amount d. If the plate has thickness t, it is easy to see that

$$d \approx (1 - 1/n) \, it \tag{2.60}$$
$$\approx it/3$$

for glass of refractive index $n \approx 1.5$. With a displacement of one third of the thickness times the angle of tilt, the glass plate is a very convenient device for obtaining small transverse displacements. The tilt-sensitive afocal system described in section 2.2.5 behaves like a very thick tilting plate. A similar analysis for the optical wedge (Fig. 2.27(b)) shows that the angular deviation D is

$$D \approx (n-1) A$$
$$\approx A/2. \tag{2.61}$$

In the rifle barrel system, two such parallel-sided plates were placed in the beam from a 2 mW helium-neon laser in a cylindrical housing. The beam was then projected

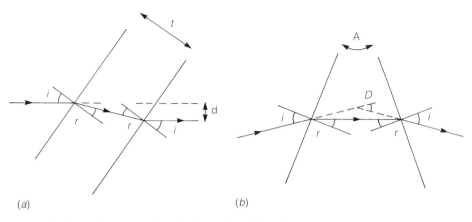

Fig. 2.27 Beam-directing optics: (a) tilting plate, (b) wedge prism.

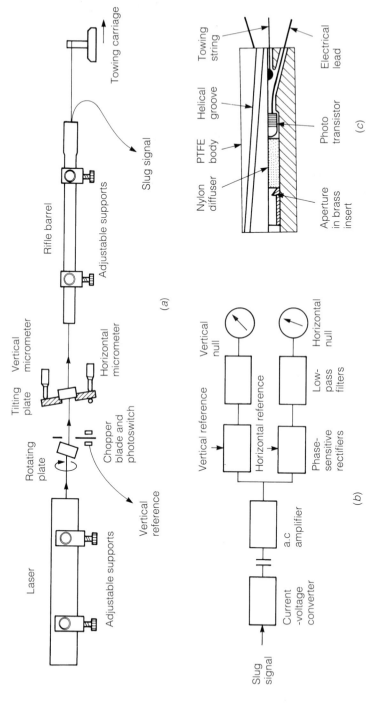

Fig. 2.28 Measuring system for rifle barrels: (*a*) schematic diagram, (*b*) electronic layout, (*c*) detector slug.

down the centre of the barrel to act as a straightness reference (Fig. 2.28(a)). The first plate was held with adjustable tilt in a rotating mount driven by a small motor, the axis of rotation coinciding with the axis of the incident beam. This caused the emergent beam to sweep out a cylinder whose radius was made comparable with the beam radius. The second plate enabled null settings of the beam position to be made, being tilted in two orthogonal directions by micrometers bearing on lever arms.

A Teflon slug (Fig. 2.28(c)) was machined carefully to be a close fit in the barrel, with four helical grooves cut to match the rifling. The slug could be pulled through the barrel by a cord, slowly rotating with the rifling. It contained a centred aperture with a diameter of 1 mm, comparable with the size of the laser beam. Behind the aperture was a small phototransistor, and between them was a piece of nylon acting as a multiply scattering diffuser. This ensured that light from a particular part of the aperture reached all parts of the detector surface in a variety of directions, so that the light was thoroughly scrambled. Without this precaution the phototransistor signals were badly distorted, due to the highly asymmetric geometry of the phototransistor surface. A singly scattering diffuser in the form of a small piece of ground glass placed immediately behind the hole was found to be inadequate, still giving pronouncedly irregular signals.

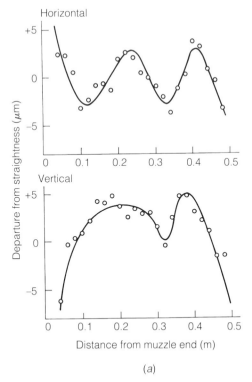

Fig. 2.29 Rifle barrel results; (a) barrel straightness.

56 Laser beam geometry and its applications

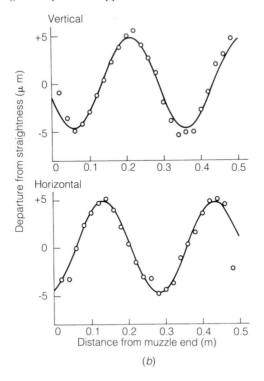

Fig. 2.29 Rifle barrel results; (b) instrumental straightness.

The signal from the phototransistor was fed to two phase-sensitive rectifiers operating in phase quadrature (Fig. 2.28(b)). Their references were generated by a semi-circular blade attached to the rotating mount which passed through two photoswitches positioned 90° apart.

The principal source of error was imperfect centring of the aperture in the slug. On this account, the effective aperture centre executed a helical motion as the slug rotated with the rifling, and the errors produced were the projections of the helix in the horizontal and vertical directions. These were sinusoids with amplitudes equal to the centring error and wavelengths equal to the pitch of the rifling, having a quadrature phase relationship for the horizontal and vertical directions.

The errors were eliminated by means of the reversal principle explained in section 2.2.1, the barrel being traversed twice and inverted between the two traverses. Figure 2.29(a) shows the straightness errors given by half the sum of the two sets of readings, while Fig. 2.29(b) shows the eccentricity errors given by half the difference. The departures of the latter results from the sinusoidal curves drawn through them demonstrate that the accuracy of the system was about 1 μm, corresponding to one thousandth of the diameter of the laser beam. This provides some confirmation that a real imperfect laser beam can provide a reference line which is remarkably straight.

2.4.4 Quadrant detectors

The arrangements described thus far can in principle employ any type of photo-detector. If the ultimate in signal-to-noise ratio is required in combination with a large frequency bandwidth, then photomultipliers can be used. However, the solid-state photodiode is usually the most convenient form of detector, and specialized devices of this type are available for the detection of image position. The commonest of these is the quadrant photodetector, in which the sensitive area is divided in a cruciform fashion into four adjoining sectors. There are four separate leads for each sector and a common ground lead.

To detect the two orthogonal components of image movement, the quadrant cell can be used in two distinct ways. The simplest is to position the dividing lines at 45° to the movement components (Fig. 2.30(a)). One can then take the two differences between opposite pairs of quadrants, obtaining null signals when the image is centred as when using pairs of apertures and discrete detectors. The other possibility is to position the dividing lines along the movement directions. For each direction, it is then necessary to sum the outputs from two pairs of adjacent quadrants before taking the difference. Figure 2.30(b) shows the circuit layout for the vertical direction. The cell becomes equivalent to two separate bipartite ones, and the arrangement has the advantage that quite large movements in one direction scarcely affect the signal for the other direction. Suitable electronic units to accompany quadrant cells mounted on cross-slides are available commercially.

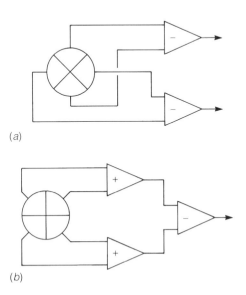

Fig. 2.30 Quadrant photocell configurations: (a) axes bisecting measurement directions, (b) axes aligned with measurement directions.

The quadrant cell is a type of detector which only gives a changing signal for image movements from the null which are small compared with the image size. The exact behaviour of the signal has been calculated for a Gaussian image profile (Bennett and Gates, 1970). To accommodate larger movements, a possibility is again to make the image centre execute a circular motion, which allows any movement from the null which is less than the radius of the circle to be detected. One can either take the time-averaged signals from the quadrants and use them in the usual way, or measure the amount of time spent in each quadrant by starting and stopping a timing circuit when the corresponding signal exceeds or ceases to exceed that from the adjacent quadrant.

2.4.5 Axial position detection

If a beam is converging fairly strongly, it is possible to use a quadrant cell to locate the longitudinal position of the focus along the optical axis as well as its lateral position. The technique has been applied to video disc and compact disc players, and also to the images formed by concave reflectors in a prototype system defining points in space for the calibration of volumetric measuring machines. A cylindrical lens is placed at a short distance from the focus, so that the converging spherical wavefront is converted to an astigmatic ellipsoidal wavefront with different curvatures in two orthogonal directions. This produces two orthogonal line foci F_1 and F_2 with a longitudinal separation, as shown in Fig. 2.31. Between them the light patch has an elliptical shape with varying eccentricity, becoming circular at the mid-position. Figure 2.31 shows this falling on the quadrant cell at three different positions.

The quadrant cell is orientated with its dividing lines at 45° to the directions of the line foci, which are also the directions of the ellipse axes. In this case the signals from opposite quadrants are added, and the two resultant sums are subtracted. As the cell is moved along the optical axis, the final signal passes through zero when the light patch becomes circular and the major and minor ellipse axes interchange. The transverse movements of the light patch can be read out at the same time by also combining the same cell signals in one of the ways described above.

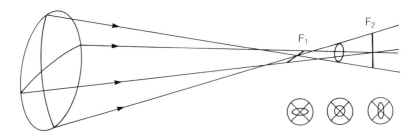

Fig. 2.31 Longitudinal focus detection using astigmatic optics.

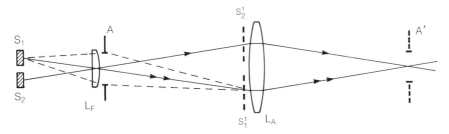

Fig. 2.32 Longitudinal focus detection using two sources.

By tracing rays from the edge of the converging wavefront to the edge of the circular image, it can be seen that the latter is a laterally inverted geometrical image of the former. It follows that non-uniformity of the irradiance over the area of the converging wavefront will change the relative values of the quadrant signals and alter the position of the null. In experiments with the system for volumetric machines, it was found that the longitudinal reading was particularly sensitive to this effect. An alternative was therefore sought which would be immune to changes in the reflectance distribution of the surface reflecting the spherical wavefront to the image detector.

A solution, developed as a form of proximity gauge (Williams, 1978), is illustrated in Fig. 2.32. It is arranged that the light patch A′ on the detector is the image formed by the lens L_A of a small aperture A. The aperture is transilluminated in two different directions by two diffuse sources S_1 and S_2 placed side-by-side, so that the light from the sources forms opposite halves of the beam converging to the image. The diagram also shows an example of a very useful device, namely a **field lens**. The lens L_F forms images S_1' and S_2' of the sources which coincide with the clear aperture of the lens L_A. This ensures that all of the light flux which passes through the aperture A also passes through the lens L_A; without the field lens L_F there would be a considerable light loss. The procedure also tends to reduce the distortions produced by the lens aberrations.

The sources are energized alternately; another equivalent procedure uses a single source and a vibrating slit placed in front of it. When the aperture image is in focus on the detector surface, both sources illuminate the same image area. When the image is out of focus, however, the different directions of the beams from the two sources cause them to illuminate different areas with a transverse separation, which reverses on the opposite side of the focus position. It follows that the difference signal from a bipartite photocell, or a quadrant cell connected as such, will give an a.c. signal at the source alternation frequency whose amplitude is proportional to the distance from focus. This can be phase-sensitively rectified to give a null signal at the focus in the manner already described.

If the light from the sources is diffused thoroughly before passing through the source aperture in this arrangement, light from a particular place on the imaging optic is spread over the whole illuminated area of the detector. Thus obstruction of part of the optic does not alter the position of the longitudinal null setting, although it does reduce the sensitivity. Instead of employing two sources and one detector, it is also possible to use one source and two detectors (Hanke, 1990), the two halves of the beam converging towards the image being directed by mirrors to two separate bipartite or quadrant detectors. The difference signals from the two detectors are then subtracted to obtain the focus signal. The source does not have to be modulated in this case.

To null the image position transversely as well as longitudinally in the dual-detector case, one can simply add the difference signals instead of subtracting them. In the dual-source case, the mean d.c. signal levels from the detector could be used. To obtain an equivalent phase-sensitive a.c. system, the two sources could be switched in phase with each other instead of in antiphase. It might appear that it is not possible to switch the sources in phase and in antiphase simultaneously to combine the lateral and longitudinal functions, but this can in effect be done by switching them in phase quadrature.

Suppose that the sources are modulated sinusoidally in phase quadrature, their intensities as a function of time t being given by

$$I_1 = I_M + I_A \sin(\omega t + \pi/4) = I_M + I_A (\cos \omega t + \sin \omega t)/\sqrt{2},$$
$$I_2 = I_M + I_A \cos(\omega t + \pi/4) = I_M + I_A (\cos \omega t - \sin \omega t)/\sqrt{2}. \qquad (2.62)$$

The two detector signals which measure the longitudinal focal position z and the lateral component of position x will be of the form

$$S_1 = (\alpha + \beta x)(I_1 + I_2) + \gamma z (I_1 - I_2),$$
$$S_2 = (\alpha - \beta x)(I_1 + I_2) - \gamma z (I_1 - I_2), \qquad (2.63)$$

where α, β, γ are constants of proportionality, so that the difference is

$$S = S_1 - S_2$$
$$= 4 I_M \beta x + \sqrt{2} I_A (\beta x \cos \omega t + \gamma z \sin \omega t). \qquad (2.64)$$

Thus phase-sensitive rectifiers with references in quadrature can be used to separate the x and z components.

2.4.6 Photopotentiometers

For the measurement of image movements which are larger than the image diameter, a nulling or scanning system is often an inconvenient complication, and a detector which is able to respond directly to relatively large movements is highly desirable. Such detectors are available which are purely analogue in operation.

The analogue type of detector is sometimes termed a **photopotentiometer**. For a two-dimensional device, a large-area silicon photodetector has an additional resistive layer incorporated at the back of the semiconductor slice. Pick-off electrodes are provided on four sides of the layer, the front electrode being common to all four of the others. When a spot of light generates charge carriers at a particular place, the currents flowing from the four pick-off electrodes are inversely proportional to the corresponding resistive paths through the layer. Thus the difference in the currents from two opposite electrodes is a measure of the position of the light spot in the direction joining them.

The difference current is also proportional to the light intensity, and may be affected slightly by the currents to the other electrode pair, but these effects can readily be compensated for by dividing the difference of the currents by their sum using an analogue divider module. The relationship between difference current and spot position is usually non-linear, particularly for large excursions from the centre, but the nonlinearity can be reduced by suitable design of the electrode geometry. An improved performance has been obtained by using two resistive layers at the front and rear of the silicon slice, one for x-movement and the other for y-movement. The relationship between difference current and position can also be irregular due to inhomogeneities in the semiconductor material, but this effect has been reduced by improvements in processing technology.

Considerably better linearity is usually obtained from a single-axis device made with a long and narrow strip of silicon, and single-axis devices can be obtained to accommodate larger movements than the dual-axis ones. If movements in the orthogonal direction cause the image to move off the detector surface, these can be accommodated by means of a long cylindrical lens placed about one focal length from the surface.

This type of detector basically measures the position of the centroid of the energy distribution over its whole sensitive area. The devices therefore perform better with small images. In the case of a zone plate image, where a significant proportion of the energy is contained in the ring structure surrounding the main image, the apparent position could be considerably in error, particularly when the image is off-centre so that the contributing ring structure is cut off asymmetrically at the detector edges. In this case it is much better to use the detector only as a null sensor.

2.4.7 Intensity-modulation methods

Because the detectors have relatively large areas, unwanted background light may be particularly significant. For laser light against a daylight background, a combination of dyed glass filters can give a fivefold improvement and a narrow-band interference filter a fiftyfold improvement. However, essentially complete isolation of the wanted signal can be achieved by intensity modulation of the light at the source. A simple method would be to apply an a.c. variation to the laser drive current, but a good depth of modulation cannot easily be obtained in this way with gas lasers, although it can with diode lasers. Acousto-optic or electro-optic modulators can be used, but they are comparatively expensive.

The most practical method is often a mechanical chopper in the form of a rotating radial grating with alternating bars and spaces. This was introduced in section 2.4.1 as a means of actual beam-position measurement, but here it is used just as an intensity modulator. For a reasonable speed of rotation, say 3000 rev min^{-1}, frequencies of a few kilohertz can be achieved if the laser beam is focused to a small diameter and the grating lines are correspondingly fine. Such frequencies are above the dominant noise frequencies caused by atmospheric turbulence, and are readily handled by basic electronics.

Phase-sensitive rectification can be applied to the difference between a.c. currents from opposite electrodes of a quadrant cell or photopotentiometer in the same way as for a single detector with a scanned aperture (New, 1974). The reference signal can be derived from the chopper, possibly by means of a separate small source and detector, but for a simple system this requires a wire link between the main source and detector units. For a photopotentiometer, the sum of the currents from all the electrodes remains virtually constant for any image position, so that this sum can be used as the reference with no link to the source.

Rather than using the noisy signal itself, it is preferable to use a square wave of the same frequency derived from a separate oscillator. The phase of this square wave must be kept in step with that of the sum current, this normally being achieved by means of a **phase-locked loop**. The technique requires a square wave in phase quadrature with the reference one, and a simple way to derive both accurately is to use an oscillator operating at four times the frequency required. A particular point in the oscillator cycle is then used to generate all the required square wave edges by means of divide-by-two circuits. If the sum current is passed to a phase-sensitive rectifier with the quadrature square wave as reference, the output will be zero if the phase relationship is correct. If the phase of the current begins to lead or lag, the output will become positive or negative, so that it can be used to control the frequency of the oscillator and maintain the correct condition.

It is possible to use several modulated sources simultaneously if the frequencies are different and incoherent. A system of this kind using a number of light-emitting diodes has been used to monitor human body movements in medical investigations and for time and motion studies.

As well as the specialized devices which have been described, one also has the option of using general-purpose image sensors such as vidicons and photodiode arrays. Despite the increased complexity of the processing electronics, its ready availability makes this a good option in many cases. Accurate calculation of the image centres is discussed in section 4.5.3. Advantages include the ability to locate multiple images and, for diode arrays, good geometrical accuracy and linearity. A particular advantage is the ability to set a signal threshold level and thereby discriminate against background light and structure surrounding the main image. A useful device would be the optical equivalent, namely a non-linear medium which only transmitted light above a certain intensity, but such a material is not available at present. The main disadvantage is the inability to follow fast image movements, the frequency response often being limited to values below the standard video frame rate of 50 Hz.

It is also possible to arrange that a change in position of a beam passing through a suitable wedge prism alters its polarization state in a way which can be translated into an intensity change (King and Raine, 1981).

2.4.8 Measurement of large movements

In some civil engineering applications of laser monitoring systems, such as the sway of suspension bridges, the image movements may be much larger than the range of available detectors, and too large to be reduced by conventional telescopic optics. A large diffusing screen viewed by a camera would give an unacceptable light loss. In this situation, reducing optics can be constructed using moulded acrylic Fresnel lenses. These consist of a series of concentric annuli with triangular profiles which act like segments of an ordinary lens. However, the performance of such an element is inevitably relatively poor, so that it forms a rather blurred image with a pronounced radial flare.

Fresnel lenses can nevertheless be used to obtain accurate imaging by using them only as field lenses in planes where the image is brought to a focus. Figure 2.33 shows a two-stage reduction system, which was used to transfer an image area 40 cm across to a detector D 12 mm across. The lenses L_1 and L_2 form the reduced images, the Fresnel lenses Z_1 and Z_2 ensuring that most of the light passes through their apertures. Light scattered in any direction from S is imaged to S_1 provided only that it passes through the aperture of the transfer lens L_1. The second Fresnel lens is followed by a ground glass G; once the Fresnel lens has directed the light towards the centre of the aperture of the transfer lens L_2, the ground glass diffuses the light into a cone of directions sufficient to fill the aperture fairly uniformly.

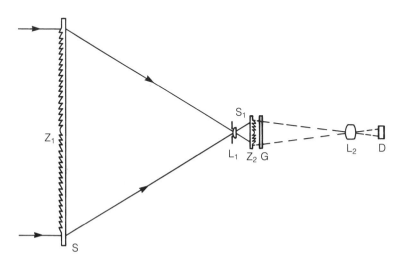

Fig. 2.33 Fresnel lenses used as field lenses.

This principle is used in instrumentation designed specifically for laser guidance of tunnelling machines (Doyle, 1978). The guiding beam falls on a large Fresnel lens with a diffusing surface immediately behind it which is imaged on to the detection system. To avoid the complexity of a two-dimensional detector array, two linear arrays are positioned horizontally and vertically in the same transverse plane, a little away from the central axis of the system. The Fresnel field lens with its screen is imaged on to them by cylindrical transfer lenses with their generator axes perpendicular to those of the detectors. The sampling of the resultant lines of light by the detectors gives inefficient collection, but ensures that the vertical detector receives a similar signal for all horizontal positions.

This system is further able to sense the direction of the laser beam as well as its position. The diffuser incorporates a pattern of closely spaced holes so that parts of the laser beam pass through to fall on a further diffusing screen. This lies in the focal plane of the Fresnel lens near the position detectors, which are sufficiently far off axis not to block it. Being in the focal plane, the patch of light formed indicates direction rather than position, and it is re-imaged for directional detection in the same way as for position. Figure 2.34 shows one of nine such target units in use during the construction of the Channel Tunnel (Johnston, 1991).

Fig. 2.34 Laser guidance system in use during construction of the Channel Tunnel. (Courtesy of Zed Instruments.)

2.5 PROBLEM SOLVING WITH OPTICAL DEVICES

Laser-based techniques are being used increasingly in industry for applications such as dimensional gauging, and many of the most effective techniques exploit the properties of lenses and mirrors in ways which are in principle very simple. This final section reviews briefly some existing developments and describes some further relevant optical devices and their applications, aiming to show the principles on which new developments to solve particular problems can be based.

2.5.1 Scanning and triangulation techniques

If a rotating mirror is placed at the focus of a lens and a laser beam is directed on to the mirror surface, the beam can be scanned across the lens aperture. After passing through the lens, the beam will remain parallel to itself as it scans. It can then be used to gauge the diameter of a component by measuring the time interval over which the light is blocked. This measurement will be independent of the precise position of the component, and a high repetition rate can be achieved by using a multifaceted polygon instead of a single mirror. It is convenient to place a second lens beyond the component, so that the scanning beam is always directed to a particular point in the focal plane where a detector can be placed. Such laser scanners are used for gauging wire, cable extrusions, glass tubing, rolled steel billets, etc. (Hanke, 1990).

Similar types of polygon scanner, without the collimating lens, are used for in-process gauging of float glass sheet (Slabodsky, 1990). By comparing momentary reductions in the intensity of the transmitted scanning beam with corresponding increases in the scattered light picked up by other detectors, various types of defect can be separately identified and quantified.

A very common type of sensor exploiting beam geometry and position-sensitive detection to measure the position of a surface uses the simple principle of triangulation. The light beam is projected on to the surface in a particular direction, and the spot of light formed is viewed from a different direction by a camera in which it is imaged on to the detector. Any movement of the surface out of its own plane then alters the position of the image. If the incident beam is normal to the surface, out-of-plane movement does not alter the position at which it is sampled; if the incident beam direction and camera axis are disposed symmetrically about the surface normal, one can take advantage of any specular component of reflection.

Several commercial heads based on this principle are available for in-process dimensional gauging, offering a wide choice of range and corresponding accuracy. Small units are now being used on three-axis measuring machines, replacing the well-established mechanical touch probes (Fig. 2.35). The great advantage of a non-contacting probe is that the null position can be sensed while the machine is in motion, so that it does not have to be halted at each measured point. On a larger scale, the triangulation principle is being used to monitor tunnel profiles. The laser and detector units are mounted on the centreline of the tunnel and rotated to scan

Fig. 2.35 Laser triangulation probe for coordinate measuring machines. (Courtesy of Renishaw.)

the circumference and obtain data in real time (Clarke, 1990; Clarke and Lindsay, 1990).

By introducing a scanning mirror or polygon, component geometry along a line can be inspected by triangulation, the deviations from straightness in the image of the projected line revealing the surface shape. One such system incorporates a conical lens, as described in section 2.2.6, so that a narrow line is obtained for a large range of depths (Häusler and Heckel, 1988). Rather than a scanned spot, a continuous line can be projected by means of cylindrical optics (Cheng *et al.*, 1991). By scanning a laser beam in raster fashion over a complete surface using orthogonal rotating mirrors, rapid whole-field gauging has been demonstrated; the system uses a photopotentiometer supplemented by a fast photodiode for spot-intensity compensation. The principle is equivalent to that used in projection moiré (section 9.1.2), which offers whole-field capability with parallel rather than serial data acquisition.

The use of two or more theodolites for measurement of spatial coordinates on large engineering structures is outlined in section 4.5.2. One of the theodolites can be fitted with a laser eyepiece so that the beam is projected to the sighted point, while the other theodolite incorporates a diode array viewing the image. The technique then becomes a generalized form of the laser triangulation principle.

2.5.2 Two-mirror and related devices

Several optical devices which are simple basically but also very useful in metrology if applied in sound ways have already been introduced in previous sections. These include the parallel-sided plate and the optical wedge for lateral and angular displacement of a beam (section 2.4.3), and the double lens for obtaining a variable wedge (section 2.2.4). Another valuable group of devices uses two mirror surfaces which form a single unit and are arranged so that a ray of light is reflected from each surface in turn, with the input, output and intermediate paths in a common plane. The

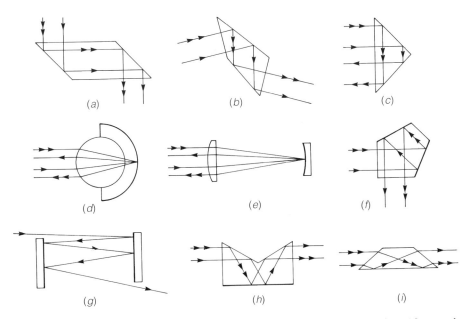

Fig. 2.36 Optical devices: (*a*) periscope prism, (*b*) deflector prism, (*c*) Porro prism, (*d*) capped sphere retroreflector, (*e*) cat's eye retroreflector, (*f*) pentagonal prism, (*g*) nearly parallel mirrors, (*h*) reflecting beam inverter, (*i*) Dove prism beam inverter.

simplest of these is the periscope, for which the mirror surfaces lie in parallel planes, so that input and output rays are parallel to each other but laterally displaced (Fig. 2.36(a)).

It is a general property of this family of devices that rotation of the input ray and therefore of the device about an axis perpendicular to the plane containing the rays produces no change in the angle between input and output rays. If the input ray rotates clockwise, its reflection from the first mirror will rotate anticlockwise by the same amount, and this change of sense is again reversed to clockwise by reflection at the second mirror. The output ray direction is also unaffected by any small lateral movement of the mirror combination, and in the special case of the periscope by rotation about any other axis. This enables a particular beam deviation to be maintained without precise maintenance of the mechanical location of the unit.

The device can be constructed in hollow form using a pair of surface-coated mirrors in a suitably rigid mounting. However, if the size required is not prohibitively large it is usual to employ a solid block of glass. This requires adequately homogeneous material and the faces must be polished adequately flat, but maintenance of the defined angle is virtually guaranteed. Figure 2.36 includes some particularly useful prisms in solid form. In all cases the input and output faces are nominally normal to the light beam, so that they cause no further deviation. More importantly, the small prismatic deviations which will occur when the block is rotated through small angles will cancel each other, producing no change in the total deviation. It is usually desirable to introduce a deliberate small angular offset, so that the secondary beams which are partially reflected from the input and output faces are slightly deviated with respect to the main beams. The accuracy of figure required for the transmitting faces is only about one-sixth of that required for the reflecting faces with refractive index $n \approx 1.5$; this is because an error in surface height e produces an optical path change of $2ne$ at an internally reflecting surface compared with $(n - 1)e$ at a transmitting face.

In general, the angle between the input and output rays will be twice the angle between the reflecting surfaces. A prism giving a deviation of about 24° (Fig. 2.36(b)) was used to adapt the three-point alignment system for use on an arched dam (Harrison, 1978). If the places to be measured lie on a circular arc between the ends of the line, the prism when placed adjacent to the zone plate will always redirect the light from the source to the detector. A glass wedge can be used for the same purpose. If two wedges are placed in series in hollow bearings so that they can be rotated independently about the beam axis, the angle of bending and its azimuth can be made adjustable. Units of this kind are available commercially, and were used in the Channel Tunnel project to carry the survey lines round curves.

For an angle of 90° between the surfaces, the output ray is reversed in direction relative to the input ray. This gives rise to the Porro prism (Fig. 2.36(c)), which has the property that an incident beam is retroreflected back in its own direction, for any orientation of the prism. This only applies when the beam is perpendicular to the roof edge; in the azimuth parallel to the roof edge the prism behaves like an ordinary mirror. A family of applications for which a commercially packaged Porro prism is available is described in section 2.5.6.

To achieve retroreflection for all directions of incidence, reflection from a third surface at right angles to the other two is necessary. This gives the cube corner reflector, which is particularly valuable for laser interferometry (section 5.1.4 and Fig. 5.2). The cube corner in solid form can be visualized as being a corner cut from a solid glass cube; the plane of cut, which is perpendicular to the line joining the corner to the centre of the cube, forms the entrance and exit face of the device. For a hollow cube corner, the effective plane from which a beam is retroreflected always passes through the apex, where the three surfaces intersect. For a solid cube corner, the equivalent point is the apparent position of the apex as seen through the glass front surface, which is distant t/n from the surface, where t is the distance to the real apex and n is the refractive index of the glass.

The range of directions over which a cube corner can retroreflect is limited, and a useful alternative is the capped sphere (Fig. 2.36(d)). The capped sphere can be regarded as a variant of the cat's-eye retroreflector (Fig. 2.36(e)), in which a mirror is placed in the focal plane of a lens. Then, recalling the principle introduced in section 2.1.3, a beam will always be returned parallel to itself. To return oblique beams correctly, the mirror should be spherical with its centre of curvature at the lens. In the capped sphere, the front surface acts as a lens with its focus coinciding with the back surface, so that a beam is always returned parallel to itself. For a material with a refractive index of 2, a plain sphere could be used, but for realistic glasses with index near 1.5 the focus is further back and a cap is necessary. The device is able to return a beam with little distortion over nearly a full hemisphere of directions.

In a technique which is similar in principle to the use of two theodolites, two or more laser beams can each be scanned in space by a pair of tilting mirrors which operate in azimuth and elevation. Each beam is directed at a retroreflector, and the returned beam is again reflected by the mirrors and then directed to a quadrant photocell. As the retroreflector moves in space, it can then be tracked automatically. This scheme is appropriate for checking the movements of a robot arm, and possibly also the motion of the probe of a three-coordinate measuring machine. If the distance to the retroreflector is also measured by interferometry using the same laser beam, only one scanning unit is necessary. Figure 2.37 shows a system of this kind which is available commercially.

Collections of small spheres or cube corners are to be found in retroreflecting paint or paper, as used for traffic signs. This can be very useful in some circumstances; the light from an optical image formed on the surface will be scattered through a small range of angles and can be reimaged quite accurately and efficiently by a lens. The automatic inspection process for sheet glass production mentioned in the previous section has been improved by introducing such a retroreflecting screen. After the scanning laser beam has passed through the glass, the screen reflects the light back in the reverse direction; it can then be detected efficiently at the polygon scanner for all positions of the laser beam across the sheet (Claridge, 1991).

70 *Laser beam geometry and its applications*

Fig. 2.37 Laser beam total survey station. (Courtesy of Leica.)

An angle of 45° between the two mirror surfaces gives a right-angle between the input and output beams; this is the basis of the pentagonal prism or **optical square** (Fig. 2.36(*f*)). In this case the angle of incidence on the reflecting faces is 22.5°, which is less than the critical angle for total internal reflection (substitution of $n = 1.5$ and $i = 90°$ in equation (2.59) yields $r = 41°$ for the critical angle). These faces therefore have to be given a metallic coating to obtain a high reflectance. As we have seen, rotation of the prism about an axis perpendicular to input and output rays has no effect on the nominal 90° angle between them. Rotation about the input ray has

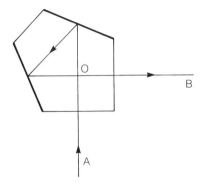

Fig. 2.38 Pentagonal prism misalignment analysis.

no effect on the 90° angle, but causes the output ray to sweep out a plane which is perpendicular to the input ray; it is this property which makes the prism particularly useful. Lateral translation in any direction also has no effect on the deflection angle if the surfaces are adequately flat. Movement in the sixth degree of freedom, namely rotation about the output ray direction, is the only movement which does slightly affect the deviation angle, and this effect can be analysed in a simple way.

Referring to Fig. 2.38, suppose that the input ray AO rotates through an angle θ with the line OB as axis. Relative to the point O, the point A moves out of the paper a distance $OA.\theta$. By the laws of reflection, the output ray OB then rotates through the same angle θ into the paper, becoming further below the paper at B than at O by a distance $OB.\theta$. Now suppose that the input ray and the prism are rotated together through the same angle θ in the opposite direction, so that the input ray is restored to its original position and the prism has been rotated by the angle θ about the original line of OB. During this process the output ray will rotate sideways about its original position OB, the lateral upward movement of the point B relative to the new position of the ray OB at O being $(OB.\theta).\theta = OB.\theta^2$. Thus the angle AOB increases by θ^2, the square of the prism rotation angle in radians. If the prism is misaligned by a milliradian, the resultant increase in the 90° deviation angle will be a microradian. The refraction at the glass-to-air interfaces affects input and output beams equally and therefore does not alter the validity of the argument.

Commercial units incorporating a pentagonal prism to turn the line of sight of an alignment telescope through a right angle in any azimuth have been available for many years, and are discussed more fully in section 3.1.1(a). More recently, laser units with a rotating pentagonal prism to define points in a horizontal or vertical plane have become widely used, particularly in the construction industry; these are described in detail in Price and Uren (1989). Figure 2.39 shows a typical application. A precision system of this type for engineering applications at shorter ranges uses a rotating wedge to trim the prism deviation angle to precisely 90°.

72 *Laser beam geometry and its applications*

Fig. 2.39 Ground levelling by laser. (Courtesy of Nevill Long.)

This concept has been developed further to produce a prototype total station which gives full information at a mobile survey staff about its position relative to the fixed laser transmitter (Gorham, 1988). The transmitter and staff units are shown in Fig. 2.40. Two laser beams are swept round in a horizontal plane by a continuously rotating pentagonal prism unit, which incorporates strong cylindrical lenses to spread each beam into a fan. The fans slope at 45° to the horizontal and at rightangles to each other, so that they form a V at any distance from the projector. They are received by at least two vertically separated detectors on the staff, and the angular positions of the rotating unit for which the beams cross the detectors can be used to deduce the azimuth and distance of the staff from the unit as well as the height.

The final interesting case with two mirror surfaces is when there is an angle of 180° between them so that they face each other. A laser beam can then be reflected many times from one to the other (Fig. 2.36(g)), if it is sufficiently narrow to avoid overlaps. If one of the mirrors tilts through a small angle, the corresponding tilt of the beam is this angle multiplied by the number of reflections, so that a striking

Fig. 2.40 'Laserfix' total survey station. (Courtesy of University of East London.)

amplification can be obtained. The arrangement is also useful for simulating large distances in the laboratory.

Since a system of two mirrors, and therefore any even number of mirrors, only alters the direction of a beam, it follows that any odd number of mirrors will produce a lateral inversion similar to that produced by a single mirror. In particular, three reflecting surfaces can be arranged as in Fig. 2.36(*h*) so that the beam is not deviated, and it can be seen that there is a lateral inversion in the plane of the paper but not

74 Laser beam geometry and its applications

perpendicular to it. A consequence of this is that if the mirror system is rotated through a certain angle about the beam axis, the beam rotates about the axis through twice that angle. Thus rotation through 90° provides a convenient method of inverting a beam to eliminate systematic errors, and since two positions at 180° both give a beam inversion, asymmetry in the system itself can also be eliminated. There are other equivalent devices, in both hollow and solid forms (Swift, 1972); Fig. 2.36(*i*) shows the commonest version, the Dove prism.

2.5.3 Surface profilometer

A laser optical technique will now be described which involved both the pentagonal prism and the periscope during its development, as well as other devices introduced earlier in the chapter (Virdee, 1988). When examining the flatness of optical surfaces, the established method is to use a Fizeau or Twyman–Green interferometer, similar to that shown in Fig. 5.7. Here, wavefronts reflected from the surface under test and from a reference surface are combined with a slight wedge angle between them so that a pattern of nominally straight, parallel and equi-spaced interference fringes is formed. Departures from straightness or periodicity of the fringes then contour the departures from flatness of the specimen, as in Fig. 10.3(*a*).

The interference method has several limitations. It requires a reference surface of similar shape to that being tested, it is not readily amenable to automation and the fringe positions have to be interpolated to obtain results on a regular grid. More significantly, there are some surface shapes for which it cannot easily be adapted.

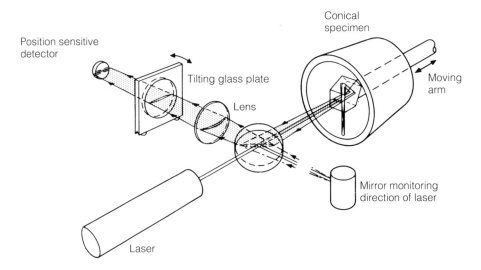

Fig. 2.41 Slope integration profilometer.

Problem solving with optical devices 75

The technique to be described was developed initially for application to X-ray microscope mirrors, the reflecting surface being the interior of a small cylinder with a slightly conical taper, as seen in Fig. 2.41.

The beam from a helium-neon laser is projected into the cone parallel to the surface, and reflected on to it at nominally normal incidence by a small pentagonal prism. The pentagonal prism is attached to an arm which is in turn attached to a linear stepper motor stage, so that the beam can be tracked along the surface. After reflection by the surface, the beam retraces its path through the pentagonal prism and a proportion of it is then deflected by an oblique partially reflecting surface to a telescope lens. As the specimen surface is scanned, changes in its slope along the direction of motion produce proportional changes in the direction of the reflected beam, and hence of the position of the spot formed in the focal plane of the telescope lens. For a cylindrical surface, the spot of light is drawn out into an astigmatic line, but this does not affect the measurement principle.

Here the property of a lens introduced in section 2.1.3 is used in reverse. A ray passing through the optical centre of the lens at a small angle θ to the optical axis is not deviated, so its intersection with the focal plane is distant $\theta.f$ from the axis, where f is the focal length. Any other parallel ray travelling in the same direction will then be directed to the same focal position. This property is exploited in the metrological instruments described in section 3.4.

Since the integral of slope is height, the surface profile along the chosen generator may be obtained by simple summation of slope values measured at a series of equal increments. A quadrant photocell is placed in the telescope focal plane to measure the spot position. For the usual reasons, null settings are made by tilting a parallel-sided glass plate placed in front of the photocell. The plate is tilted by an eccentric motorized cam operating on a lever arm, whose position is monitored by a displacement transducer giving readings proportional to the tilt angle.

The step length must be comparable with the spot size on the surface – if it is longer, so that parts of the surface are missed, the integration may not be accurate; if it is shorter so that successive portions overlap, then small-scale undulations will not be revealed. The spot size on the surface will vary slightly as it is traversed, and the beam geometry should be tailored so that the waist lies at the middle of the range. Figure 2.8 applies in this situation, so that the spot size on the detector remains constant. The beam geometry has no influence on the relationship between its direction and the resultant spot position, provided that the detector surface lies exactly in the focal plane.

It follows from the diffraction theory discussed in section 2.1.1 that if the spot reflected from the surface tilts by a certain fraction of a wavelength over its diameter, then the spot on the detector will move by a comparable fraction of its width. For example an accuracy of six nanometres over the spot diameter and step length requires the spot position on the detector to be measured to about 1% of its width. The initial beam drift mentioned in section 2.1.3 is particularly significant in this case. The stationary plane mirror in Fig. 2.41 reflecting back the second output from

76 *Laser beam geometry and its applications*

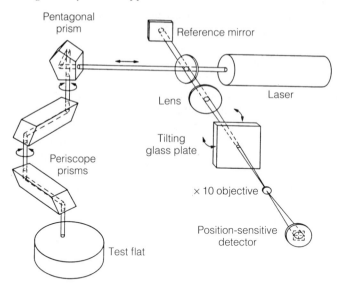

Fig. 2.42 Dual-axis profilometer using double periscope.

the partial reflector allows the laser beam direction to be monitored independently by the quadrant cell.

2.5.4 Profilometry in two dimensions

Having achieved promising results with this equipment, it was decided to extend the principle to two dimensions and make measurements over the whole area of a flat surface rather than along single generators. This again gives a good illustration of right and wrong ways to approach a measurement problem. In the approach used initially (Fig. 2.42), the pentagonal prism was kept in a fixed position and the output beam was scanned over the surface by means of two periscope prisms in series. A single prism attached to a hollow bearing at its input end enables the output beam to take any position around a circle, and a second prism similarly attached to the first enables any position within a circular area to be reached, the pair acting in combination like an articulating arm. The positioning can be achieved in Cartesian coordinates by moving the output port with a pair of orthogonal slides. In this case the image movement was magnified by a microscope objective.

Ideally the output beam should maintain a constant direction parallel to the input beam, and two orthogonal slope measurements can be obtained at each position. The x-direction slopes yield a set of x-direction profiles whose relative slopes are known but whose relative heights are unknown. Similarly, the y-direction slopes yield y-direction profiles with a known slope relation but an unknown height relation. It should then be possible to fit these two sets of orthogonal profiles together and obtain

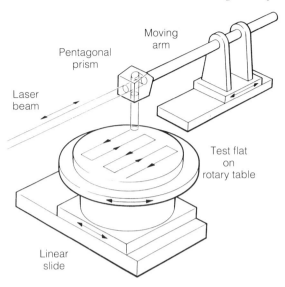

Fig. 2.43 Dual-axis profilometer using orthogonal scans.

an over-determined solution for the surface shape. In practice, it was found that the overall misfit between the two sets of profiles greatly exceeded the localized random errors. For the system to work with the same degree of accuracy as its predecessor, the laser beam must always take precisely the same path through each prism, and it is not possible in practice to adjust and maintain the positions and, more particularly, the directions of the bearing axes with the requisite precision.

A much better performance has subsequently been obtained by extending the profilometer principle in its original form with the moving pentagonal prism (Fig. 2.43). A series of parallel linear scans are made in raster fashion across the surface, incrementing its position between each of the scans in a direction at right angles to them. The specimen is then rotated through a right angle and the process is repeated. This again yields separate sets of x- and y-profiles, but in this case the slopes as well as the heights of the individual scans are unrelated. On combining the two sets of profiles a satisfactory fit is obtained, the residual differences between the x-scan and the y-scan height values at each point typically being two or three nanometres.

With the profile slopes unrelated, the surface shape is still determined uniquely, apart from any twist component for which the height is proportional to the coordinate product xy. However, the twist can readily be obtained by also taking measurements along one or preferably both diagonals of the grid. The method is then entirely analogous to the well-established **Union Jack** procedure for determining the flatness of surface tables from autocollimator slope measurements which is described in section 3.4.3. A further advantage of this approach compared with the use of periscopes is that with successive scans having unrelated slopes the system has only to remain stable for the duration of each individual scan.

78 *Laser beam geometry and its applications*

2.5.5 Profilometry refinements

Accuracies approaching a nanometre have been achieved with this equipment, provided that certain precautions are observed carefully. The outgoing and returning beams must be aligned precisely with the direction of travel of the pentagonal prism, so that the part of the prism which the light traverses is always the same. This ensures that imperfections of homogeneity or surface figure of the prism do not produce spurious changes in the beam direction. The working area must be shielded from temperature gradients and allowed to stabilize before measurements are taken, since any spurious tilting or bending of the laser beam would be interpreted incorrectly as a change in surface slope.

The initial beam-drift problem has been overcome very effectively by passing the beam through an optical fibre and placing the output end of the fibre in the focal plane of a collimating lens. Beam movements at the input end of the fibre then only affect the energy transmitted and not the effective point of divergence of the output beam.

As well as stability of the laser beam direction, the method requires that the specimen shall not tilt relative to the rest of the equipment while a scan is taking place; such tilts could again be misinterpreted. Both of these error sources can be eliminated substantially by directing a second beam derived from the same laser to a fixed position on the surface via a second pentagonal prism. Subtraction of the two position signals produced by the beams reflected through the moving and fixed prisms then enables any common-mode drift to be eliminated. Another approach is to

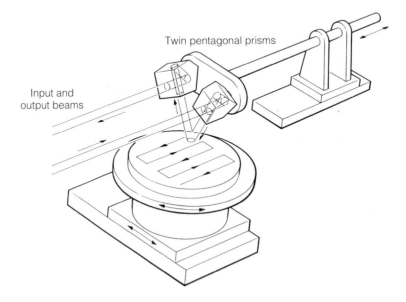

Fig. 2.44 Oblique incidence profilometer using two pentagonal prisms.

measure the difference between the directions of two adjacent laser beams, both scanned over the surface, and integrate twice rather than once (Kiyono *et al.*, 1991).

Further errors arise if the surface of a transparent parallel-sided specimen is being tested, particularly if the second surface is also polished. Part of the light reflected from this surface will find its way to the detector and produce a spurious signal. The reflected energy can be reduced considerably by contacting an absorbing surface with similar refractive index using an index-matching fluid; however, the matching of the refractive indices is bound to be imperfect, and the residual reflected energy may still be sufficient to cause significant errors when the highest accuracy is sought.

This problem can be eliminated completely by means of a modified approach. If the pentagonal prism is turned through a few degrees about the laser beam axis, as in Fig. 2.44, the beam is directly obliquely onto the surface. The reflected beam is then separated from the incident one, so that it can be picked up by a second prism placed beside the first one. The reflection from the back surface is then physically separated from the main one. The use of parallel but separated paths for the outgoing and returning beams also means that the partially reflecting beam divider, which can also give troublesome second-surface reflections, may be replaced by a fully reflecting mirror. The obliquity of the sensing beam does not reduce the angular sensitivity of the system; its only effect is to make the light patch on the specimen surface slightly elliptical. This is a literal example of lateral thinking!

A further alternative is to arrange for the obliquity to be in line with the laser beam rather than perpendicular to it; this can be achieved using two pentagonal prisms modified to give a deviation of rather more than 90° and directing the output beam in line with the input beam to the opposite side of the specimen.

2.5.6 Roller parallelism measurements

The double periscope principle has been applied more successfully to a common industrial problem, namely the alignment of the axles of large rollers in a paper processing machine (Penfold and Williams, 1983). If a plane mirror equipped with tilt adjustments is attached to the end of a roller, its plane can readily be set perpendicular to the axis of rotation by observing the direction of a reflected light beam. A circular motion will be described as the roller rotates whose radius reduces to zero when the mirror orientation is correct.

One could use a laser beam, but this condition can be observed conveniently by means of a commercial autocollimator such as that described in section 3.4.1. Essentially the light from a collimator is reflected back into it so that a pair of back-illuminated crosswires in the focal plane are imaged back into the same focal plane. Their position is then a measure of the direction of the reflected beam, as in the laser profilometry. In the example being considered, the critical rollers were arranged approximately on an arc of a circle, so it was convenient to arrange a pair of large articulating periscopes between the autocollimator and the roller ends. The position of the reflected crosswires, viewing through the periscopes, should then be the same

80 *Laser beam geometry and its applications*

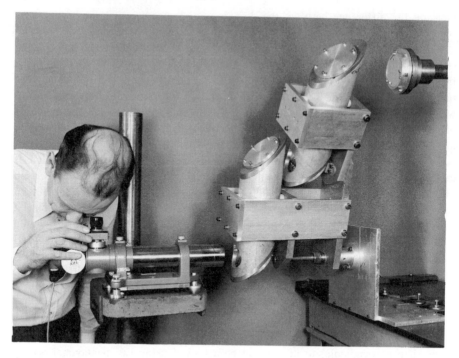

Fig. 2.45 Double periscope with autocollimator for roller alignment.

for each roller after adjustment of the mirror attached to it. Figure 2.45 shows the system being tested with a dummy mirror.

In this case the perfection of the bearings is not critical, because the beam of light is much larger and its transverse movements are small compared with its size. The periscope mirrors were set parallel by observing the transmitted fringe patterns when each periscope was placed in the measurement path of a large field interferometer. If such an instrument is not available, there are other possible procedures. One of these uses the same autocollimator and tiltable mirror, with the addition of a pentagonal prism. The beam from the autocollimator is turned horizontally through a right angle by the prism and then reflected back on itself by the mirror. If the periscope to be tested is placed in front of the mirror, the prism can be moved along the autocollimator beam until the beam is again reflected back on itself via the periscope. It is necessary to ensure that the pentagonal prism has the same orientation in both positions, which can easily be achieved by rigidly attaching to it a pair of orthogonal precision spirit levels.

There is also a more general method available for checking the parallelism of rollers, which is not confined to a particular arrangement of the rollers. This makes use of an ordinary theodolite, its telescope being fitted with a graticule illuminator so

Problem solving with optical devices 81

Fig. 2.46 Autocollimation prism for roller alignment. (Courtesy of Leica.)

that it can function as an autocollimator. Most machines have an access corridor down at least one side, allowing the theodolite to be mounted on a tripod with its telescope viewing the end of each roller in turn. Since the telescope direction is referenced to gravity when set to the horizontal on the vertical circle, it is a simple matter to check the parallelism of each roller in the vertical direction, either by viewing a suitable target placed at intervals along it or by autocollimating to a mirror mounted on the end as before.

82 *Laser beam geometry and its applications*

For the parallelism in the horizontal direction, use is made of a Porro prism in a mounting matching that of the theodolite. This accessory is available commercially and is known as an **autocollimation prism** (Fig. 2.46). The prism is set up in the corridor on a second tripod placed at one end of the machine, and the mount is rotated about its vertical standing axis until the prism roof edge is nominally parallel to the direction of the roller axes.

A reflection can now be obtained from the prism with the theodolite at any position along the corridor and at any height; it may need to be moved across the corridor until it is in line with the prism. Once set up at the end of a roller, the telescope is turned about the horizontal and vertical axes until it is pointing at the prism and the crosswires are in autocollimation. The line of sight now lies in a vertical plane perpendicular to the roof edge. The telescope is then set to the horizontal, and rotated about the standing axis through an exact right angle. This makes the line of sight parallel to the roof edge, so that the roller axis can also be set parallel to the roof edge. An equivalent but less flexible procedure is described in section 3.2.2, for which the prism is replaced by three plumb lines and the theodolite by an alignment telescope and pentagonal prism.

2.5.7 Intensity-based measurements

This concluding section gives some further examples of how a little ingenuity can enable quite simple optical principles to be used in highly effective ways. A common requirement is to compare the diameters at particular places of a series of nominally similar components, or to monitor the changes in a diameter. For this purpose, the beam-scanning system described in section 2.5.1 can be employed. Alternatively, using parallel beams of light projected past the ends of the diameter, the edges can be imaged on to one or a pair of linear diode arrays, and this kind of technique is now commonly used.

However, one can also consider simply measuring the change of total intensity of the beams with position of the edges, the sum of the intensities being a measure of the diameter which is independent of the absolute position of the component. The

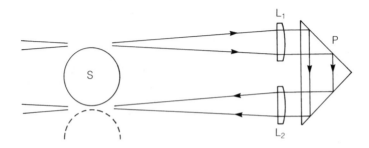

Fig. 2.47 Intensity-based diameter measurement.

fractional changes of intensity can be enhanced considerably by introducing a second fixed edge in close proximity to each component edge, so that a narrow slit of light is formed. The auxiliary masks need not be physically close to the component, but can be in the conjugate image plane of a transfer lens.

In suitable circumstances, an elegant development from this concept is effectively to sum the signals from the two ends by optical rather than electronic means. A possible arrangement is shown in Fig. 2.47. Light passing one end of the diameter is imaged on to the other end, so that the two edges of the specimen S form the two sides of the slit. This requires that the image should be of unit magnification and not inverted, which is achieved conveniently with two similar lenses L_1 and L_2 at one focal length from the object in combination with a Porro prism P. The separation of the lens centres is set to be slightly greater than the length of the diameter, and can be used to control the width of the slit so that it is always finite for the possible variation in size.

Overall movements or misalignments of the specimen along the length of the diameter being measured change the position of the slit formed but not its width. To obtain immunity to other specimen movements, the image must also be erect in the direction along the length of the slit, parallel to the roof edge, and this can be achieved using a cube corner instead, which still enables the incoming and outgoing beams to be separated. It may be advantageous to use laser light for this type of application, because ordinary incoherent light will contain wavefronts propagating in a finite cone of directions which will cause blurring in the shadow of an edge of a deep object.

A special variant of this type of technique is illustrated in Fig. 2.48. The requirement was to determine accurately the centring error of a small lens L_2, that is to determine whether the surfaces were exactly parallel at the centre of the circle

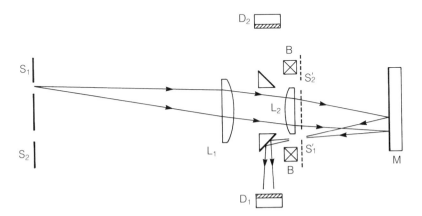

Fig. 2.48 Intensity-based lens centration measurement.

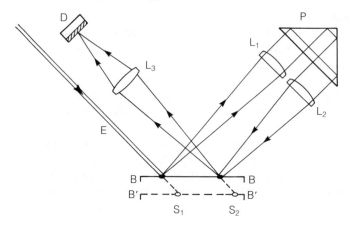

Fig. 2.49 Surface tilt measurement.

defined by the lens perimeter. The lens is mounted in a ball bearing BB, with clear spaces to allow light to pass its edge. In a similar manner to the previous case, light from a slit S_1 is collimated by another lens L_1 and then passed through the specimen lens L_2 which forms an image S_1'. By means of a mirror M placed at one half of its focal length from L_2, the image S_1' is reflected back to be formed in the plane of L_2. The slit is offset laterally to position the image so that one of the edges is redefined by the lens edge. The light passing the edge is reflected by the small prism to the detector D_1.

Rotation of the lens L_2 about its optical centre will give no movement of the image, but there will be a periodic variation in the lens edge position if there is any eccentricity, causing a periodic change in the amount of light passing the edge. The useful feature of the arrangement is that rotation about any other axis will have the same effect. This is because, with a parallel input beam, sideways movement of the lens produces an equal sideways movement of the image S_1', so that the amount of light passing the edge is unaffected. The bearing BB in which the lens is mounted can therefore have a much lower order of precision than that required from the measurement. It is convenient to add a second diametrically opposite slit S_2 and subtract the signals from the detectors D_1 and D_2, which gives the advantages discussed in section 2.4.1.

As a final example, shown in Fig. 2.49, the requirement was to measure the squareness of the side face of a small ball bearing BB to the axis of rotation. A spot of light S_1 was formed at one end of a diameter by an oblique laser beam E. The scattered light was imaged to the other end of the diameter by again using a Porro prism P together with two lenses L_1 and L_2. The secondary image S_2 was then imaged on to a position-sensing detector D by a lens L_3. The position of the final image is then unaffected by in-plane movement or, as indicated in the diagram, out-of-plane translation of the bearing surface to $B'B'$. Tilt along the diameter being examined is the only component of movement which will affect the image position; the position will thus vary sinusoidally as the bearing is rotated if the face is not square.

This introductory chapter has given only a sample of the many geometrical applications for laser beams, but it is hoped that the main principles have been presented in a way which may stimulate further developments.

REFERENCES

Arnaud, J. (1985) Representation of Gaussian beams by complex rays. *Appl. Opt.*, **24**(4), 538–43.

Bara, S., Jaroszewicz, Z., Kolodziejczyk, A. and Moreno, V. (1991) Determination of basic grids for subtractive moiré patterns. *Appl. Opt.*, **30**(10), 1258–62.

Bennett, S. J. and Gates, J. W. C. (1970) The design of detector arrays for laser alignment systems. *J. Sci. Instrum.*, **3**, 65–8.

Bloom, A. L. (1968) *Gas Lasers*. Wiley, New York.

Burch, J. M. and Williams, D. C. (1974) Afocal lens systems for straightness measurement. *Opt. Laser Technol.*, **6**(4), 166–8.

Burch, J. M. and Williams, D. C. (1977) Varifocal moiré zone plates for straightness measurement. *Appl. Opt.*, **16**(9), 2445–50.

Cheng, X., Su, X. and Guo, L. (1991) Automated measurement method for 360° profilometry of 3-D diffuse objects. *Appl. Opt.*, **30**(10), 1274–8.

Chrzanowski, A., Jarzymowski, A. and Kaspar, M. (1976) A comparison of precision alignment methods. *The Canadian Surveyor*, **30**(2), 81–96.

Claridge, J. F. (1991) Automatic glass examination; use of retroreflective laser scanning. *Sensor Review*, **11**(1), 17–21.

Clarke, T. A. (1990) The use of optical triangulation for high speed acquisition of cross sections or profiles of structures. *Photogrammetric Record*, **13**(76), 523–32.

Clarke, T. A. and Lindsay, N. E. (1990) A triangulation based profiler. *Proc. SPIE*, **1395**(5/2), 940–7.

Collins, S. A., Jr (1964) Analysis of optical resonators involving focusing elements. *Appl. Opt.*, **3**(11), 1263–75.

Doyle, K. J. (1978) British Patent No. 1 513 380.

Durnin, J. (1987) Exact solutions for nondiffracting beams. I. The scalar theory. *J. Opt. Soc. Am.* A, **4**(4), 651–4.

Dyson, J. and Noble, P. J. W. (1964) Electrical readout from optical alignment devices. *J. Sci. Instrum.*, **41**, 311–16.

Edlén, B. (1966) The refractive index of air. *Metrologia*, **2**(2), 71–80.

Gates, J. W. C. and Bennett, S. J. (1968) A confocal interferometer for pointing on coherent sources. *J. Sci. Instrum.*, **1**, 1171–4.

Gates, J. W. C. (1975) Three-dimensional location and measurement by coherent optical methods. *Photogrammetric Engineering and Remote Sensing*, **41**(11), 1349–54.

Gerrard, A. and Burch, J. M. (1975) *Introduction to Matrix Methods in Optics*, Wiley, New York.

Gorham, B. J. (1988) Measurement of spatial position using laser beams. *Land & Minerals Surveying*, **6**, 121–6.

Hanke, G. (1990) Applying optical measurement techniques. *Sensor Review*, **10**(1), 30–4.

Harrison, P. W., Tolmon, F. R. and New, B. M. (1972) The laser for long distance alignment – a practical assessment. *Proc. Instn Civ. Engnrs*, **52**, 1–24.

Harrison, P. W. (1973) A laser-based technique for alignment and deflection measurement. *Civ. Engng. Pub. Works Rev.*, **68**, 224–7.

Harrison, P. W. (1978) Measurement of dam deflection by laser. *Water Power and Dam Constr.*, **30**(4), 52.

Häusler, G. and Heckel, W. (1988) Light sectioning with large depth and high resolution. *Appl. Opt.*, **27**(24), 5165–9.

Hecht, E. (1987) *Optics*. Addison-Wesley, Reading, Massachusetts.

Herrmansfeldt, W. B., Lee, M. J., Spranza, J. J. and Trigger, K. R. (1968) Precision alignment using a system of large rectangular Fresnel lenses. *Appl. Opt.*, **7**(6), 995–1005.

Huiser, A. M. J. and Gächter, B. F. (1989) A solution to atmospherically induced problems in very high-accuracy alignment and levelling. *J. Phys. D: Appl. Phys.*, **22**(11), 1630–8.

Johnston, A. (1991) Lateral refraction in tunnels. *Survey Review*, **31**(242), 201–20.

King, R. J. and Raine, K. W. (1981) Polarimetry applied to alignment and angle measurement. *Opt. Engng*, **20**, 39.

Kiyono, S., Asakawa, Y., Odagiri, H. and Kamada, O. (1991) 3-dimensional expression by differential laser autocollimation. *Int. J. Japan Soc. Prec. Eng.*, **25**(2), 148–9.

Kogelnik, H. and Li, T. (1966) Laser beams and resonators. *Appl. Opt.*, **5**(10), 1550–67.

McLeod, J. H. (1960) Axicons and their uses. *J. Opt. Soc. Am.*, **50**, 166.

New, B. M. (1974) Versatile electrooptic alignment system for field applications. *Appl. Opt.*, **13**(4), 937–41.

Penfold, A. B. and Williams, D. C. (May 1983) Periscopic alignment of out-of-sight shafts. *Quality Today*, 21–3.

Price, W. F. and Uren, J. (1989) *Laser Surveying*. Van Nostrand Reinhold.

Slabodsky, F. B. (1990) Automatic inspection of glass. *Sensor Review*, **10**(2), 79–83.

Strohbehn, J. W., (ed.) (1978) *Laser Beam Propagation in the Atmosphere. (Topics in Applied Physics,* Vol. 25), Springer-Verlag, Berlin.

Swift, D. W. (1972) Image rotation devices – a comparative survey. *Opt. Laser Technol.*, **4**(4), 175–88.

Tillen, R. J. (1965) Optical alignment devices and their applications in engineering. *The Quality Engineer*, **29**(5), 2–16.

Turunen, J., Vasara, A. and Friberg, T. (1988) Holographic generation of diffraction-free beams. *Appl. Opt.*, **27**(19), 3959–62.

van Heel, A. C. S. (1961) Modern alignment devices. *Prog. Opt.*, **I**, 289–329.

Virdee, M. S. (1988) Nanometrology of optical flats by laser autocollimation. *Surf. Topogr.*, **1**, 415–25.

Wallis, D. (1984) Laser alignment techniques at the Wimbleball reservoir. *Water Power and Dam Constr.*, **36**(5), 33–34.

Williams, D. C. (1983) Laser alignment techniques. *Phys. Technol.*, **14**(2), 61–7.

Williams, D. C. and Penfold, A. B. (1986) Three-point alignment in non-ideal situations. *Proc. FIG XVIII Int. Congr.*, Toronto, Canada, **5**, 424.

Williams, D. C. and Kahmen, H. (1984) Two wavelength angular refraction measurement, in *Geodetic Refraction* (ed. F. K. Brunner), Ch. B, Springer-Verlag, Berlin.

Williams, T. L. (1978) A scanning gauge for measuring the form of spherical and aspherical surfaces. *Opt. Acta*, **25**(12), 1155–66.

3
Alignment metrology

B. S. PEARN

Alignment measurements which have to be made on workpieces whose size allows them to be placed on a surface table do not in general present serious problems. The reason is that the table itself provides a flat reference **plane** and any **line** across the table provides a straight reference line (Fig. 3.1). 90° lines which are **square** to the plane or line are provided by L-type or box squares or cylindrical squares. Dial test indicators, height gauges, gauge blocks or even feeler gauges are used to make the actual measurements with respect to the plane, line or square.

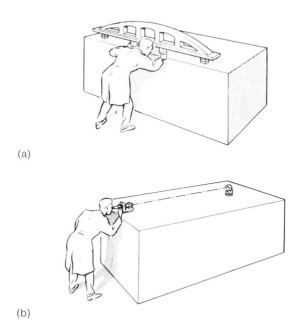

Fig. 3.1 Using (*a*) a straightedge and (*b*) a line of sight on a surface table.

Optical Methods in Engineering Metrology. Edited by D.C. Williams. Published in 1993 by Chapman & Hall, London. ISBN 0 412 39640 8

88 *Alignment metrology*

Workpieces up to about 3 or 4 m can usually be handled in this well proven manner. However, beyond these sizes the traditional methods become impractical, and it becomes necessary to create these same planes, lines and squares actually on to the work. To consider tables, straightedges or squares of sizes greater than 3 or 4 m is impractical. Their weight and size make them very difficult to handle, and their weight also causes distortion of the workpiece which will undoubtedly destroy their accuracy. The employment of optics provides the solution to this situation, which exists in many industries, for example shipbuilding and ship repair, diesel engines, petrochemical compressors, aircraft assembly jigs, paper making and printing, nuclear reactors and nuclear research, to name just a few.

3.1 INSTRUMENTATION FOR ALIGNMENT

The basic instrument is a focusing telescope, described in detail in Dagnall and Pearn (1967) which provides the straight **line**. A pentagonal prism, as described in section 2.5.2, provides the 90° **square**. Rotating the prism around the axis of the telescope generates the **plane**. The telescope and pentagonal prism combined with a range of mechanical accessories for holding and pointing the telescope provide the optical equivalent of the surface table, straightedge and square. Although constructed very rigidly, the instruments and accessories are of relatively light weight and no distortion problems are caused. The operating range of optics is in general only limited by the environment in which the alignment work must be undertaken.

3.1.1 Micro-alignment telescope

The precision steel tube of the telescope (Fig. 3.2) contains the focusing lenses and cross-line graticule which generate a line of sight which is concentric to and parallel with the tube axis, an essential feature which establishes a known relationship between the line of sight and the instrument and its mountings. In use the telescope sights targets or scales attached to the work at nominally ×30 magnification. Built-in optical micrometers of the tilting parallel plate type enable the displacement of the target relative to the cross-lines to be measured to accuracies of 0.01 mm up to 3 or 4 m, 0.05 mm at 15 m, and 0.1 mm at 30 m.

A frequently used mounting arrangement for the telescope can be seen in Fig. 3.12. The telescope barrel closely fits a precision bore in a sphere which is held in an outer cage with micrometers providing tilt in azimuth and elevation. To enable the accuracy to be checked, the telescope tube can be rotated in its mounting sleeve through 180° about its axis, so that any residual errors are reversed in direction. This is an instance of the general principle introduced in section 2.2.1.

(a) Pentagonal prisms

The five-sided solid glass prism deviates the telescope line of sight (LOS) through 90° within 1 second of arc (0.005 mm per metre). Four types of the prism are used

Fig. 3.2 Alignment telescope (photos courtesy of Rank Taylor Hobson).

for alignment work. Two of these are mounted directly on to the telescope tube, where they can be used to give a single 90° LOS or rotated 360° to generate a plane perpendicular to the telescope LOS. This line or plane can be either co-planar with the mounting sphere centre, or 4.00 in (101.6 mm) forward. The latter version is shown in Fig. 3.3. The zero-offset version is limited to providing only some 200° of plane due to the mechanical mountings obscuring the 90° LOS. These two types are always used mounted on the telescope tube, and the prisms incorporate matched glass wedges which enable a through-sighting facility for setting and validating the main telescope reference LOS. The third version is used in situations where the 90° line is required to be some distance from the telescope. This prism is normally used with an adjustable table, facilitating its setting to the work.

The fourth version of the pentagonal prism is the sweep optical square (Fig. 3.4), which is used especially for generating flat planes. The telescope itself is fitted into a special bore in the prism housing which locates it concentric to the precision bearing about which the prism is rotated. The prism is rotatable through 360°, producing an infinite number of lines all precisely at 90° to the telescope line of sight, a complete 360° sweep generating a flat plane. The flatness accuracy of the swept plane is derived from the 1" deviation accuracy of the prism, plus the concentricity of the

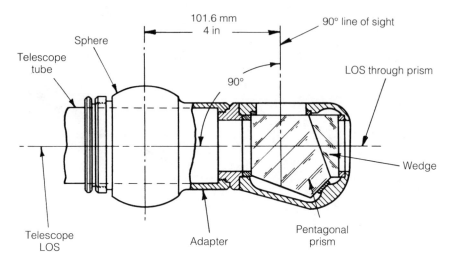

Fig. 3.3 Pentagonal prism, 4 in (101.6 mm) forward from mounting centre.

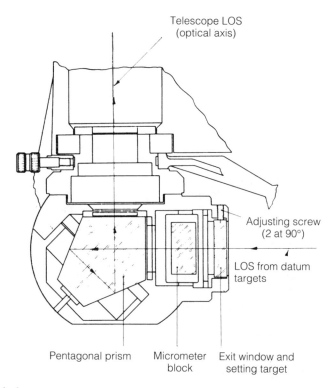

Fig. 3.4 Sweep optical square.

Instrumentation for alignment 91

prism bearing to the telescope line of sight. Typically the accuracy on a 10 metre plane is 1" on 5 m = 0.025 mm plus concentricity of 0.01 mm, giving a total of 0.035 mm.

(b) Targets

Sighting targets consist in the main of 2.25 in (57 mm) and 1.5 in (38 mm) diameter glass discs printed with a pattern of centre dot and circles which are precisely concentric to the perimeter (Fig. 3.5). In general targets have to be centred into bores using specially made adapter plates, although for bores greater than 180 mm target centring spiders are used. The centre dot and its adjacent circle are the most used. However, the telescope micrometer measuring range is restricted to ±1.2 mm, and there are many situations where errors or even design features require larger displacements to be measured. Accordingly, most targets incorporate extra concentric circles graduated out to 20 mm radius.

Since the optical magnification of the telescope is nominally ×30 throughout its focusing range, the images of target patterns reduce in size as operating distances increase. To compensate, special target patterns are used for distances greater than 25 m. For distances up to 3 or 4 m the centre dot is employed, the telescope micrometers being used to centre the dot precisely to the cross-lines. The resultant measurement is read directly from the micrometer drum graduations. At this operating distance a measuring accuracy of <0.01 mm is achievable. Beyond 4 or 5 m the centre dot disappears. The inside of the first circle is then used, the micrometers being used as before to centralize the circle to the cross-lines, the four small segments being symmetrical. In the range 4 to 15 m an accuracy of 0.02 mm should be achievable. This sighting technique is used for all other operating distances and should yield accuracies roughly in proportion to the distances, for example at 30 m an accuracy of 0.1 mm should be possible.

For situations where it is necessary to sight distant targets through intermediate ones, special high-quality optical glass targets are used. These targets are polished

Fig. 3.5 Sighting targets for alignment telescope.

92 *Alignment metrology*

Fig. 3.6 Scales for optical sighting.

parallel within 2" in order to minimize any deviations to the reference lines of sight. They use an identical dot and circles pattern, and are easily identified from the normal type by the glass colour. White glass should only be used for through-sighting situations, the green-tinted type being only used as the end target of any line.

(c) Scales

Graduated scales or rules of almost any type can be employed and sighted, and subject to the actual scale accuracy can be used to determine the offset of any surface or face with respect to the telescope line of sight. Special scales for sighting by optical instruments are available (Fig. 3.6); these scales have a white space representing each 2 mm or 0.1 in, which enables the black cross-lines of the telescope to be viewed for centrality between black intermediate lines. The accuracy of measurement obtained from these special scales is comparable with targets. The scales also have line pairs with special separations for use according to operators' preferences in conjunction with the different operating distances.

(d) Autoreflection

The telescope provides another squareness facility which uses a reflector, and it is applied to the setting of squareness of small surfaces, in general not larger than 500 mm. The autoreflection facility consists of a target pattern printed on the front cover-glass of the telescope. The target is concentric to the line of sight, and illuminated from within the telescope. A flat reflector is placed on the surface to be measured, and the telescope focused to twice the reflector distance. As shown in Fig. 3.7, the telescope will now sight the mirror image of its cover glass target, the position of this image with respect to the cross-lines indicating whether the reflector is square to the line of sight. Displacement of the autoreflection image from the cross-lines indicates the amount out of square. The target circles are graduated and are used to quantify the squareness by applying a simple formula.

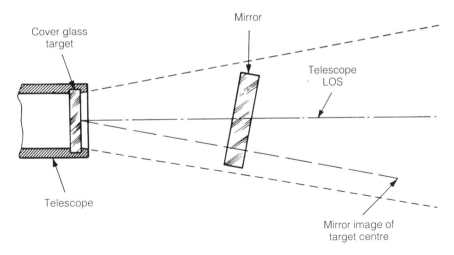

Fig. 3.7 Autoreflection.

3.1.2 Lasers

Everything that has been said about the basic geometry of alignment can in general be applied to lasers, with the exception of the target and its measurement. The laser is a projector of coherent energy, and the target therefore becomes a photoelectric detector system; this results in the measurements being made at the target station. Lasers are therefore often thought to be the ideal alignment tool, with no focusing optics and photoelectric rather than visual detection, to name just two advantages over optical telescopes.

However, unless special techniques such as those given in section 2.1 and 2.2 are employed, lasers have to be applied in exactly the same geometric manner to that already described, and this is often their practical limitation. Like telescopes they are physically single-ended devices, and the final accuracy of the alignment operation is dependent on the accuracy to which the basic reference lines can be positioned and also maintained. To explain further, telescope and laser instruments are both about 250 mm long over their mounting length, and in order to achieve an alignment accuracy of 0.04 mm over a workpiece 10 m long would require an initial pointing accuracy of $<0.8''$ or 0.001 mm over the instrument length. Furthermore, this pointing accuracy must be maintainable and stable throughout the measuring time.

Environmental stability is the most difficult to control. Floor movements, perhaps due to overhead cranes, and thermal movements caused by draughts or space heaters, all affect the initial pointing accuracy and destroy the geometrical references essential to this type of metrology. The inherent instability of the laser beam energy centre with respect to the instrument mounting tube is a constant source of trouble, especially during warming-up periods. All these instabilities manifest themselves as

94 *Alignment metrology*

angular changes and are therefore most troublesome as operating distances increase. These are fundamental difficulties and apply to any single-ended aligning instruments.

However, telescopes with their built-in measuring facilities are equipped to handle these difficulties in a very practical way. Because the operator has to be situated at the telescope, it only requires a change of focus to enable him to verify the accuracy of the reference line of sight. Any angular corrections can be effected immediately, being facilitated by the spherical mounting and the adjusting bracket horizontal and vertical screws. The situation is not so practical with the laser because the detection of any reference-line errors takes place remote from the instrument mounting, necessitating remote correction adjustments. Until these problems are resolved, laser alignment must be used with considerable caution.

3.2 TELESCOPE APPLICATIONS

3.2.1 Crankshaft bearings

A typical application for the alignment telescope is the measurement of the straightness of a line of bores for the crankshaft bearings of a diesel engine. The basic principle of this application is to compare the straight line of sight of the telescope with that of the line of bores. The centre of each bore is defined by a target, with the telescope mounted to sight along the line of bores and focused to sight on to the target as it is moved to each bore. With the telescope cross-lines as reference, the target image position is compared, and any displacements are then measured in x and y directions with the telescope micrometers.

In this application the work took place in a ship's engine room, where the centreline of the bores in the bed-plate was almost at floor level. As there was no suitable stable structure available on which the telescope could be mounted, the **attached** mounting method was used. Accordingly, the telescope was mounted directly into the No. 1 bore using a bore fixture. This mounting locates the telescope concentrically using a spherical adapter which permits the pointing adjustments to the telescope to be executed without loss of concentricity. The centre of the last bore was defined by its target, and the telescope was adjusted angularly to point precisely into its centre, establishing the reference line of sight position passing centrally through the end bores of the engine bed-plate. The end target is left in position always to be available as the line of sight validator.

A second target is now centred into each intermediate bore in turn and its centrality measured with respect to the telescope line of sight. A graph of the resultant set of measurements (Fig. 3.10(*a*)) displays the bore alignment with respect to zero readings from the first and last bores. The graph shows a typical set of readings, which suggest that the intermediate bores have 'sagged' with respect to the end bores. It is important to appreciate that the graph can be interpreted in different ways; the bed-plate could have lifted at its ends! Measurements and graphs

Fig. 3.8 Target spider for bores larger than 180 mm in diameter. T: target in rotating bearing, C: centring screws, D: dial test indicator for centring.

must be analysed having due regard to the method of measurement and the reference datum employed.

The attached method of mounting requires that the instrument mounting locates the telescope and its line of sight precisely through the centre of the bore. In practice, however, tolerances of bore diameters and concentricity of mountings invariably can produce some errors. The alternative mounting method **detaches** the telescope from the bed-plate and places it on a rigid piece of floor adjacent to the No.1 bearing bore. Targets are centred in the first and last bores by spider fixtures (Fig. 3.8), and the telescope is now set up and adjusted to sight zero through the two end targets. Figure 3.9 shows an identical bed-plate undergoing bore-alignment tests in a workshop environment. The procedure for measuring the intermediate bores is identical to that used with the attached method.

The rigidity of the instrument mountings and of the floor is of paramount importance with this method and the operator would be well advised to monitor the readings of the end target frequently; the end target should be left in position throughout the measuring operation. The detached method is often preferred because it allows target concentricity to be verified, especially at the No.1 bore position, by rotation of target adapters whilst being sighted from the telescope. There are some situations where it is difficult or too time-consuming to adjust the zero-zero datum; the detached method allows the measuring operation to proceed with the end targets not at zero.

Figure 3.10(*b*) shows a set of readings made from the same bed-plate. Note particularly that whilst none of the readings are the same, the misalignments of the bores to each other are exactly the same. The presentation of results in graphical

96 *Alignment metrology*

Fig. 3.9 Bore alignment on diesel engine bed-plate. T: telescope with right-angle eyepiece mounted detached from bed-plate, E: reference targets in spider fixtures in first and last bores, I: intermediate target in spider fixture.

form is extremely valuable because it presents very clearly the shape of a line of bores. A further advantage is that alternative datum lines can be drawn in using different reference points (Fig. 3.10(c)); this is an option which is extremely useful when assessing the possible effects of corrective adjustments.

The recording of alignment information together with other relevant data is an essential part of the operation. This applies especially to the direction of errors. It is good practice to record the eyepiece view, make a sketch marking the cross-lines, and then add a dot and circle in the quadrant where the target appears. Always do this first, then make the measurement, and then add the vertical and horizontal readings. It is ambiguous and dangerous to use + or − to indicate a target position; always use the L U R D convention (left, up, right, down). Always record the target position with respect to the cross-lines, never vice versa. It is also very valuable to record by sketch the position of the telescope relative to the workpiece. Add North, South, East and West, Port or Starboard, date and time of test, air condition and temperature. All this information can often be extremely valuable for some later investigation of the alignment condition.

Illumination behind the target is of vital importance for the obtaining of accurate measurements. This light should be chosen carefully, and where possible the instrument manufacturer's product should be used. In cases where temporary

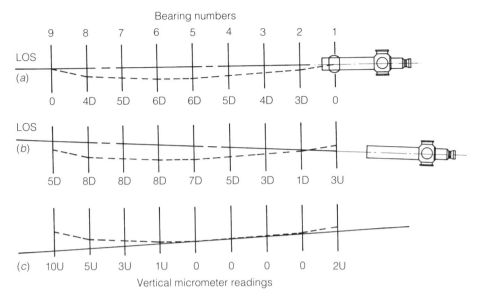

Fig. 3.10 Bearing alignment measurements; (*a*) with telescope attached, (*b*) with telescope detached, and (*c*) adjusted to inner bearings.

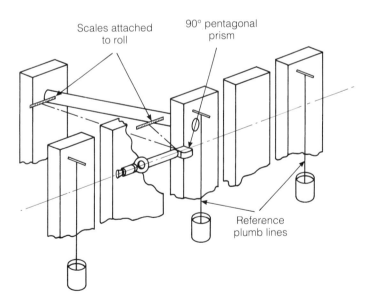

Fig. 3.11 Alignment of machine rollers.

98 *Alignment metrology*

illumination has to be employed, always diffuse the light in some way and avoid very strong white. Pale greens or yellows are good. If hand lamps or torches are used, ensure that they are fixed to some rigid structure and not hand held. In cases where only natural illumination is available, arrange to place a white paper or card angled at 45° behind the target. Some care and experimentation will be well rewarded in this important area.

3.2.2 Transport and process rollers

In machines used in the paper-making and printing industries it is essential that the transport and process rollers be parallel to each other. These critical rolls usually are considerable distances apart and may also be at various heights down the plant. It is possible to use an alignment telescope for this task, as an alternative to the specialized methods described in section 2.5.6. Most installations are by proportion long and narrow, and whilst parallelism between rolls is of prime importance, they have also to be square to the line of the plant. The basic principle of the alignment method is to establish a reference line down one side of the machine which is parallel to the machine centreline. Lines of sight set off at 90° to this master line sight into the machine adjacent to the rolls and are used to set or measure the roll parallelism (Fig. 3.11). All rolls can then be set parallel to each other, and also square to this master line.

The master line of sight is established parallel to the machine centreline by setting up the telescope to sight equal readings off graduated scales stood off from the machine side frames. Only nominal accuracy is required at this stage. Three plumb lines are now hung from the machine frame and adjusted to intersect the telescope master line of sight. Plumb lines are usually nylon fishing line of a suitable diameter compatible with the sighting distances, for example 2 mm diameter is suitable for 30 m sights, reducing to 0.25 mm for short distances. All plumb lines must have their bob weights immersed in cans of oil in order to damp out gyratory movements. In outdoor conditions, or where severe draughts are present, heavier weights may be required and possibly a more viscous damping oil.

The pentagonal prism (optical square) is now fitted to the telescope and the instrument repositioned to sight at 90° into the machine adjacent to a roll. Resetting of the master line of sight is achieved by sighting straight through the prism and adjusting the telescope to sight coincident with any two of the three plumb lines. A graduated scale is then attached to each end of the roll and sighted from the telescope 90° line, as shown in Fig. 3.11. Differences between the scale readings give directly the squareness error with respect to the master line, and more importantly represent the parallelism error to the 90° line of sight. Any number of rolls can be measured or set by repeating this procedure at any position along the machine. Plumb lines draped over each end of a roll can be sighted from the 90° line of sight. This technique is particularly useful where some rolls are high up in the machine, provided that the edge of the roll projects beyond the machine side frames, or sighting apertures are available.

3.2.3 Coaxial spindles

Squareness and parallelism can be related, for example two 90° angles added together equal 180° and can be considered as two parallel lines. An example of how parallelism is derived from measuring squareness is described in the following application, which was to align two opposed spindles spaced 2 m apart. Neither spindle was hollow, so a four-target method could not be used. The solution to the problem consisted of positioning the telescope between the spindles and sighting it coaxially to the fixed spindle, then reversing the telescope in its mounting to sight the second spindle, which was then adjusted to be coaxial.

The setting of coaxiality is achieved by mounting an adjustable mirror target holder on to the end of each spindle. Starting with the fixed spindle, the concentricity of the mirror target is first set by sighting from the telescope and rotating the spindle 180°, adjusting radially until concentricity is achieved. The mirror target is then set for perpendicularity using autoreflection, the spindle being rotated 180° as before and the mirror being adjusted in its spherical seating until perpendicular. Finally, the telescope was adjusted to sight perpendicularly to the reflector and also central to the target. To enable the necessary adjustments to be made, a universal mounting for the telescope was used. This provides horizontal and vertical transverse adjustments in addition to directional adjustments for azimuth and elevation. The telescope was now set coaxially to the fixed spindle.

Utilizing the precise parallelism and concentricity of the telescope line of sight to the mounting tube axis, the telescope was removed carefully from its mounting and then turned 180° and replaced, pointing now at the second spindle. This spindle was then rotated 180° and the mirror target adjusted for perpendicularity and concentricity. Finally, the whole spindle assembly was adjusted into position to centralize and square the mirror target to the telescope. The second spindle was now coaxial with the first. The principle of using a reflective target to establish a spindle or shaft axis is applied widely to machine tool spindles, engine crankshafts, and gear boxes. The method is very accurate because the reflector and target accuracy are determined by generation from the spindle rotation, all fixture and coupling flange errors being eliminated.

3.3 FLATNESS OR LEVEL

It is important to understand and appreciate the difference between a straight line and a level line, similarly a flat plane and a level plane. Also, a line at 90° to a straight line and a vertical line each have their own unique specification. There are many cases where alignment requirements are incorrect or specified too loosely. For example, the request to 'level a machine bed' means literally that the bed must be horizontal. Whilst this would undoubtedly achieve the specification, there would be unnecessary work done in striving for the horizontal. The essential requirement of machine tool beds is that they be flat, and it is a second-order requirement that they be level. In all but some exceptional cases it is only necessary for the bed to be

100 *Alignment metrology*

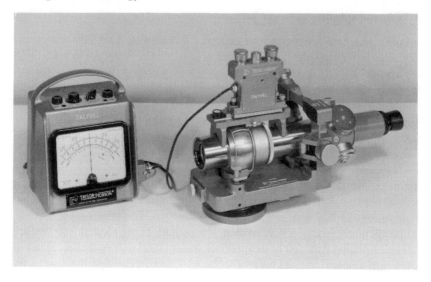

Fig. 3.12 Alignment telescope fitted with stride level.

nominally parallel to the floor. Vertical slides on machine columns are essentially required to be at 90° to the bed slideway and not necessarily vertical.

The importance of getting these requirements exactly correct is that it changes the equipment and often also the method. Where level is really needed, the telescope would be required to sight a horizontal reference. This requires that a precision level be used to set the telescope horizontal. The 'Talyvel' electronic level is used for this operation and is shown in Fig. 3.12 mounted on a stride base incorporating adjustment facilities which enable calibration to absolute horizontal. When the system is required to generate a horizontal plane the telescope, complete with mounting sphere, horizontal base and adjusting bracket would be rotated around the vertical axis of the fixed mounting cup, relevelling to zero before each azimuth sighting. The reason for the very common looseness of specification is that prior to the use of optical instruments, most engineers used gravity-referenced devices such as bubble levels, water levels in the form of water-filled pipes with sighting glasses, or even troughs of water or oil as their horizontal reference. These methods automatically achieve level as well as flatness.

3.3.1 Flatness on-board ship

A very common requirement in ship-building and repair, and also on offshore oil and gas platforms, is to measure the flatness of diesel engine bed-plates, gas compressors and machinery bases. Ships and offshore platforms must be recognized as dynamic structures. They are moving constantly from the effects of wind and tide, which makes the employment of gravity-referenced devices extremely difficult and even impossible.

Flatness or level

Since there cannot be any level on floating structures, any flatness reference must be related to the structure itself. Measuring the flatness of a diesel engine bed-plate installed in a ship requires the use of a reference plane which will at all times be related to the ship. The sweep optical square was used to generate this reference plane for measuring the bed-plate of a 16-cylinder twin-crankshaft diesel engine.

The sweep optical square was mounted on a rigid but temporary platform made from a steel plate attached to four studs, thus enabling the sighting plane to be clear of obstructions. A graduated scale on a magnetic base was positioned adjacent to each bearing pocket and sighted from the rotating prism unit, the height of the bed-plate being measured from the scale at each position. In an ideal situation the sweep square plane would have been adjusted to sight the same height on scales placed at three of the bed-plate corners. This would have enabled all the intermediate scale measurements to be related directly to the plane; indeed, it would have been necessary had the measuring operation included adjusting the bed-plate. However, in the harsh world of reality the work had to be done with only one scale, and as time was short the sweep square reference plane was only set parallel to within about 0.5 mm.

All scale positions were measured and recorded. However, due to the large setting error the flatness is difficult to assess, so to improve this situation the figures were manipulated for clearer presentation. We refer to the four corners of the bed as A, B, C, D, as in Fig. 3.17. The first stage was to subtract 0.5 mm from every reading to lower the plane and bring corner D to zero. The second adjustment was to tilt the plane about line AD until corner C also came to zero, taking 0.2 mm off all values along the line BC and a proportionate amount off each intermediate reading back to the line AD. Finally the plane was tilted about line DC to bring corner A to zero, treating the readings in a similar manner. The final readings can now be related clearly to a plane defined by the zero values at corners ACD.

3.3.2 Alignment on an offshore platform

An interesting combination of a straight line and perpendicular planes was used to set up a portable milling machine on a mud pump on an offshore drilling platform, in order to remachine a badly distorted cylinder mounting face. Mud pumps are of rugged yet simple design and construction. The three-throw crankshaft operates connecting rods through cross heads and extensions to the pistons reciprocating in separate cylinder blocks bolted on to the mounting plate. The distorted face caused the cylinder bores to be out of parallel to the cross-head line, resulting in excessive wear of pistons, rings and cylinder liners.

The reference chosen was the stroke of the cross head, and the telescope was set up to sight parallel to its top and bottom dead-centre positions. The optical square mounted on the telescope and rotated 360° then generated a flat plane perpendicular to the cross head. Scales attached to the face were sighted and the whole face was surveyed, the differences between each scale position showing directly the amount of machining required to restore the face to its original perpendicularity.

Fig. 3.13 Cylinder mounting face of mud pump. T: telescope, Q: optical square, S: scales on magnetic bases.

The telescope complete with optical square was positioned about 200 mm away from the cylinder mounting plate. It was adjusted to sight parallel to the stroke of the cross head as it was moved from top dead centre to bottom dead centre, the target being mounted on an angle bracket bolted on to the cross head. Figure 3.13 shows the graduated scales attached magnetically to the cylinder mounting face.

The portable milling machine was now installed in front of the cylinder mounting face and the telescope equipment repositioned and reset parallel to the cross head stroke line. Graduated scales were then attached to each end of the machine bed, and a third scale to the vertical slide containing the milling spindle. The telescope and optical square were now rotated to sight the LH and RH bed scales, the bed being adjusted until both scale readings were equal. The telescope and prism were then rotated to sight the scale on the vertical column, placed first at the top and then moved to the bottom, the machine bed again being adjusted until both readings were equal. The cutting plane of the milling machine was now correctly positioned perpendicular to the line of the cross head.

The final stage of this task was to ensure that the axis of the cutting spindle was perpendicular to the cutting plane, otherwise concavity and stepping would result on the finished surface. Autoreflection was used to set the cutting spindle parallel to the cross head line (perpendicular to the cutting plane) by placing a flat reflector on to the spindle shoulder, the whole column being adjusted to bring the reflector square to the telescope. This was first done by rotating around its vertical axis to eliminate cutting concave, and then about a horizontal axis to avoid cutting steps.

This example is an application which would appear to use a mixture of good and bad metrology. In particular, the original datum for all the work was limited to the 0.4 m stroke of the cross head, even though the face to be machined was 2×0.6 m and the milling machine had a bed length of 6 m and a column height of 1.5 m. However, the situation offered no alternative reference datum, and furthermore in the pump functional geometry, the piston stroke is equal to the cross head travel. Therefore, the final parallelism error between piston and cylinder would never exceed any datum-setting error.

When planning alignment work and procedure it is essential that the machine design and the function of the components is understood fully. When dealing with overhaul and repair work it is essential to be aware of the machining procedure that was employed for the original manufacture, and from this knowledge choose the reference or datum points for optical alignment. It is regrettably often the case that insufficient attention is given to this aspect of alignment by plant engineers and designers. All too often the words 'align it optically' are spoken without enough thought being given to the details of how it is to be achieved.

It must not be forgotten that optical instruments do no more than provide highly accurate references, weightless and unrestricted for size. The success of optical alignment lies entirely in the hands of the engineers who plan the operations. Manufacturers themselves could contribute greatly to this situation by providing reference faces or holes put in during machining stages; these could then be available for re-establishing reference lines of sight. These comments apply particularly to the industries in which it is accepted practice to have regular overhauls when alignment is checked and maintained.

3.4 AUTOCOLLIMATION

Whilst the accuracy of telescope equipment is quite adequate for the majority of applications, it is not good enough for the calibration of such items as machine tools, measuring machines, surface tables and straightedges. In these industries, accuracy requirements are subtly different, the most significant difference being that the precision straightness or flatness is built into one piece, like the bed or column or table. The position of one accurate piece with respect to another accurate part does not have to be precise. This unique situation is the prime characteristic which enables the autocollimator to be applied.

3.4.1 Straightness by autocollimation

The autocollimator is a telescope permanently focused at infinity, and it is specially designed to measure the tilt of a reflector. Autocollimators cannot be used to measure targets or scales, and must never be confused with focusing telescopes. Their

104 Alignment metrology

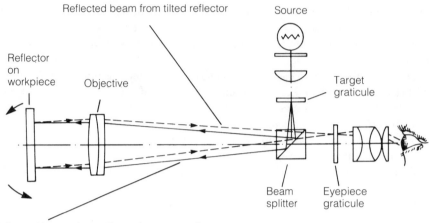

Fig. 3.14 Optical arrangement of an autocollimator.

uniqueness lies in their ability to measure angular changes of reflectors to tenths of seconds. Figure 3.14 shows a diagrammatic illustration of an autocollimator and reflector. Autocollimators measure straightness by measuring the change of angle of a reflector carriage as it is moved along the surface in increments of the spacing of its locating feet. Assuming that the pitch of the feet is 200 mm, and the angle of the reflector is changed by 1", this would have been caused by a surface imperfection of 0.001 mm under one of the feet. The process is analogous to the use of a laser beam reflected from the surface itself as described in section 2.5.3.

The procedure for measuring straightness along a machine slideway would be first to position the reflector carriage close to the autocollimator and adjust either to give a near zero reading, then record the reading. The reflector carriage would then be moved along the bed by a distance equal to the pitch of the feet. This would place the carriage front foot over the position previously occupied by the back foot, any lack of straightness would lift or lower the back foot in its new position.

In straightness measurement with the autocollimator, one always subtracts the first reading from every other reading. Cumulative addition of successive values then converts slope to height. It is usual to find that the graph is inclined, and it is normal practice to correct this inclination in order to present the shape in an undistorted form. Operators who are recording and plotting results manually must be sure of the numerical sign associated with the direction of tilt of the reflector; an incorrect sign would produce a hollow contour instead of a rounded one. To check this out it is always good practice to insert a cigarette paper under one foot and observe the direction of movement of the image.

3.4.2 Electronic autocollimators

Electronic autocollimators work in exactly the same way as visual instruments, but using photodiodes to detect the position of the reflected image. Using a rectangular array of diodes, position and hence magnitude of reflector angles can be measured in both x- and y-directions. This enables straightness of the top and side faces of a workpiece to be measured at one single traverse of the reflector carriage, without the need to touch the autocollimator throughout the measuring operation. For machine tool makers this represents a considerable reduction in measuring time, and coupling a printer into the system enables the graphical results to be available a couple of minutes after the measuring operation has been completed.

Figure 3.15 shows a dual-axis autocollimator as it would be set up to measure the top and side edges of a machine bed. Note particularly the two sets of locating feet on the reflector carriage, adjustable between 25 and 200 mm spacing. Figure 3.16 shows diagrammatically the principle of the photoelectric autocollimator. An infrared diode supplies the measuring optics, and a green LED supplies the light for the visual setting up of the instrument to the reflector.

Computerized measurement of straightness produces a slightly different format to that of the manually drawn exercise. The graphs for the top and side edges of the workpiece (x and y measuring axes of the autocollimator) are drawn with both their end points at zero. The actual inclination of the line through the end points is described as the slope and is printed out in millimetres per metre. Note there is a

Fig. 3.15 Photoelectric autocollimator measuring slideway straightness.

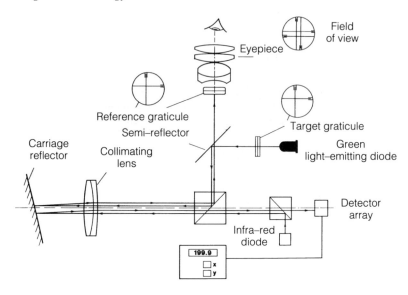

Fig. 3.16 Optical arrangement of photoelectric autocollimator.

special use for this piece of data when measuring squareness between a bed and a column as described in the next paragraph. Note also that other data are made available, like peaks and valleys and also the peak-to-valley magnitude, which is the total deviation from the straight line joining the end points. The peak-to-valley magnitude is extremely valuable and time/cost saving when adjusting machine beds, indicating immediately changes brought about by the adjustments. Also, most computer programs utilize the auto-ranging facility, which always fills the printing paper whether the magnitude is tenths or tens of millimetres.

Squareness of machine columns with respect to their base slideways are measured using an almost identical procedure, with a pentagonal prism to deviate the autocollimator reference axis precisely 90°. The base or reference slideway must first be measured. Then, without disturbing the autocollimator, the reflector carriage is transferred to the column, stepping according to the foot pitch as before. It is advisable to use two tabulations and treat the base and column readings as having separate origins. The graphs can be drawn side by side, or end on to each other as though they were in line, the error of the 90° being determined by comparing the two slope values. When checking squareness it is not possible to use the graph attitude correction technique, as this would invalidate the 90° relationship.

3.4.3 Flatness of surface tables

Flatness measurement for the calibration of surface tables is simply an extension of the straightness routine used eight times around and across a table in the form of a

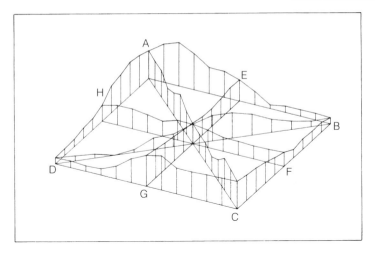

Fig. 3.17 Isometric plot of Union Jack method of flatness measurement.

Union Jack flag (Fig. 3.17). Each straight line can be measured without reference to any other, and they can be measured in any order, the only essential requirement being that each generator line must contain an even number of measuring positions.

The final assessment and presentation of flatness is done by utilizing the unique position of the crossover of the two diagonal lines AC and DB. Line AC is laid over line DB and balanced so that its ends A and C are at equal heights above or below points D and B. The straightness plot of line AB is then placed with A on A and B on B, similarly the other straightness plots are placed into their appropriate positions. Lines HF, EG are plotted in according to their places on lines AD, BC and AB, DC. Any mismatch of the heights at the central crossover point is considered to be an error in the original straightness data, and is often the result of dirt or dust getting between the feet of the reflector carriage and the table. This method of measuring the flatness of surface tables has been available for many years but due to the tedious nature of analysing and plotting the data it was only used by dedicated metrologists. However, with the present day availability of small portable computers, the method is now widely applied.

3.4.4. Angle comparison

Autocollimators also excel in the comparison of angles, in particular the calibration of dividing heads and circular tables. Because the autocollimator can only measure small angles, typically 10 or 15 minutes, it is necessary for it to be used as a comparator. Polygons are used as master angle standards. They are normally available with 4, 6, 8 or 12 sides giving angles of 30, 45, 60 and 90 degrees. Any other angle can be obtained, but would have to be specially made and would therefore be

108 Alignment metrology

expensive. Polygons are normally polished to 5" accuracy and supplied with a calibration certificate stating angular accuracy to within 0.5 or 1".

To measure a dividing table, the polygon is mounted nominally central (within 5 mm) on to the table top. The autocollimator is then rigidly mounted as close as possible and pointing to the dividing table centre. The table is set to zero degrees, and the polygon and autocollimator adjusted to give a zero or near-zero reading. As an example, using a 12-sided polygon the table is then rotated precisely 30°, which should have indexed the next polygon facet into exactly the same position as the first one. Any discrepancy would be measured directly by the autocollimator, the procedure being repeated 11 times until the table has been rotated 360°, where the original start reading should be repeated proving validity of the measurements. The use of polygons for angular divisions limits the method to specific angles, which because of their repetitiveness could possibly miss a periodic error. For this reason it is often good practice not to start at the table zero, but to start at some random angle such as 3° 27' 13".

In cases where specific angles have to be measured, a set of angle gauges can be used. Eight polished steel angles make up a set, which can be wrung together and finally wrung on to a four-sided polygon placed on the rotary table. These sets of angle gauges are used to make up any angle in increments of 5 minutes. Angle gauges are accurate to 2", the one drawback being that the method is not quick. A method which enables random angles to be checked quite quickly is to use another indexing table of known accuracy to generate the master angle, still leaving the autocollimator to make the comparison measurement. A precision clinometer can be used to measure vertical angular indexing. Accuracy to better than 5" is easily achievable with a standard clinometer. However, this can be improved by having the clinometer calibrated. For the ultimate accuracy a precision multi-tooth indexing table can be used. Such tables have 1440 teeth and are capable of generating any angle in 15' increments accurate within less than 0.5".

3.5 PRACTICAL PROBLEMS

3.5.1 Environmental problems

All optical instruments are affected to some degree by the air through which they operate. Thermal changes are the most common causes of trouble and in telescopes and visual autocollimators the trouble is visible immediately in the eyepiece as an unstable image shifting about, predominantly in the vertical direction but also horizontally, and in very bad cases there is an inability to obtain sharp focus. Lasers are affected similarly, displaying unstable readings on their readout panels, and photoelectric autocollimators also show unstable displays. Factory heaters, open doors, electric motors, sunlight and recently welded areas are the most common offenders. Temperature-controlled environments including standard rooms have been found to cause trouble; the air exchange and circulating systems produce thermal changes in the air of quite short duration, but nevertheless sufficient to cause instability in readings and affect measurements.

Temperature changes in the air change the density and so the velocity of light is increased or slowed. The velocity change in itself is not really the problem, as this only causes focus changes which are only experienced in extremely bad cases. The real problem is the differing velocities of light rays across the line of sight. These are changed quite rapidly by the generally upward movements of the air, resulting in an angular deviation of the line of sight, as analysed in section 2.3.2. Thermal disturbances situated near to the telescope, autocollimator or laser will cause more trouble than thermal changes occurring near the target or laser detector. The air temperature itself is generally not the cause of the problems, it is the variation of temperature. For example, a temperature change of 0.5°C about a norm of 20°C is likely to cause more trouble than 0.5°C about 30°C. This phenomenon is often experienced when working on steam turbines. Targets or scales placed close to very hot casings often result in acceptably good measurements, the air being so hot that it does not cool across the line of sight.

It must not be forgotten also to account for the effects of thermal expansion of materials on the overall metrology. It is fortunate that, in general, thermal problems rarely produce misleading results. This is because the symptoms are recognized easily in the form of disruption to sighting conditions. Thermal conditions are almost impossible to eradicate and each situation must be experimented upon to find a satisfactory solution. In general it is necessary to first remove the reasons for the thermals. Heaters should be turned off, doors closed, draughts eliminated, and welded areas allowed to cool or removed.

Sometimes it is possible to erect screening, particularly where fan heaters are operating. Sunlight is a source of heat, and must be screened off from the measuring area. Whilst it is very important to remove turbulence, it should be realized that it is a warning of danger, and screening and protection methods could create a situation where a temperature gradient exists across the line of sight. This could be extremely dangerous; temperature gradients of only 1 or 2°C have been known to produce deviations amounting to many millimetres over sights 20 m long, with no danger signals.

In situations where it is not possible to remove the source of heat, for example in hot countries, the only solution is to defer the work until the cool of the night, the hours between 2 and 4 o'clock in the morning being the most unsociable but the best. This is particularly important in the shipbuilding and repair industries, on account of the distortions to a ship's hull caused by the sun's heat. In general the alignment engineer is given little choice of the alignment situation; however, where possible some forward planning must be adopted if good and reliable results are to be achieved.

3.5.2 Obstructed sight line

Situations where sights have to be taken through small holes should be avoided; in general holes less than 25 mm in diameter are likely to cause trouble. Whilst targets

110 *Alignment metrology*

can be seen through smaller holes, considerable difficulty can be experienced with low lighting contrasts. Should holes also be off line then asymmetrical focusing problems will show. The target image will appear to be badly off centre in slightly off-focus conditions, and whilst perfect focusing is necessary to eliminate parallax the problem is nevertheless very disconcerting. It is sometimes necessary to introduce a design change in order to enable alignment to be done through a larger hole by incorporating a concentric adapter plate to accommodate the small hole. Such action eliminates completely the small hole problem.

Asymmetrical focusing can be experienced if there is a partial obstruction to the telescope field of view. The same condition will also be present if the target illumination is off centre, a problem often experienced when using temporary illumination like hand lamps. One should never attempt to sight through windows, since ordinary window glass is rarely flat enough to permit focusing, neither is it parallel-sided enough not to deviate the line of sight. In cases where it is essential to sight through windows of environmental test chambers, special high-quality windows must be arranged.

3.5.3 Parallax

Whilst parallax is not induced by any environmental conditions, it is nevertheless a phenomenon which causes considerable problems, especially to the less experienced user. It is a condition which accompanies all focusing types of instrument. Parallax is caused by incorrect focusing of the eyepiece or objective lenses, resulting in the

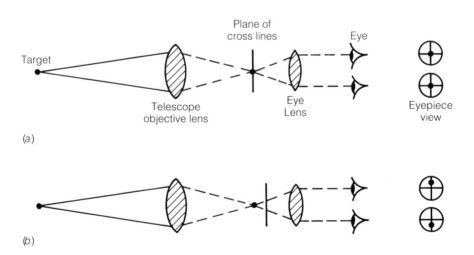

Fig. 3.18 (*a*) Eyepiece correctly focused – no parallax, (*b*) eyepiece incorrectly focused – parallax introduced.

target image being positioned in a slightly different plane to that of the graticule cross lines (Fig. 3.18). Parallax manifests itself in the operator's inability to obtain repeatable measurements of the target image position with respect to the cross-lines.

All instruments have eyepieces which are adjustable for focusing the lines of the reference graticule. Most eyepieces cover a range of at least 5 dioptres, enough to accommodate the normal variations in operator's eyesight. Any incorrect focusing of the eyepiece results in the operator's eye looking at a position either in front of or just behind the plane of the cross-lines. This situation results in an apparent shift sideways or up/down of the cross-lines as the operator's eye moves sideways or up/down, as shown in Fig. 3.18(b). This eye movement is unavoidable and is caused by body movement, including breathing.

The eyepiece setting will remain constant for quite long periods, but the target focus has to be reset for each and every sighting distance. As with the eyepiece, any incorrect focusing will compound the inability to obtain consistent measurements between target and cross-lines. Both these incorrect focus conditions are visually evident as unsharp and less black target or cross-lines. Unfortunately, the condition can be easily ignored by less experienced operators. A parallax condition can only be experienced by the individual operator; no other person can see the same condition, because no two people have exactly the same eye defects.

Parallax must be detected and corrected by the operator. The procedure for focusing a telescope consists of six stages which are as follows:

1. Position illumination behind the target.
2. Defocus the telescope.
3. Adjust the eyepiece until the cross-lines are focused sharply and appear as dense and black as possible.
4. Focus the telescope to view the target and adjust for an image that is as sharp and black and dense as possible.
5. Move eye or head up/down or sideways, and if there is any evidence of any relative movement between cross-lines and target, then parallax is present.
6. Repeat stages 2, 3, and 4 until no movement between cross-lines and target can be observed.

Do not attempt to make any measurements until all traces of parallax are eliminated. Never attempt to verify settings or make measurements from another person's focus setting; always adjust focus to suit your own eyesight. When focusing and when using optical instruments it is good practice to keep both eyes open, as eye fatigue will be reduced. Do not stare into an eyepiece for long periods. If experiencing trouble with a focus setting it is essential to look away and relax the eye for a few seconds and then try again. Wearers of spectacles should keep them on when using optical instruments. This applies particularly to people with astigmatic defects, as failure to wear astigmatic compensating glasses will result in the inability to focus both vertical and horizontal cross-lines simultaneously.

3.5.4 Conclusion

In all alignment work, the stability of the instrument mounting is of prime importance. Remembering that telescopes and, of course, alignment lasers are single-ended measuring devices, a tilt of 1 μm of the instrument mounting length will result in an error of 0.04 mm over a 10 m workpiece. Direct mounting (attached method) on to the workpiece is the most reliable, but will usually require the provision of special mounting adaptors or brackets. Detached mountings should be chosen with care. In particular, look out for unstable floors, cracked concrete, and beware of the movement due to heavy workpieces being moved whilst measurements are in progress.

Beware also of tidal situations; these often cause floors and even machines to move and tilt. Structures like ships also expand and contract and bend from the effect of the sun's heat. A submarine, for example, will bend many millimetres during the sun-up, sun-down period, caused by the upper surface expanding from the sun's heat whilst the underside remains constant due to the cooling effect of the water or just the shade. In every application it is good engineering practice to provide a reference target which can be used to validate the instrument line of sight. As a final comment, avoid placing any tools or note pad, etc., on or near the instrument mounting.

REFERENCE

Dagnall, R. H. and Pearn, B. S. (1967) *Optical Alignment*, Hutchinson, London.

4
Photogrammetry in industrial measurement

R. A. HUNT

Crown Copyright

The word **photogrammetry** originally applied to any form of measurement using light, being equivalent to what we might now call optical metrology, but it rapidly became virtually synonymous with 'air survey'. With the advent of high-speed computers it became practical to generalize the methods developed in air survey to produce accurate estimates of the shape of relatively small and close objects, typically in the size range from one to one hundred metres. In order to make the distinction between this generalized form of the subject and conventional air survey, it became known as close-range photogrammetry which is taken to include industrial, medical and architectural applications.

Recently, the word has become used in a broader sense to include methods which are not based on conventional photography or even conventional image formation (see, for example, Gruen and Baltsavias, 1990). All these methods have in common the concept of recording two-dimensional projections of the three-dimensional world essentially in parallel and the use of some form of triangulation method to reconstruct objects of interest from the two-dimensional records.

4.1 THE TECHNIQUE OF PHOTOGRAMMETRY

The basic principles of conventional photogrammetry are very simple. Let us assume that we have two cameras as shown in Fig. 4.1. If the positions and orientations of the cameras are known together with their basic internal geometry, then the image-forming pencils of rays from some unknown point P may be reconstructed using measurements of the photographic records of the images of P from the two cameras. The intersection of these reconstructed pencils then gives an estimate of the position of the point P at the instant when the photographs were taken.

The simplest form of photogrammetry is stereo photogrammetry, where only two cameras are used. It is usual to explain stereo photogrammetry using human vision as a model, as illustrated in Fig. 4.2. In order to introduce some symbols, let us write down the angle θ between the pencils of rays emanating from the candle in typical photogrammetric notation. The position of the effective centre of a camera lens is referred to in photogrammetry as the perspective centre of the camera (this concept

Optical Methods in Engineering Metrology. Edited by D.C. Williams. Published in 1993 by Chapman & Hall, London. ISBN 0 412 39640 8

114 *Photogrammetry in industrial measurement*

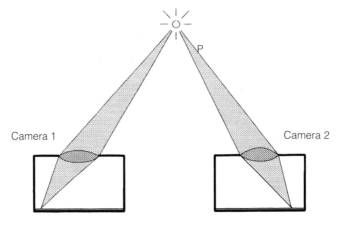

Fig. 4.1 Two cameras imaging a self-luminous point P.

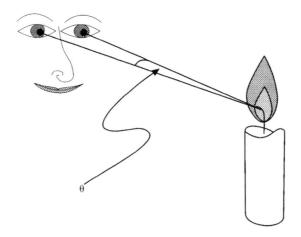

Fig. 4.2 Human vision as a model of photogrammetry.

will be defined more carefully in section 4.3). Let the perspective centres of the two eyes be at positions defined by the vectors $X_0 + \Delta X$ and $X_0 - \Delta X$ and the candle at X. Then the angle θ is given by the normalized scalar product

$$\cos \theta = (X_0 + \Delta X - X).(X_0 - \Delta X - X) / |X_0 + \Delta X - X|.|X_0 - \Delta X - X|. \tag{4.1}$$

If the subject turns to look at the object then $\Delta X.(X_0 - X) = 0$ and if the object is at a reasonable focusing distance $|X_0 - X| \gg |\Delta X|$. With these conditions satisfied equation (4.1) becomes

$$\theta \approx 2|\Delta X| / |X_0 - X|. \tag{4.2}$$

The technique of photogrammetry 115

This example is not as arbitrary as it seems, for the technique of photogrammetry is much older than the digital computer. Even today much stereo mapping is done by presenting pairs of photographs to human operators and letting them use their brain-based stereo vision software to judge the apparent height of an object in the field of view in comparison with an artificially produced floating mark. The accuracy of this process is limited by the convergence angles which may sensibly be interpreted by the human vision system.

In this context it is often said that human beings in fact perceive the three-dimensional nature of the world using the stereoscopic effect. This seems much more dubious. Certainly this effect can be used to create the illusion of a three-dimensional scene but so can perspective and subtle use of colour in a strictly monoscopic image. No less certainly persons with monocular vision have a completely three-dimensional view of the world.

Nevertheless, stereoscopic vision can play an important part in human perception of depth at relatively short distances. The distance between the pupils of my eyes, which is somewhat smaller than average, is about 60 mm and I can detect a significant difference in depth perception when one eye is covered for objects between approximately 0.2 m and 6 m in front of me. This corresponds to angular differences between 0.5 and 17 degrees. The reader can easily verify this by attempting to play a simple ball game with and without one eye covered and noting the distances at which depth perception is degraded. Experimental notes: 1. play in a good light where the improved noise reduction due to using two eyes is unimportant and 2. use a soft ball!

Thus the eyes form a fairly poor photogrammetric system and our relatively accurate three-dimensional perception of the world comes from our ability to include many different types of information gathered at different times and in different locations (patients with some memory disorders cannot maintain an accurate three-dimensional picture of their environment).

Yet many aspects of the stereoscopic use of the eyes exactly parallel the use of cameras in close-range photogrammetry. It is natural, in stereo photogrammetry, to associate the eyes with the cameras and the retina with the photographic plate. Unfortunately, this implies that the elegant transfer of the preprocessed images to the brain is replaced with the cumbersome business of measuring up the plates and analysing the results on a digital computer. Just like the photogrammetric cameras, the eyes must be calibrated in some way so that angular differences, however they are perceived, between the apparent positions of correlated patches of the visual field are associated with the distances to corresponding physical objects. To continue in this vein would lead us quickly into a mathematical description of imaging by eyes and cameras and a discussion of machine vision but these considerations are deferred to sections 4.3 and 4.5, respectively.

In practice, close-range photogrammetric surveys very seldom use only two photographs of the object. There are two main reasons for this. First, with only two cameras the position of each new point is estimated on the basis of the data on two ray pencils emanating from that point. If for some reason there were an error in the data associated with one of these pencils little could be done to retrieve the co-ordinates of the point. This is also reflected in the statistical analysis, so that only a very poor estimate of the uncertainty of the co-ordinates is obtained. Second, many engineering objects are truly three-dimensional and their entire surface cannot be shown on two overlapping photographs; because the photographs must overlap to obtain estimates of the point positions on the surface, such objects can only be analysed in terms of more than two photographs.

Sometimes surveys with many photographs are still analysed in terms of sets of stereo pairs, as in conventional air survey, the advantage being that the pairs may still be presented to a human operator to exploit his or her free brain-based software to identify and correlate target points in the two views and associate them with an apparent position in space. The difficulty with this procedure is that only relatively small angles of convergence may be accommodated by the human visual system. Our estimate above gave $\theta \approx 17°$ for the maximum angle of convergence at which it is still possible to apparently fuse different views of an object into a single three-dimensional image. This can be improved by cunning design of the stereo comparator but it still falls short of the 90° necessary for equal uncertainties in all three co-ordinate directions.

For the most accurate work, it is usual to place measuring marks on the points of interest in order to facilitate the correlation of these points between views from different camera positions. Once this has been done, there is much less point in giving oneself a headache trying to fuse different perspective projections of target images, as only the coordinates of the image centres are needed by the software to locate the points in space using any number of views in parallel.

4.2 HOW TO MAKE A MEASUREMENT

There now follows a description of an 'ideal' experiment. There are basically two forms of experiment: 1. the leap in the dark ('I wonder what will happen if . . . ?'); and 2. the careful estimation of some set of parameters believed to represent the process being investigated. A photogrammetric survey, or any other metrological experiment for that matter, falls firmly into the second category.

Set out below are what the author believes to be the optimum set of operations which should be gone through in order to perform a successful metrological experiment. It should be noted that the emphasis is on the estimation of uncertainties and the evaluation of parameters, for these are indeed the criteria upon which good metrology should be judged.

1. Think of a method.
2. Form a mathematical model.
3. Design the experiment.
4. Use 2 and 3 to optimize the accuracy.
5. If the accuracy is still not good enough go to 1.
6. Perform the experiment.

7. Estimate the actual accuracy of parameter estimation.
8. If the final accuracy is not good enough go to 1.
9. End.

Let us consider these stages in turn.

4.2.1 Choice of method

In this chapter it is assumed implicitly that the method chosen for the measurement task in hand is a photogrammetric one. Nevertheless it is always advisable, as has been seen in earlier chapters, to consider alternative measurement strategies for attaining the same end.

An obvious alternative to photogrammetry in many types of work is theodolite survey. A theodolite is essentially a telescope, containing a cross hair or other aiming mark, mounted rigidly with respect to the object of interest via two accurate angular scales. Usually the axis of rotation of one of the scales is horizontal and other vertical (see Fig. 4.14 and 4.15) although this need not always be the case. Similar basic geometrical information about the object is obtained from both methods. The difference is that in conventional photogrammetry data about many points on the object are collected in parallel but processed later, whereas in a conventional theodolite survey data are gathered sequentially but are capable of being processed on line. Thus transient occurrences such as distortion during loading are best recorded photogrammetrically, whereas surveys of essentially static objects where the results are needed quickly, e.g. on-site monitoring of construction work, are best suited to theodolite survey. With the advent of accurate methods for the extraction of metric information from video cameras, these two techniques show some signs of converging in an updated form of an old instrument called the **phototheodolite**, which is effectively a theodolite with a camera on top. If the camera is replaced by a CCD video camera, then the resulting instrument could be described as an example of real-time photogrammetry or as an image-processing theodolite. At least two automatic theodolite systems based on this principle already exist and they will be discussed briefly in section 4.5.2.

These considerations of the appropriateness of the technology may be constrained by the usual economic considerations which include the expertise of staff and the nature of the equipment most readily available.

4.2.2 Mathematical modelling

It is a true if much neglected fact that that no experiment, however accurate, can ever prove a theory to be correct; much less choose the appropriate theory for you. Conversely, any experiment can potentially prove a theory wrong, at least to within some experimental uncertainty, and indeed some of the most fundamental experiments are designed to do just that. Thus experimental physicists and engineers who claim to develop general empirical relations between cause and effect devoid of any theoretical framework have either uttered a 'terminological inexactitude' or they are in error. Even a mapmaker constructing a digital terrain model using spline functions is making a theoretical, if unjustified, assertion about the analytic form of the contours.

In our case the parameters that concern us are the co-ordinates of the points of interest, but we are obliged unfortunately to calculate the positions and orientations of the cameras as a means to an end. The directly observed quantities are the two-dimensional co-ordinates of the image centres on the photographic plates and our theory must express the latter in terms of the former, as will be done explicitly in section 4.3.

4.2.3 Experimental design

The formal design of the experiment is a step that is all too often left out. In our case this involves getting a fair assessment of the size and shape of the object to be measured and its immediate environment. An assessment must also be made of the relative importance of the various parameters to be evaluated. Often this is only possible if some indication is given of the use to which the experimentally estimated parameter values will be put. For example in an 'as-built' survey of an engineering structure required before an additional part may be fitted, the surfaces actually in contact with the proposed new part are of overriding importance. The camera positions for the various views can then be concentrated on this important area.

4.2.4 Prediction of uncertainties

The experimental design plus the mathematical model and statistical estimates of the uncertainties in the values that will be obtained for the measurable quantities, e.g. plate co-ordinates, should be sufficient to produce simple statistical estimates of the expected uncertainties in the estimated parameter values. Initially this analysis may be used to try and optimize the experimental design with due regard to the relative importance of the parameter values.

4.2.5 Assessment of the method

Once the design has been optimized the same analysis may show that the chosen experimental method should be rejected and another must be thought of. The advantage of this methodology is that it is much cheaper to discover an inappropriate method on paper than to commit resources to it only to find that it was not capable of the required accuracy.

4.2.6 The experiment

A conventional photogrammetric experiment can be represented in terms of the sequence of operations shown below.

1. On arrival on site the target points must be marked on the object and the cameras positioned according to the proposed experimental design.
2. The photographs are then taken and developed on site. There are two reasons for this: first the latent image on the plates fades during the first few hours after exposure; and, second, if a plate has been exposed incorrectly, or otherwise rendered useless, there is a chance to retake the photograph before the object changes in any significant way.
3. The positions of the target images on the photographs are measured in some co-ordinate system based on fiducial marks on the photographic plates which serve to relate the plate co-ordinates to the internal co-ordinate system of the camera. Occasionally, these fiducial marks are mechanical indentations made in the emulsion at the time the photograph is taken but are more usually images of small geometrically stable reference light sources within the camera.

The importance of care and accuracy on the actual day of the experiment cannot be over emphasized; no amount of statistical analysis can get good results from bad data! Sometimes interesting inferences can be drawn from data sets that were spoiled accidentally or from historical records not originally intended for analysis – good examples of this may be found in Thompson (1962) and Robertson (1990) – but the power of the analysis should never be used as an excuse for poor experimental technique.

Once the raw data, i.e. the plate co-ordinates and any additional data on the size and shape of the object, have been gathered, ordered and put into machine-readable form, the computer can begin to estimate the co-ordinates of the unknown points. More practical details of the experiments will be given in sections 4.4 and 4.6.

4.2.7 Estimation of experimental uncertainties

Once the experiment has been performed, the software which was used to predict the probable uncertainty of the estimated parameter values before the experiment may also be used to estimate the uncertainty of the experimental estimates of the parameter values. In this case experimental estimates of the uncertainties are used rather than the theoretical estimates used in 4.2.4. Thus these estimates are more likely to be correct but if, contrary to expectation, the uncertainties are greater than is acceptable then either the experiment needs to be repeated, the method must be modified or the experiment abandoned.

4.3 MATHEMATICAL FORMULATION

As mentioned in section 4.2, an essential prerequisite of setting up any metrological experiment is to set up an appropriate mathematical model of that experiment. In the case of a photogrammetric survey the theory is that of the imaging of a photo-grammetric camera. To get a grasp of the physics of the situation, let us consider a simple pinhole camera as illustrated in diagrammatic form in Fig. 4.3.

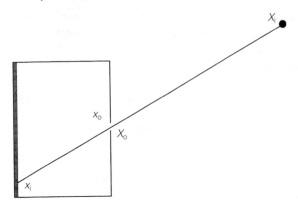

Fig. 4.3 Diagram of a pinhole camera.

In an elementary theory of imaging, the camera may be considered as an opaque box with a pinhole in the middle of one side and a photosensitive surface on the inside surface opposite. The function of the box is to divide space into two parts: the inside of the box and the outside. It is customary to label points on the outside of the box with upper-case letters and points on the inside with lower-case letters. Thus the inside may be referred to as the image space (x, y, z) and the outside as the object space (X, Y, Z).

The position of any point may then be expressed in terms of either set of co-ordinates. In particular, let the pinhole be at x_0 in the image space and at X_0 in the object space. In addition, let us suppose that a target at X_i in the object space is imaged at x_i on the photosensitive surface in the image space. Then we can define new vectors u and U as

$$u = x_i - x_0$$

and

$$U = R^{-1}(X_i - X_0),$$

where R is a rotation matrix chosen in such a way that U_x, U_y and U_z are parallel to u_x, u_y and u_z.

It is usual to define R in terms of three rotations about the co-ordinate axes. In particular the notation inherited from aerial survey suggests a rotation of ω about the X-axis, ϕ about the Y-axis and κ about the Z-axis. Nevertheless, this notation should be treated with caution as the senses of the rotations sometimes differ.

By similar triangles

$$u/u_z = U/U_z, \qquad (4.3)$$

where the z-axis has been chosen perpendicular to the photosensitive surface of the box. It is traditional to write u_z as $-f$, in which case (4.3) becomes

$$u = -fU/U_z. \qquad (4.4)$$

In photogrammetry, f is called the principal distance, and the foot of the line perpendicular to the photosensitive surface which passes through x_0 is known as the principal point. These definitions are not those used in mainstream optics, but they will be used in the rest of this chapter for compatibility with the photogrammetric literature.

The simple equation (4.4) contains the essence of photogrammetric survey and, incidentally, the fundamentals of the perspective painting and drawing of the renaissance.

In painting it is usual to omit the negative sign in order to obtain a positive rather than a negative image. There is only one position from which to view a perspective projection in order to recreate the implied three-dimensional scene and, as the two eyes can hardly be put in the same place, this is best done using one eye only. This situation is illustrated in Fig. 4.4.

It is seen that the best position for the eye is at x_0 corresponding to the imagined position of the painter's eye at X_0 in the original scene. The practical difficulty is to find x_0 in the art gallery and put one's eye there. The key to this question is that parallel lines on the object appear to come from a single point in the perspective drawing usually referred to as the **vanishing point**. We can see this formally for a family of parallel lines given by

$$U = U_0 + tn,$$

where t varies along the length of the line and different vectors U_0 distinguish different members of the family of lines. If we assume that $n_z \neq 0$ we restrict

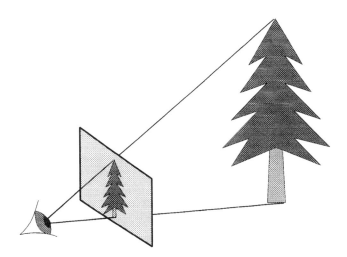

Fig. 4.4 Perspective drawing.

ourselves to lines which do not lie in planes parallel to the artist's canvas. Using equation (4.4) with the sign modified for perspective drawing we obtain

$$u_I = f(U_0 + t\mathbf{n})/(U_0 + t\mathbf{n})_z. \tag{4.5}$$

As n_z is not zero we can always choose the origin of t such that $(U_0)_z = 0$, in which case equation (4.5) becomes

$$u_I = f\{\mathbf{n}/n_z + (1/t)(U_0/n_z)\}, \tag{4.6}$$

which states that the image of a straight line is a straight line. This is also true when n_z is zero and $(U_0)_z \neq 0$ as the reader may easily verify. If we assume that n_z is positive, i.e. that the line proceeds away from the artist as t increases, then as t goes to infinity equation (4.6) becomes

$$u_\infty = f\mathbf{n}/n_z$$

or

$$x_\infty = x_0 + f\mathbf{n}/n_z \tag{4.7}$$

and it is seen that the projected lines converge to a single point x_∞ independent of U_0 which is the 'vanishing point'.

If a vanishing point may be located in the picture which corresponds to known directions $\mathbf{n}$ in the object space, then a vector with direction $-\mathbf{n}$, i.e. coming towards the viewer, starting from the corresponding vanishing point x_∞ will, from equation (4.7), pass through the perspective centre x_0. Further, if two or more such points may be located on the picture then the perspective centre x_0 may be estimated and the eye placed there for viewing the picture. There are two minor points to observe during this experiment; make *mental* estimates of the vanishing points, as drawing on the face of the painting is often frowned upon, and watch out for the inevitable aspidistra carefully placed at x_0 by the ever helpful curator.

It must also be noted that in real paintings these rules are not always followed exactly, as artists may find it necessary to distort the picture in various ways in order to emphasize some artistic, philosophical or religious point. The beauty of this technique is that it is not necessary to understand it intellectually either to do it or to appreciate it, as the firmware of the brain decodes the message whether we understand the mechanism or not.

In the photogrammetric community it is usual to use the identity $\mathbf{R}^T = \mathbf{R}^{-1}$ to write equation (4.4) in the longer form

$$x_i = x_0 - f[\mathbf{R}^T(X_i - X_0)]_x / [\mathbf{R}^T(X_i - X_0)]_z \tag{4.8}$$

and

$$y_i = y_0 - f[\mathbf{R}^T(X_i - X_0)]_y / [\mathbf{R}^T(X_i - X_0)]_z. \tag{4.9}$$

In this form equations (4.8) and (4.9) are known as the collinearity equations, presumably due to the assumption that the object, the image and the pinhole lie in a straight line.

4.3.1 The bundle solution

Equations (4.8) and (4.9) express the measured quantities x_i and y_i as functions of the parameters x_0, y_0, f, $R(\omega, \phi, \kappa)$, X_0 and X_i. In a typical survey we only wish to know the X_i corresponding to the points of interest. However, we are obliged to compute parameters x_0, y_0 and f which are internal to the camera and $R(\omega, \phi, \kappa)$ and X_0 which correspond to the camera but are external to it. In fact there must be at least two cameras imaging each point X_i, but this and other practical constraints on the solution will be discussed towards the end of the section.

The bundle solution is a global least-squares solution for all the parameters in terms of the measured values, the expression 'bundle' presumably indicating that entire bundles of 'rays' are used simultaneously in the derivation of the solution.

4.3.2 Least squares revisited

In this section X and y will be used with a different meaning to that in the rest of this chapter; this potentially confusing decision has been taken to ensure uniformity of notation with Rao (1973). Any linear least squares problem may be written in terms of two matrix equations

$$E(y) = X\beta \qquad (4.10)$$

and

$$D(y) = I_n \sigma^2, \qquad (4.11)$$

where the vector y represents the n observed quantities and β the set of m parameters in terms of which they are to be described. Equation (4.10) then gives the mean or expected values of the observed quantities $E(y)$ in terms of some linear combination of the parameters β written in terms of the matrix X as $X\beta$. Thus (4.10) contains all the information about the supposed theory of the experiment and the experimental design. All the statistical information is contained in equation (4.11), where the dispersion or variance-covariance matrix $D(y)$ is set equal to the unit matrix I_n multiplied by the variance σ^2 of a single observation. This equation asserts that each row of equation (4.10) has equal weight and that the observed quantities, i.e. the elements of y, are statistically independent. If it is suspected that (4.11) is not true then one may use a more complicated theory or change variables to make it true; in photogrammetry it is always possible to do the latter. The significance of (4.11) is that if it is true then the least-squares estimate of the parameters β may be shown to be the linear estimate with the minimum variance.

The formal definitions of $E(y)$ and $D(y)$ are

$$[E(y)]_i = \int \ldots \int y_i P(y_1 \ldots y_n) dy_1 \ldots dy_n$$

and

$$[D(y)]_{ij} = \int \ldots \int (y_i - [E(y)]_i)(y_j - [E(y)]_j) P(y_i \ldots y_n) dy_1 \ldots dy_n$$

124 *Photogrammetry in industrial measurement*

where $P(y_1...y_n)\delta y_1...\delta y_n$ is the probability of finding the value of y_1 in the interval y_1 to $y_1 + \delta y_1$, etc., and the integrals are taken over the ranges of the corresponding variables.

The value of putting any experimental results in the standard form given in equations (4.10) and (4.11) is that various standard procedures are available immediately. The formal least-squares solution to (4.10) and (4.11) for β is $\bar\beta$, where

$$\bar\beta = (X^TX)^{-1}X^Ty \qquad (4.12)$$

and the variance-covariance matrix for $\bar\beta$ is given by

$$D(\bar\beta) = (X^TX)^{-1}\sigma^2, \qquad (4.13)$$

where the value of σ^2 may be estimated from

$$\sigma^2 \approx (y - X\bar\beta)^2/(n - m). \qquad (4.14)$$

The difficulty in photogrammetry is that the equations (4.8) and (4.9) which we wish to use to construct (4.10) are highly nonlinear.

4.3.3 The Newton–Raphson method

In the following no attempt will be made to give the detail of the numerical analysis but rather an idea will be given of how it is done in principle, in order to discuss the solution. To simplify the notation let (4.8) be written as

$$x_i = F_i(x_0, f, \omega, \phi, \kappa, X_0, X_i).$$

Let $x_0 = \bar{x}_0 + \Delta x_0$, etc. Then, expanding to first order in Δx_0,

$$\xi_i \equiv x_i - F_i(\bar{x}_0, \bar{f}, \bar\omega, \bar\phi, \bar\kappa, \bar{X}_0, \bar{X}_i) \approx (\partial F_i/\partial x_0)\Delta x_0 + \ldots . \qquad (4.15)$$

This then gives an approximate linear equation for the new variables ξ_i and equation (4.9) for y_i may be linearized in a similar manner.

Suppose that suitable initial estimates of the parameters x_0, etc., may be found such that the least-squares solution of the set of equations like (4.15) yields values of Δx_0 which when added to the initial estimates $\bar{x}_0$ of x_0 give better estimates of x_0 and the same is true for all the other parameters. Then, as long as this remains true, repeated solution of the linearized equations and refinement of the estimates will converge towards the least-squares solution of the nonlinear equations (4.8) and (4.9). This basic system of solution is known as the Newton–Raphson method. A useful discussion of this type of problem from a numerical point of view is given by Press *et al.* (1988).

One thing is certain; if only the equations (4.8) and (4.9) are used then even the linearized least-squares problem has no solution. This is because seven constraints are needed to fix the origin (three constraints), the orientation (three constraints) and the overall scale (one constraint) of the output co-ordinate system. These seven minimal constraints do not change the shape of the object but only its size and

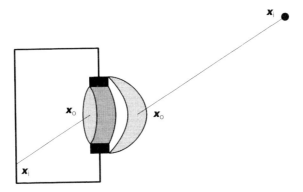

Fig. 4.5 Model of a conventional camera.

position in space. Additional constraints may start to distort the shape. A more detailed study of such questions is given in Cooper (1987).

4.3.4 Treatment of distortions

The difficulty with using a pinhole camera to conduct a real survey is that it has poor light-gathering ability. The usual solution is to use a lens system to increase the light-gathering power of the camera. The penalty for this is that the imaging equations become somewhat more complicated. A model for a conventional camera is shown in Fig. 4.5.

It is seen from the figure that for the model of the conventional camera x_0 does not coincide with X_0 as it did for the pinhole camera. In fact the inner and outer perspective centres x_0 and X_0 correspond to the nodal points of the lens. If the camera system is cylindrically symmetrical but has radial distortions, then this has the effect of changing f and U_z in equation (4.4) to $f + \Delta f(\alpha)$ and $U_z + \Delta U_z(\alpha)$, where α is the angle that the imaging pencil of rays makes with the axis of the system. This angle may be expressed in terms of the interior or the exterior co-ordinate systems and here common photogrammetric practice (see, for example, Granshaw (1980)) will be deviated from by using the exterior set so that

$$\alpha \equiv \tan^{-1}[\sqrt{(U_x^2 + U_y^2)}/U_z].$$

The reason for this is to simplify the statistical analysis of the results by ensuring that only measured values appear on the left hand side of the equations for x_i, etc., and only parameter values on the right.

With these definitions (4.4) becomes

$$\left. \begin{aligned} u_x &= -[f + \Delta f(\alpha)]\, U_x/[U_z + \Delta U_z(\alpha)], \\ u_y &= -[f + \Delta f(\alpha)]\, U_y/[U_z + \Delta U_z(\alpha)] \end{aligned} \right\} \quad (4.16)$$

and

$$u_z = -f.$$

126 Photogrammetry in industrial measurement

If it is further assumed that $\Delta f(\alpha)$ and $\Delta U_z(\alpha)$ are single valued, continuous and differentiable, then they are even functions and may be approximated by an even power series in α.

Let

$$\left. \begin{aligned} s &\equiv -fU/U_z, \\ k(s^2) &\equiv \Delta f(\alpha)/f, \\ q(s^2) &\equiv \Delta U_z(\alpha)/U_z \end{aligned} \right\} \quad (4.17)$$

and
where
$$s^2 \equiv (s_x)^2 + (s_y)^2.$$

Then for a cylindrically symmetrical camera system we have from the definition of **u**, (4.16) and (4.17)

$$x_i = x_0 + s_x[1 + k(s^2)]/[1 + q(s^2)]$$

and

$$y_i = y_0 + s_y[1 + k(s^2)]/[1 + q(s^2)].$$

Most real camera systems, with an exception for some video systems, are very nearly cylindrically symmetrical but any small asymmetrical terms may be approximated by

$$x_i = x_0 + s_x[1 + k(s^2)]/[1 + q(s^2)] + T_1 s^2 + P_1(s_x^2 - s_y^2) + 2P_2 s_x s_y \quad (4.18)$$

and

$$y_i = y_0 + s_y[1 + k(s^2)]/[1 + q(s^2)] + T_2 s^2 + P_2(s_x^2 - s_y^2) + 2P_1 s_x s_y, \quad (4.19)$$

where T_1, T_2, P_1 and P_2 are extra parameters.

These equations (4.18) and (4.19) which are used at the author's institution are the equivalent of (4.8) and (4.9) for a real conventional camera. Similar equations have been developed elsewhere for the same purpose. These equations fall into two broad classes; (4.18) and (4.19) typify the class which attempts to produce a suitable parameterization by appealing to the physical properties of photogrammetric lenses. The alternative approach is to fit any deviation from a perfect pinhole image with a general two-dimensional mapping such as a Fourier series. Surprisingly, these two very different methods lead to similar results for air survey work where film is used rather that glass plates. This is because the camera parameters are usually evaluated in parallel with other parameters after the photography has been taken. Thus they include such effects as the flatness of the film when the photography was taken, in-plane distortion of the film due to the camera mechanism and the properties of the measuring machine used to evaluate the target co-ordinates x_i and y_i which can dominate their apparent values.

4.3.5 Error ellipsoids: fact or fiction?

It is always possible to find a set of co-ordinate axes such that the 3×3 submatrix of the variance-covariance matrix corresponding to the three components X_i, Y_i and Z_i

Mathematical formulation

of X_i can be put into diagonal form. This being the case, it is often supposed that the new co-ordinate components $\tilde{X}_i$, $\tilde{Y}_i$ and $\tilde{Z}_i$ are statistically independent. If this were true, then, assuming the input data were distributed in a Gaussian fashion, the probability density for finding the point x_i, could be written as

$$P(\tilde{x}_i, \tilde{y}_i, \tilde{z}_i) = C \exp\left[-(\lambda_1 \tilde{x}_i^2 + \lambda_2 \tilde{y}_i^2 + \lambda_3 \tilde{z}_i^2)/\sigma^2\right], \quad (4.20)$$

where σ^2/λ_1, etc., are the eigenvalues of the 3×3 submatrix.

It is seen from (4.20) that the surfaces of constant probability density are ellipsoids. The error ellipsoid is then that surface of constant probability density that encloses say 95% of $\int P(\tilde{X}_i, \tilde{Y}_i, \tilde{Z}_i) \, d\tilde{x}_i d\tilde{y}_i d\tilde{z}_i$. It is then easy to show that this ellipsoid is the minimum volume with this property.

The difficulty with the above argument is that the 3×3 submatrix cannot be treated in isolation and therefore the new co-ordinates are not necessarily statistically independent. If this were so the matrix $(X^T X)^{-1}$ would be block diagonal and the bundle solution would be unnecessary. In fact, the estimates of the positions of the points of interest interact via the camera positions and orientations and the constraints used to fix the co-ordinate system.

Nevertheless, the error ellipsoid is often quoted as a measure of the accuracy to which the point positions have been determined and there are two possible interpretations of this. The first is the unusual case where the camera positions are indeed considered to be known exactly, i.e. to better than any accuracy required of the experiment. For the second, it is considered that integration has taken place over all possible values of the parameters other than the co-ordinates of the point of interest. Then the error ellipsoid does indeed give the minimum volume in which there is 95% probability of finding the point of interest. It is always best to find out what particular measures of the object are of interest to the end user of the measurements and evaluate confidence estimates for those measures.

4.3.6 Singularities in the solution

If, by cunning placement of the cameras or otherwise, we manage to set the computer an impossible problem, then the matrix $X^T X$ will be singular so that the correction vector β cannot be computed. This sometimes happens if self-calibration parameters to describe the imaging properties of the camera have been included in the model but cannot be evaluated from the current experiment.

An example of this is encountered in air survey for, when flying over flat terrain, f cannot be determined. Looking vertically on to a flat landscape, changes in f are indistinguishable from changes in flying height, i.e. changes in X_0. This effect can readily be seen in television coverage of cricket matches, in which views of flat objects such as the score board appear quite natural, whereas views of the wicket using the same telephoto lens appear unreasonably foreshortened. In both cases our eyes are not at the correct perspective centre but in the former case it is easy for the brain to suppose that the image on the television is at some appropriate distance. This involves some intellectual suppression of evidence from stereo vision about the position of the set but appears to be quite painless and is possibly acquired by avid viewing. The view of the wicket on the other hand cannot be rationalized by placing the action at any one distance since even the monocular image cannot be seen by an unaided human eye at any position on the field. This is because the relative magnifications of figures at different distances depends on X_0 but the angular size of the image depends on f. Similarly, it is much more difficult to rationalize any scene with nontrivial depth unless the effective value of f taking into account the size of the screen and the distance of the viewer from it correspond to some possible position of the unaided eye. No doubt a good cameraman exploits these effects to the full, like the painter mentioned earlier in the section.

These observations underline the fact that it is always advisable to include explicit constraints to link the internal parameters of the cameras to known calibrated values (i.e. values obtained from a separate calibration experiment) with suitable weighting factors proportional to the inverse of the standard deviation of the calibration. Some may say that to allow such constraints is to leave the way open for unscrupulous persons to cheat on the results and obtain apparently accurate parameters from a poor survey but it could also be said that this method removes unnecessary singularities and consequent waste of time and holds no dangers for the honest experimenter. If there be any dishonest workers in the field then nothing written here will alter their point of view and no doubt they will reap their reward in due time.

4.3.7 Design of experiments

It may be noted that in equation (4.13) the term $(X^TX)^{-1}$ only depends on the experimental design and the parameterization of the imaging device, whereas from (4.14) the estimate of σ^2 depends only on the accuracy of the measurements. In fact, a poor parameterization may affect the apparent experimental estimate of σ^2 but this ought to be detected when the cameras are calibrated. Assuming that the parameterization is good, the fact that $(X^TX)^{-1}$ is independent of the experimental results provides an excellent method of computing the expected uncertainties of the estimated parameters using different experimental designs.

Ideal imagery is obtained by computing the images of a proposed array of points X_i using the calibration values of the imaging parameters and the proposed positions X_0 and orientations (ω, ϕ, κ) of the cameras. A data file is then made up for the ideal experiment and the matrix $(X^TX)^{-1}$ computed using the bundle solution software to be used to analyse the real experiment. This matrix is the variance-covariance matrix but for a factor of σ^2. The small value of σ^2 computed by the software for the ideal experiment is a reflection of nothing but the numerical accuracy of the computer and the software but a more sensible value may be obtained from calibration data on the plate-measuring machine, etc. This may then be used to multiply $(X^TX)^{-1}$ and form an *a priori* estimate of the variance-covariance matrix from which statistical estimates of the uncertainties of some critical measure of the worth of the experiment may be computed. If all else fails even the dreaded error ellipsoids will do, with due reference to the warnings in section 4.3.5.

4.3.8 Calculation of initial values

A major disadvantage of the iterative scheme outlined above for the solution of a nonlinear least-squares problem is the need for initial estimates of all the parameters. In principle, the parameters could just be guessed for the positions of the points and the cameras and for a simple geometry this may well be good enough to start the iteration. In many industrial experiments the positions of the points of interest may be well enough known from drawings or previous surveys both to carry out the design analysis mentioned in section 4.3.7 and to provide initial data for the real calculation, given that the proposed design was followed faithfully in the real experiment.

In more complicated cases where the proposed design might have been modified on site due to unforeseen events, it may be necessary to resort to the so-called **space resection** to place the cameras in the object space, followed by an intersection to find the approximate positions of the points of interest. In the space resection, initial values are assigned to at least four points which may be identified on each camera and these values are used together with the calibrated imaging parameters to calculate the position of the perspective centre X_0 and the orientation $R(\omega, \phi, \kappa)$ of each camera in turn. An intersection program then computes the remaining X_i by intersecting the ray pencils projected from the image points x_i into the object space, again using the calibrated values of the imaging parameters. One method for doing this is outlined by Hunt (1984).

In the worst case, where photography has to be analysed without a detailed knowledge of the experiment, e.g. historical photography (Thompson, 1962), unplanned records of accidents (Robertson, 1990) or plain bad experiments, the computation of the initial values is a formidable task in itself. The saving grace of the method is that if the initial values, however they are obtained, are good enough to start the iteration the final result should be independent of them.

4.4 IMAGING DEVICES

The first and most important property of any photogrammetric imaging device is that it should produce readily identifiable images of target points. The co-ordinates of some reference point on the image, e.g. the centre, should be easy to measure on a suitable measuring machine. The device should be described in terms of some internal parameter set $\{p\}$ and the co-ordinates of the image centres should be related to $\{p\}$, X_0, R and X_i by a known simple formula.

In particular, there is no reason why the image should be in any sense pictorial, although most of them are, or that the formula for computing the co-ordinates of the image centres should bear any resemblance to a pinhole camera as described in section 4.3.

130 *Photogrammetry in industrial measurement*

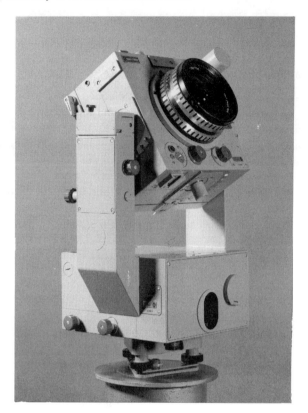

Fig. 4.6 The UMK10 metric camera.

4.4.1 Metric and semi-metric cameras

Metric cameras are expensive devices which form images as close as possible to the equivalent pinhole images. Yet they have similar light-gathering capacity to a normal 35 mm camera and produce a very good quality pictorial photograph. A specific example of this type of camera is the UMK10 camera obtained from Zeiss in Germany, shown in Fig. 4.6.

It can be seen from the figure that such a camera is a substantial piece of experimental equipment. They originated in an era when computing power was strictly limited and it was therefore necessary to eliminate distortion by careful lens design rather than to compensate for it in the software. This was particularly true if the images were being analysed by presenting them to the human vision system. In the case of the UMK10 camera the calibration certificate suggests that at the optimum focus setting the image deviates from the perfect pinhole image by at most 5 μm over a plate area of 150 mm × 180 mm, i.e. one part in 3×10^4. However, using

Imaging devices 131

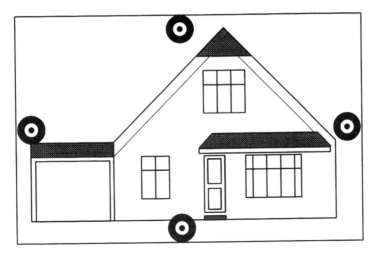

Fig. 4.7 A diagram of a metric camera image showing fiducial marks.

a holographic emulsion like Agfa 10E75 to eliminate grain size and a high-precision measuring microscope, it is possible to locate the image to 1 μm, which shows that it is worth compensating for distortion even in the best metric cameras. The internal co-ordinate system is defined by so-called fiducial marks mentioned in section 4.2.6 which are produced on the photographic plate, often using an internal light source, immediately before the main exposure. An illustration of the image format is given in Fig. 4.7.

Semi-metric cameras are good modified medium-format ($\approx$ 60 mm square) cameras, usually using film rather than glass plates. It is common in such cameras to have a reseau grid etched on a glass plate within the camera. This grid is measured accurately before use and, when the image co-ordinates are measured from the photographs, the image of the grid is used to estimate and eliminate any distortion of the film. This grid may be considered as an extension of the fiducial marks as it defines the internal co-ordinate system, but in this case, with a much larger number of points, the co-ordinates are defined locally as well as globally. An illustration of a typical semi-metric camera image is shown in Fig. 4.8. The camera is always calibrated to enable computer correction of lens distortions which are usually much larger than in metric cameras. Despite all this it is very difficult indeed to get the same accuracy as with a metric camera, yet a good semi-metric system is not much cheaper.

The chief advantage of the semi-metric camera is its comparatively light weight and ease of handling. The controls, etc., are similar to those of a domestic camera and therefore they can be operated by nonspecialist staff, a good example of this being the Rollei system used by the German police force to record traffic accidents and other scenes where three-dimensional information is important.

132 *Photogrammetry in industrial measurement*

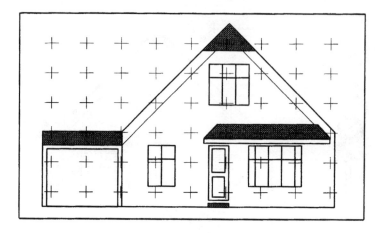

Fig. 4.8 A diagram of a semi-metric camera image showing resean marks.

Historically, cameras were developed to produce visually satisfying records of people, places and events. This tradition has carried over into the production of cameras for metric work where it is not strictly relevant. The disadvantages of this approach may be listed as follows.

1. The lens distortion may have to be modelled by several parameters.
2. Some images may be unclear due to finite depth of field.
3. Results depend critically on capturing geometrical information on a physical recording medium.
4. The images do not necessarily have a characteristic signature.
5. There are vast amounts of unwanted information in regions of the plate containing no target images.

Nevertheless, conventional cameras are capable of giving clear and easily interpreted pictures which are very useful as archival material.

4.4.2 The NPL monocentric axicon camera

The NPL monocentric axicon camera (Burch and Forno, 1982, 1984; Forno, 1990) exemplifies a different approach to photogrammetric imaging. The monocentric axicon lens consists of two concentric spheres of refracting material; in the prototype the inner sphere was liquid but current examples have both spheres made of glass. The camera is designed to image point sources of light, so if the lens were hanging freely in space the imaging pencil of rays would be symmetrical about the line joining the centre of the lens to the centre of the source, just as it is for a simple pinhole. In practice the lens is ground cylindrical at the equator to enable it to fit in the camera and an opaque layer ensures that only rays passing through both spheres

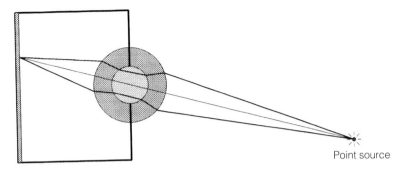

Fig. 4.9 A diagram of the NPL monocentric axicon camera.

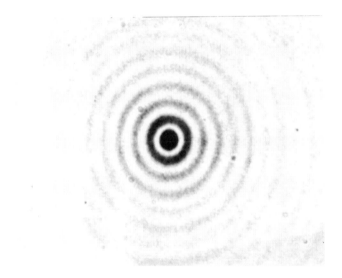

Fig. 4.10 An image from the NPL monocentric axicon camera.

are allowed to reach the image plane, as shown in Fig. 4.9. Thus the perfect symmetry is broken but it turns out that the lens can be made sufficiently well that the deviation of the centre of the image from the position expected for a perfect pinhole image cannot be detected.

The image produced by the camera is a set of concentric diffraction rings as shown in Fig. 4.10. They are formed by an annular region of the aperture over which the phase is stationary and are therefore analogous to the ring fringes discussed in section 2.2.6. As the camera is not focused in the usual sense the depth of field is very large indeed; the current model with a principal distance of approximately 50 mm produces metric images of suitable point sources from 300 mm to infinity. A photogrammetric survey of a three-dimensional test object which was also measured using a very accurate three-co-ordinate measuring machine gave agreement between the two methods to two parts in 10^6, which is at the limit of both techniques.

4.4.3 The role of the measuring machine

No matter how good your cameras or how cunning your experimental set-up, all is in vain if your measuring machine gives you the wrong image co-ordinates. One of the major difficulties with photogrammetry is that nobody has yet designed and marketed a cheap, accurate two-dimensional measuring machine. Expensive accurate measuring machines we have already! In particular NPL has at least three of them with a nominal accuracy of better than 1 μm in x and y. Smaller accurate measuring machines are coming on to the market but they are still expensive.

An obvious design difficulty with most measuring machines is that they measure the wrong thing. The photogrammetrist is only interested in the positions of the image centres with respect to the principal point in some consistent set of plate co-ordinates or the equivalent angular offsets, yet it is usual to measure the fiducial marks in the co-ordinate system of the measuring machine and then to measure the target images in the same system and difference the results. Obviously, a more direct method should prove to be more accurate and robust.

4.4.4 Finding the image centres

With a conventional camera and a typical high-contrast target, it is usual for the operator to set a cross-hair or a ring graticule on the point that appears to be the image centre when viewing the plate with a microscope or on a video screen. This is a classical ergonomic problem which can be analysed in terms of the psychology of visual perception and work is still going on in this field. The gross conclusion appears to be that, for high-contrast targets, the best target to set upon is one which has very nearly the inverse intensity distribution of the graticule or measuring mark. At this level the differences between human vision and its computer equivalent become apparent. Severn (1991), who has recently studied this problem at NPL in connection with measuring machine design, suggests that a simple target like those shown in Fig. 4.11 is less confusing to the visual system than more elaborate targets favoured by machines.

Fig. 4.11 Optimum targets for visual setting.

In the case of highly structured images, it is sometimes possible to construct specialized apparatus to find the image centres which is peculiar to those images. A simple example of this is to fit a special graticule in the microscope designed to optimize visual setting for a particular job but a more instructive example comes from the NPL monocentric axicon camera. The camera gives the very distinctive images shown in Fig. 4.10, which have a clearly defined ring structure. Using point sources of white light, the exposure and development of the photographic plate can be so arranged that the first clear ring of the image is visible over the entire field of view of the camera for a range of object distances.

A laser system has been developed by Burch and others which causes the beam to track round a circle on the photographic plate exactly matching the first clear ring of the target image. The total light intensity passing though the plate is monitored using a photosensitive detector. When the circle described by the laser beam is nearly concentric with the target image, the residual mismatch of the laser track and the first clear ring modulates the light intensity passing through the plate. The phase of the modulation relative to the signal driving the beam gives the direction of the offset and the magnitude of the modulation gives its size. This is similar to the technique for measuring bore straightness described in section 2.4.3. A phase sensitive detector is used to display this information for the operator in the form of a Lissajous figure on an oscilloscope. The operator then moves the stage until the figure collapses to a point at the centre of the screen, indicating that the target image has been centred. Using this method, which is described in detail by Forno (1990), it was possible to set on axicon images with a repeatability of better than 50 nm in x and y on a 50 mm × 50 mm plate.

Currently interest is growing in digital image processing techniques to locate the centres of target images. Such techniques can easily compensate for perspective projection and shading which are difficult for human operators. The disadvantage is that the computer lacks the human operator's intuition and background knowledge which he or she uses to reject large portions of the visual field when they are partially obscured or damaged. A more detailed discussion of these techniques will be included in the next section.

4.5 TOWARDS REAL TIME

Real time photogrammetry is one of the most interesting and fastest-moving aspects of the subject and it would require a whole chapter to do justice to the advances made in the last five years but in this section an attempt will be made to give a flavour of what is going on.

In a sense all experiments are done in real time; certainly very few are done in imaginary time! Nevertheless, the term does have some meaning. A typical photogrammetric survey might require between four hours and a day to target the object, obtain the photographs and develop them. Then it might take between one and three days to measure the plates and analyse the results. Thus, even though the data may be recorded on the plates in as little as 1/500 of a second, the whole survey may occupy several days. In a so called real-time photogrammetric survey the data are

136 *Photogrammetry in industrial measurement*

captured in some digital form and processed by an on-line computer. The aim is to return three-dimensional co-ordinates at video frame rate, i.e. updated every 1/50 of a second or so, although very few systems ever achieve this at present.

4.5.1 The coming of computer vision

It took the advent of very fast analogue to digital converters to enable computers to 'grab' single video frames from cameras originally designed for broadcast television and load them into the memory of a computer. Once this development had occurred, the prices of the components were driven down by their use in nonmetric applications such as television commercials, amateur video and machine vision for schools. These apparently trivial events have placed a precise new measuring tool in the hands of the metrologist.

Any measurement of this type which can reasonably be included in a chapter on photogrammetry has some form of video camera as the primary data-input device and may be represented schematically as in Fig. 4.12. This figure shows a rather stylized form of the apparatus needed to set up a computer vision system. The performance of any such system depends on the detailed choice of components. This is no place to recommend any particular manufacturer but some general comments on the contents of the boxes may be of help in setting up such a system.

The camera shown in Fig. 4.12 is supposed to represent a good solid-state camera adapted for metric work. In practice it is best to obtain a raw CCD camera with no anti-aliasing devices, preferably no interlace and no gamma correction but with a pixel clock output.

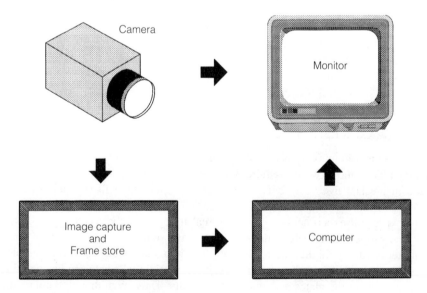

Fig. 4.12 Schematic layout of a computer vision system.

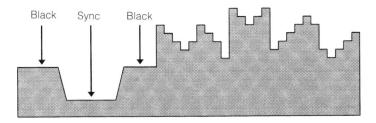

Fig. 4.13 One line of monochrome analogue video.

At the time of writing, a good CCD camera for metric work has an array of about 500 × 600 photosensitive elements so that the picture is compatible with a nominal 512 × 512 frame store and typically produces an analogue output suitable for driving some form of video monitor. The form of such a signal is shown schematically in Fig. 4.13. It is seen from the figure that the line of video starts with the black reference level (the front porch), gives the line sync pulse, goes back to the black level (the back porch) and then starts on the video data. In the figure, which is very stylized, exaggeration is given to the slope on the edges of the sync pulse which is a consequence of the finite bandwidth of the signal.

Most frame stores currently available transfer the pixels into the frame store using their own pixel clock which is triggered by the line sync pulse. Line-to-line variations in the line signal alter the difference between the trigger voltage and the black level. This effectively moves the trigger level up and down the slope between the bottom of the sync pulse and the black level, resulting in a jitter in the pixel clock timing which is observed as a jitter in the spatial registration of the lines in the frame store. The frequency of the frame store clock is locked to the frequency of the camera clock (or vice versa) and this electronic locking implies that the one frequency wanders about the other. These variations lead to errors in the spatial registration of the frame store within a line. In addition to these problems inherent in trying to synchronize one pixel clock to another, there is the more obvious effect of the frame store format just being different from the physical distribution of photosensitive elements in the camera. Thus a 500 × 582 camera chip cannot be mapped directly on to a 512 × 512 frame store; either camera pixels must be discarded and blanks left in the store or the picture must be interpolated, leading to periodic variations in apparent intensity due to sampling.

The solution to this problem is to use the camera's own pixel clock output to transfer the pixel values into the frame store. Interpolation must not be used and pixels should be ignored or left blank to ensure an exact match between the usable area of the frame store and the usable area of the CCD array. With such a digitization system it is claimed that the positions of extended images can be located accurately to about 0.01 of the period of the CCD array.

Thus measurements of image points on the CCD array may be made with a best relative accuracy of 1 in 5×10^4, comparing quite reasonably with a metric camera which achieves a best relative accuracy of 1 in 10^5. At present relatively few workers

138 *Photogrammetry in industrial measurement*

achieve this accuracy but the use of video processing in metric work is steadily increasing.

4.5.2 Video theodolites

The early work using video images to locate metric targets was limited by the various effects mentioned above and one could seldom interpolate to better than 0.5 pixels. This combined with relatively small 256 × 256 frame stores gave a relative accuracy of about 1 in 500, much worse than any metric camera.

One way around this problem is to mount the camera on top of a theodolite. The theodolite then establishes the exterior co-ordinate system and gives very accurate angular orientation of the telescope, while the video camera can establish the presence of the target in the centre of the field of view. Such a device would usually have motorized drives on the theodolites so that both the image processing and the driving of the theodolites could be performed by the same host computer.

A diagram of such a setup is shown in Fig. 4.14. Let the effective rotation centre of the first theodolite be at $X_0^{(1)}$, the centre of the second theodolite be at $X_0^{(2)}$ and the point of interest be at X_i. Then for either theodolite the horizontal bearing θ_i and the vertical bearing ϕ_i, illustrated in Fig. 4.15, are given by

$$\left. \begin{array}{l} \theta_i = \tan^{-1}(U_y/U_x) \\ \phi_i = \tan^{-1}[\sqrt{(U_x^2 + U_y^2)}/U_z] \\ U = R^{-1}(X_i - X_0). \end{array} \right\} \quad (4.21)$$

and
where

The rotation matrix R has the same significance as the rotation matrix used to define U in equations (4.3) and (4.4), in that it ensures that the vector U is expressed in terms of the interior co-ordinate system of the theodolite. It is seen that both the metric camera and the theodolite may be considered as angle-measuring devices, the theodolite measuring the angles sequentially and the camera recording the angular data for several points in parallel. The analysis of both proceeds in a similar manner,

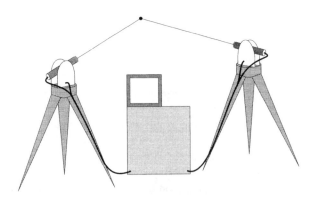

Fig. 4.14 An automatic video theodolite system.

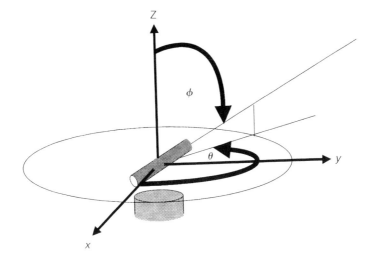

Fig. 4.15 Theodolite bearings.

although it is more usual to use theodolites from known fixed (or defined) positions. For example, $X_0^{(1)}$ might be taken as the origin and $X_0^{(2)}$ may be considered to be at $(X_{12}, 0, Z_{12})$. Then, if the Z-axes of both theodolites and the exterior co-ordinate system are aligned with the gravitational vertical, pointing the theodolites at one another gives Z_{12} in terms of X_{12} and serves to set the zeros on the horizontal scales such that the X- and Y-axes of all three co-ordinate systems also coincide. In this simple situation R is just the unit matrix and the only unknown in the setup is the overall scale, which may be fixed by including at least one known length in the survey. This form of the analysis is the one most commonly used in field work.

Such a system has the advantage that a reasonable location of the target image in the video frame results in very good estimates of θ_i and ϕ_i, because the entire field of view of the theodolite telescope only represents about one and a half degrees. Let us assume that some engineering artefact subtends about 60° at the theodolite in both θ and ϕ. Then in order to obtain an angular accuracy of 1 in 10^5, corresponding to a good metric camera, the telescope needs to be aligned to about two seconds of arc. This corresponds to about 1 in 2500 of the video field or about 0.2 pixels on a 512×512 element frame store, as opposed to centring target images to 0.005 pixels which would be necessary to use the video camera directly as a metric camera. This analysis may be unduly pessimistic as it is based on the magnification of a known theodolite. It would be perfectly possible to increase the magnification and therefore decrease the required acuity of the image centring in terms of pixels.

4.5.3 Location of images

It is all very well to require that a video system should estimate a target image position to some fraction of a pixel but it is another thing entirely to do it. In order to

140 *Photogrammetry in industrial measurement*

weigh one method of location against another, it is necessary to use some statistical analysis of the expected results and this in turn requires us to construct some model of the process.

Least-squares analysis provides us with a suitable tool for the job. Let us suppose that the image of a target formed on the light-sensitive surface of the camera is given by some function $f(x - x_0)$, where x_0 is the centre of the image and x is the position on the surface. Each effective pixel, i.e. each address in the frame store, will contain a value given by the convolution of $f(x - x_0)$ with some pupil function characteristic of the camera and the frame store. Here we will make the simplest assumption that this pupil function is proportional to a delta function. This, together with the assumption that the photosensitive surface lies in the x–y plane, gives the pixel value for the ith row and jth column as P_{ij}, where

$$P_{ij} = f(x_i - x_0, y_j - y_0) \tag{4.22}$$

and x_i and y_j are the effective co-ordinates of the pixel on the photosensitive surface.

In order to progress further it is usual to linearize (4.22) to obtain

$$P_{ij} \approx f(x_i - \tilde{x}_0, y_j - \tilde{y}_0) + (\partial f/\partial x_0)\Delta x_0 + (\partial f/\partial y_0)\Delta y_0 \tag{4.23}$$

with

$$x_0 = \tilde{x}_0 + \Delta x_0$$

and

$$y_0 = \tilde{y}_0 + \Delta y_0,$$

where $\tilde{x}_0$ and $\tilde{y}_0$ are the initial estimates of the co-ordinates of the image centre.

There are various other methods of estimating the centres of target images currently in use, namely centres of gravity, edge following and template matching but, unless new information is introduced about the nature of the image, the least-squares method may be shown to give the minimum variance linear estimator of x_0 based on the P_{ij} and in that sense it is the best that can be done.

With this simple model, the sum of the squares of the differences between the pixel values predicted by equation (4.23) and the measured values may be minimized directly with respect to Δx_0 and Δy_0 to give their least squares values $\langle\Delta x_0\rangle$ and $\langle\Delta y_0\rangle$. Thus

$$\left.\begin{aligned}\langle\Delta x_0\rangle &= \{\Sigma(f_y^2)\Sigma[f_x(P_{ij}-f)] - \Sigma(f_x f_y)\Sigma[f_y(P_{ij}-f)]\}/[\Sigma(f_x^2)\Sigma(f_y^2) - (\Sigma f_x f_y)^2] \\ \langle\Delta y_0\rangle &= \{\Sigma(f_x^2)\Sigma[f_y(P_{ij}-f)] - \Sigma(f_x f_y)\Sigma[f_x(P_{ij}-f)]\}/[\Sigma(f_x^2)\Sigma(f_y^2) - (\Sigma f_x f_y)^2]\end{aligned}\right\} \tag{4.24}$$

where
$$f = f(x_i - \tilde{x}_0, y_j - \tilde{y}_0),$$
$$f_x = (\partial f/\partial x_0),$$
$$f_y = (\partial f/\partial y_0)$$

and the sums are taken over all i and j.

Similarly, if it is assumed that the pixel values all have the same variance $\sigma^2(P)$, the expected variances of the estimates $\langle\Delta x_0\rangle$ and $\langle\Delta y_0\rangle$ may be written as

$$\sigma^2(\langle\Delta x_0\rangle) = \sigma^2(P)\Sigma(f_y^2)/[\Sigma(f_x^2)\Sigma(f_y^2) - (\Sigma f_x f_y)^2]$$
$$\sigma^2(\langle\Delta y_0\rangle) = \sigma^2(P)\Sigma(f_x^2)/[\Sigma(f_x^2)\Sigma(f_y^2) - (\Sigma f_x f_y)^2].$$

During any given iteration $\tilde{x}_0$ is a constant and therefore the variance of x_0 is equal to the variance of Δx_0. Thus if the image is symmetrical in the sense that $\Sigma f_x f_y = 0$, then the variances simplify to

$$\sigma^2(\langle x_0\rangle) = \sigma^2(P)/[\Sigma(f_x^2)] \qquad (4.25)$$
$$\sigma^2(\langle y_0\rangle) = \sigma^2(P)/[\Sigma(f_y^2)]. \qquad (4.26)$$

These equations show that the accuracy to which x_0 and y_0 may be determined depends essentially on two things, namely the accuracy of the individual pixel values characterized by $\sigma^2(P)$ and the mean-square gradients of the image distribution function f. In particular, it is manifestly counterproductive to focus the image down to a sharp spot on the array which is less than one or two pixels in diameter. In fact the image should be distributed over as many pixels as possible consistent with being able to distinguish images of different object points. Similarly, it is useful to have as bright an image as possible without saturating the camera or the digitizer in order to minimize the contributions of thermal noise and the number of available grey levels to the effective signal-to-noise ratios σ/f_x and σ/f_y. In addition to these effects, the clocking of the frame into the digitizer affects the overall registration of the target image in the frame; this is taken to limit systems which do not explicitly use the pixel clock to an accuracy of about a tenth of a pixel in x_0 and y_0. Averaging over several frames may reduce this effect as long as it is random from frame to frame. However, in many cases this would defeat the object of acquiring the data at video rate. Systems explicitly using the pixel clock are claimed to obtain accuracies of better than a hundredth of a pixel spacing.

4.5.4 True video photogrammetry

A few years ago the author was involved in setting up a demonstration of video photogrammetry using two inexpensive vidicon cameras and a BBC microcomputer. This was a very salutary exercise and he counsels anyone who is intending to take up photogrammetric survey seriously to do the same. This subsection begins by describing the steps involved in putting together the demonstration from scratch and ends by mentioning some references to more sophisticated applications of the technique. The basic apparatus necessary for the demonstration is a small microcomputer, a frame grabber, two video cameras, a video monitor and a joystick or mouse.

The first thing to do is to calibrate the cameras in some way. With the relatively low accuracy of location of the video images, it is remarkably difficult to calibrate the cameras from scratch during a survey using the least-squares software, as is often done with metric cameras. Thus an optical experiment must be performed to get at

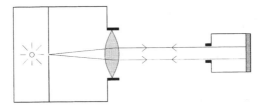

Fig. 4.16 Optical method for principal point.

least initial estimates of the principal distance f in pixel units, the effective position of the principal point in the pixel array and the ratio of vertical to horizontal pixel sizes.

The definition of the principal point gives a clue to a straightforward way of finding it. Let us imagine the camera to be set up on an optical bench as shown in Fig. 4.16 facing an autocollimator (section 3.4). The lens is removed from the camera and the camera rotated until the front surface of the photosensitive array is perpendicular to the axis defined by the autocollimator. At this stage the parallel light from the autocollimator is falling normally on the photosensitive surface, so replacing the camera lens brings it to a focus at the principal point. The centre of this image as detected by the host computer in the frame store gives the effective co-ordinates of the principal point. A good theodolite can play the part of the autocollimator but if neither is available a suitable substitute can be constructed using a pinhole in a white screen and a lens as shown in Fig. 4.16. The photosensitive surface is then normal to the beam when the image of the pinhole coincides with the pinhole itself.

The principal distance may be deduced simply by imaging two points of known separation S at various distances from the camera, taking care that the line joining the points is parallel to the line scan of the camera, its image passes through the principal point (which could be marked at this stage with a video overlay that also indicated the line scan direction) and the images of the points lie symmetrically either side of it (facilitated by marking a point mid-way between them on some supporting structure such as a ruler).

Suppose that this is done twice with the midpoint at distances l_1 and l_2 from the perspective centre, giving two sets of images d_1 and d_2 apart as shown in Fig. 4.17. Then, using similar triangles,

$$S/l_1 = d_1/f \text{ and } S/l_2 = d_2/f,$$

where S, l_1 and l_2 are in millimetres but f, d_1 and d_2 are in horizontal pixel units so that

$$f = d_1 d_2 (l_1 - l_2)/[S(d_2 - d_1)].$$

This subtraction method is recommended because it avoids having to estimate the position of the perspective centre at this stage. In a similar way, the points can be replaced at a distance l_1 from the camera and the support structure rotated about its

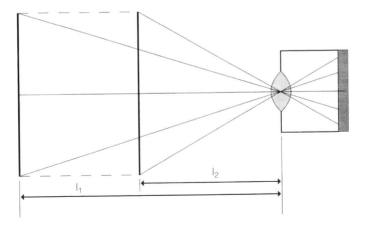

Fig. 4.17 Calibration of a video camera.

midpoint to bring the line joining their images perpendicular to the line scan while the midpoint remains in its l_1 position. If this gives images d_v apart then

$$a/b = d_v/d_1,$$

where a and b are the horizontal and vertical pixel spacings.

Obviously more sophisticated methods may be used to get a more accurate calibration of the video cameras but the simple method above proved sufficient to get the demonstration going.

In a demonstration it is simplest to involve the 'audience' by enabling them to locate the images on a monitor display with a suitable cross-hair injected into it, using a video overlay directed round the screen by means of the joystick or a mouse. This greatly simplifies the software and, if anything, increases the educational value of the demonstration. The necessary software consists of two distinct parts: first, a package to calibrate the system when viewing a known artefact which has been constructed for the purpose, and second a package to survey an unknown object using the calibration data. In traditional photogrammetry using metric cameras these operations are often done simultaneously, removing any worry that the set-up may wander out of calibration during the experiment. In the demonstration it seemed more important to obtain answers quickly using fairly modest computing power. This situation will be changed radically by the transfer of the software to a much faster computer such as the Acorn Archimedes. A similar separation is used in some medical and robotic applications of real-time photogrammetry where speed of unknown point co-ordination is of utmost importance.

If it is assumed that the cameras have been calibrated with respect to their principal points, principal distances, a/b ratios and any other internal distortions which are considered important, then each camera still has six parameters associated with it. Three of these give the position of the perspective centre of the camera and three give its orientation with respect to the co-ordinate axes. As each camera measures two internal co-ordinates for each target image, it must therefore see at least three known points, i.e. points with known external co-ordinates, in order to compute its position and orientation. In fact, it is usual to produce an artefact with four known points on it which do not lie in a plane to enable the system to choose between possible multiple solutions. There are several algorithms which may be used to do the calculation, including the least-squares method outlined in Section 4.2 above. The difficulty with the least-squares method is that it may be frustrated by the relatively poor data. The actual algorithm used is outlined in Hunt (1984).

Once both cameras have been oriented and located, it is comparatively easy to find the co-ordinates of unknown object points by extrapolating the imaging pencil of rays from the measured image out through the perspective centre to the object point. If this is done for more than one camera, the intersection of the ray pencils gives the position of the point. Once again this is slightly more difficult in practice and the actual method used is given in Hunt (1984).

The calibration object used in the demonstration was an equilateral triangle of mild steel with targets some 200 mm apart near the corners. The fourth target was placed on a steel pillar at the centre of the triangle. The object was then measured independently and mounted kinematically on the base board of the demonstration.

Each time the demonstration was set up, the calibration object was placed in position and the target positions located on the video frames with the joystick as described above. The calibration program was then run to establish the positions and orientations of the cameras, which took about 15 min on a BBC microcomputer, and the demonstration was ready to locate small objects in three dimensions such as marks on a ruler. Even this very crude system could measure to about 1 mm in a volume of about 200 mm cubed, which was sufficient for demonstration purposes. It may be noted that this corresponds well with an accuracy of about one pixel on the 256×256 frame store that was used.

More sophisticated systems with pixel clocking and subpixel interpolation have been shown to achieve 1/100 of a pixel on a 512×512 frame store, which with the same simple geometry should be able to give results to about 5 μm in a 200 mm cubed volume. Detailed descriptions of some such high accuracy applications are given in the conference proceedings from Interlaken (Gruen *et al.*, 1987), Vienna (Gruen and Kahmen, 1989) and Zurich (Gruen and Baltsavias, 1990).

4.6 APPLICATIONS FOR PHOTOGRAMMETRY

Like any other technique, photogrammetry has its strengths and its weaknesses. Its strengths lie in its portability, parallel data collection and relative robustness; its weakness is the time delay between obtaining the data and computing the results, although it is sometimes useful to be able to retire to a safe or stable environment to measure up the plates. In conventional (i.e. not video) photogrammetry a week of work may be needed for measuring the plates and feeding the image co-ordinates into computer programs.

Applications for photogrammetry 145

The direct competition to photogrammetric survey comes from theodolites and rangefinders, both of which look at object points sequentially. Three typical uses of photogrammetry will now be described with comments on the advantages and disadvantages of alternative techniques.

4.6.1 Survey of an infrared telescope

Any candidate for precise survey work of this kind must be something which depends for its function on its three-dimensional shape but is of such a size or is situated in such a location as to hinder the use of laboratory-based techniques. A typical example of such a device is a radar antenna, which has its performance intimately linked with its physical shape and is necessarily situated in an exposed location. For these reasons, if there is a problem with such an antenna thought to be due to distortion of the dish, the geometry is often checked using a photogrammetric survey. The experimental layout for such a survey is shown in Fig. 4.18 (see also Oldfield, 1984). Although the sizes of antennae vary according to the application, usually between 500 and 1000 target points are sufficient to define the shape of the dish, while four photographs should give an accuracy of about 1/100 000 of the overall size of the dish.

Fig. 4.18 Photogrammetric determination of the shape of a dish antenna.

146 *Photogrammetry in industrial measurement*

The James Clerk Maxwell telescope in Hawaii is intended to look at infrared radiation in the wavelength range 0.3–1.0 mm. This was translated into a requirement to know the surface topography of the 15 m dish to an accuracy of 0.05 mm measured along the normal to the theoretical paraboloid. As with any large dish of this kind, the distortion of the dish under gravity had to be considered, as the telescope changed its attitude to examine objects near the zenith and objects close to the horizon. A special interferometer had been devised to measure the figure of the dish when it was aimed directly at the zenith but it was impossible to use this technique when the telescope was aiming at points closer to the horizon. This was where the photogrammetric technique became valuable.

The telescope itself had been cunningly designed to maintain a parabolic shape even when it was tilted. Thus as the telescope changed its elevation its dish was intended to change from one paraboloid to another, while maintaining the same focus and therefore having minimal effect on the image quality. As mentioned above, the figure could be checked accurately when pointing at the zenith and we were concerned to measure the deviation from the design paraboloid as the telescope was brought to bear on objects nearer the horizon.

In this particular case the dish was made up of 276 panels which were assumed to be of correct geometrical form within themselves. Each of the panels was mounted

Fig. 4.19 The James Clerk Maxwell Telescope in Hawaii.

kinematically on a subframe via three computer-controlled screw adjusters, giving the surface 3×276 degrees of freedom. Fortunately, we were asked to collaborate in this project before the final assembly of the telescope on Hawaii and we were therefore able to have three retroreflecting targets placed on each panel as the dish was being assembled, thus relieving us of the very considerable task of trying to attach the targets to the completed telescope.

There were two main difficulties from our point of view: 1. the high accuracy required, and 2. the altitude of 4200 m above sea level on the volcanic mountain of Mauna Kea, which made it impossible to make tea. A photograph of the dish is shown in Fig. 4.19. While we were submitting our proposal for the job, we carried out a numerical simulation of the experiment which revealed that the required accuracy should be achievable if eight photographs were used for each orientation of the dish and the photographic plates could be measured to ± 1 μm. During this simulation a file was generated containing all the expected three-dimensional co-ordinates of the targets.

In the fullness of time two members of staff braved the rigours of altitude sickness and lack of tea to take photographs of the telescope. In order to fit in with other work on the installation and to use the most thermally stable conditions they worked mainly at night. Figure 4.20 is a flash photograph of the dish which shows the retroreflecting targets to advantage. These targets have come into their own recently with the advent of automatic target recognition as, with properly arranged illumination,

Fig. 4.20 A flash photograph of the targeted dish.

148 *Photogrammetry in industrial measurement*

they appear on the (negative) plates much darker than any other objects in the field of view, thus simplifying the task of the image analysis program.

The team returned home with a very large number of plates to be measured. In our experimental design most plates included all of the points. A typical plate therefore contained some 828 points to be measured together with four fiducial marks to establish the orientation of the plate on the measuring machine. The design had also called for eight views of the dish in each orientation, giving 6656 points to be measured to ± 1 μm in x and y. It is little wonder therefore that soon afterwards our thoughts turned towards automatic methods of target recognition and plate measurement.

The data analysis itself was done in the manner outlined in Section 4.3. The most noticeable difference when treating real experimental data is the detection of so-called gross errors, such as the wrong assignment of serial numbers to observed images. Errors of this kind are usually due to operator fatigue and are reduced in proportion to the extent to which the measurement system is automated.

4.6.2 The destruction of a bridge

Often, when some civil engineering structure has finished its useful life, it is just destroyed wantonly, but this is not always the case. It is sometimes valuable to test the redundant artefact to destruction in a controlled manner and so learn something of the ultimate strength of similar structures still in service. Such was the case when a disused Victorian canal bridge had to be removed to make way for a new road. Figure 4.21 shows an idealized view of the experiment. This omits for clarity the pouring rain, the local 'Save our Bridge' committee and the charging herd of cows. Suffice it to say that nothing is ever as simple as it appears when you write the proposal.

In the previous example, all the eight views of a given orientation of the dish were taken using the same camera and retroreflecting targets were used to give clearer images for subsequent precision measurement. This was partly because the structure

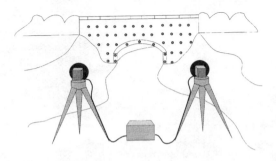

Fig. 4.21 An idealized view of a bridge test.

was assumed to be so stable that it would not have moved significantly between photographs and partly because it is hard work lugging metric cameras up mountains. In the case of the bridge the converse was true. Two cameras had to be used with synchronized flash guns in order to freeze any motion of the bridge under the pressure of the hydraulic ram used to load it and retroreflecting targets were necessary to make the images visible. Used in this way, photogrammetry is much more convenient than a more conventional theodolite survey and much more flexible than purpose-built gauges.

This required the measurement of less than 50 points on each plate and the data analysis was relatively straightforward. It was also backed up by a concurrent moiré photography experiment (see section 9.4.1) which gave a continuous strain map of the bridge surface as it deformed and this assisted with the most difficult problem of defining an appropriate co-ordinate system in which to express the results. The difficulty arises from the form of the experiment itself. If you are pressing down on a bridge (and by the same token pulling up on the ground) to such an extent that the bridge is being crushed, then the surrounding countryside is bound to move a little, taking your cameras and any local 'fixed' points along with it. It is important therefore to agree with your client exactly what frame of reference is preferred. A similar experiment done in a more friendly indoor environment is described by Forno *et al.* (1991).

4.6.3 Measurement of a fuel tank

A few years ago the law was changed regarding the taxation of oil stocks, so that the taxable volume was that held in the tank farm rather than that pumped out of the tanker. This, understandably, created a great deal of interest in the volume of fuel storage tanks. In the case to be described, the 'traditional' method of measuring the volume was to use a depth gauge on the side of the tank and multiply by the appropriate cross-sectional area to generate the required volume. Corrections were made for the distortion of the tank under the weight of the fuel and for any internal structure of the tank. The client assured us that uncertainty in the tank distortion was the major cause of error. This distortion was modelled mathematically and checked from time to time by measuring the distances from the tank sides to verticals set up using optical plummets. These were normalized by comparing the lower readings with measurements of the outer circumference measured with steel tapes and referred to as 'strappings'. The purpose of the photogrammetric survey was to measure about 500 marked points on the outside of some of these tanks to provide an independent estimate of the external geometry. The accuracy required was $\pm£1000$ in the value of the product, which turned out to be ±0.1 mm normal to the surface of the tank. Some of the tanks to be measured are shown in Fig. 4.22.

The photograph was taken in the early morning light. The reasons for this are twofold: first the tank being measured was being filled in the very early morning and, second, the author feels they look better in the dark! This view is not shared by the

150 *Photogrammetry in industrial measurement*

Fig. 4.22 A photograph of jet fuel storage tanks.

oil companies, however, who must look upon the tanks as things of great beauty, because unlike the owners of the Hawaii telescope, they would not allow any permanent marking of the surface which might spoil their appearance. This problem was solved by attaching a large number of metric spheres to annular ceramic magnets and suspending them over the side of the tank, mounted on the string at one metre intervals by means of elastic bands. Then, with a deft flick of the wrist, the magnets and balls were brought in contact with the side of the tank and held there by the magnets. Figure 4.23 shows the author preparing to perform this feat on top of one of the jet fuel tanks shown in Fig. 4.22.

The reader might enquire how we managed to acquire a large number of metric spheres of sufficient quality and at reasonable cost. The answer is that we used two-star ping-pong balls which, by the nature of their use, are very accurately made. You would never believe the difficulty of persuading the public service to requisition 1000 ping-pong balls!

Once the balls had been placed on the tank the photography was carried out using a single camera and by taking photographs successively from 7 or 13 positions around the tank. A prime number of positions was always used, to reduce the possibility of random noise in the plate measurements inducing co-operative perturbations of the results. The experimental design would have been improved considerably if photo-

Applications for photogrammetry 151

Fig. 4.23 The author preparing a spherical target for a fuel tank.

graphs could have been taken from more than one horizontal plane but the very tight fire restrictions on such a site and the pipework, etc., associated with the tank prevented the use of any mobile hoist to take elevated views. In the event the results were reinforced by additional theodolite measurements on the empty tank which served to set the scale and the gravitational vertical for the photogrammetry. This method proved very satisfactory and shed considerable light on anomalies observed in the geometry obtained by other methods.

On the whole our experience at NPL with the technique of photogrammetry has been a positive one. It has certainly enabled us to measure structures in the size range from 1 m to 100 m with an accuracy of about 1 in 10^5 of the overall dimensions. In many of these cases other techniques could not compete without incurring considerable additional difficulty and expense. As always, the trick is to choose the right method for the measurement task in hand and I hope that this section will help provide the information necessary to make an informed and correct decision.

REFERENCES

Burch, J. M. and Forno, C. (1982) Preliminary assessment of a new high precision camera. *International Archives of Photogrammetry and Remote Sensing*, **24**, Part V/1, 90–9.

Burch, J. M, and Forno, C. (1984) Progress with the NPL Centrax camera system. *International Archives of Photogrammetry and Remote Sensing*, **25**, Part 5A, Commission V, 141–50.

Cooper, M. A. R. (1987) *Control Surveys in Civil Engineering*, Collins, London.

Forno, C. (1990) PhD Thesis: *A camera for high accuracy photogrammetry*, University of Surrey, Department of Mechanical Engineering.

Forno, C., Brown, S.B., Hunt, R. A. *et al.* (1991) The measurement of deformation of a bridge by moiré photography and photogrammetry, *Strain – Journal of the British Society for Strain Measurement*, **27** (33), 83–7.

Granshaw, S. I. (1980) Bundle adjustment methods in engineering photogrammetry. *Photogrammetric Record*, **10** (56), 181–207.

Gruen, A. and Baltsavias, E. P. (eds) (1990) *Close-range Photogrammetry meets Machine Vision*. Proc. SPIE **1395**.

Gruen, A. and Kahmen, H. (eds) (1989) *Optical 3-D Measurement Techniques*, Herbert Wichmann Verlag GmbH, D-7500 Karlsruhe.

Gruen, A., Beyer, H., Parsic, Z. and Steinbrückner, L. (eds) (1987) *Fast Processing of Photogrammetric Data*, Institut für Geodäsie und Photogrammetrie, ETH Zürich.

Hunt, R. A. (1984) Estimation of initial values before bundle adjustment of close range data. *International Archives of Photogrammetry and Remote Sensing*, **25**, Part A5, Commission V, 419–28.

Oldfield, S. (1984) Photogrammetric determination of the form of a 10 m diameter radio antenna. *International Archives of Photogrammetry and Remote Sensing*, **25**, Part A5, Commission V, 590–6.

Press, W. H., Flannery, B. P., Teukolsky, S. A. and Vetterling, W. T. (1988) *Numerical Recipes in C*, Cambridge University Press.

Rao, C. R. (1973) *Linear Statistical Inference and Its Applications*, 2nd edn, Wiley, London.

Robertson G. (1990) Aircraft crash analysis utilizing a photogrammetric approach. *Proc. SPIE* **1395**, 1126–33.

Severn, I.(1991) PhD Thesis: *An automated photogrammetric plate measurement system*. Brunel University, Department of Manufacturing and Engineering Systems.

Thompson, E. H. (1962) Photogrammetry in the restoration of Castle Howard. *Photogrammetric Record*, **4** (20), 94–119.

5

Laser interferometry for precision engineering metrology

P. GILL

5.1 GENERAL PRINCIPLES

Interferometric techniques for the precision measurement of length have found increasing application since the turn of the century. With the advent of compact laser systems of narrow spectral bandwidth and corresponding high coherence, laser interferometry has developed into a standard technique for the measurement of length, displacement, angle and other related dimensional quantities at the highest precision. A number of manufacturers within the high-technology optical industries provide complete stand-alone or OEM length and displacement measuring systems. Most precision engineering concerns now take advantage of the high resolution and accuracy of these devices, both in the calibration of material standards such as gauge blocks and line standards, but also through on-line incorporation of such devices into computer-controlled lathes and three-axis co-ordinate measurement machines.

5.1.1 Optical interference by amplitude division

The most common interferometer arrangement for length measurement is based on the generation of optical interference fringes by amplitude division, and has the general arrangement of a Michelson interferometer. Basic interferometer operation (Fig. 5.1) consists of division at a beamsplitter of the initial wave from the laser source into two components $E_{1,2}(t) = E_{1,2} \cos(\phi_{1,2} - \omega t)$, where E is the field amplitude, ϕ is the phase at any time t and $\omega = 2\pi\nu$ where ν is the frequency of the radiation. Components $E_{1,2}(t)$ are reflected back to the beamsplitter by stationary mirror M_1 and moveable mirror M_2, respectively, and constitute the reference and measurement arms of the interferometer. The two retroreflected beams are recombined on the beamsplitter and a photodetector placed behind the beamsplitter records a recombined beam intensity of the form

$$I = E_1^2 + E_2^2 + 2E_1E_2 \cos(\phi_2 - \phi_1), \qquad (5.1)$$

where the interference term can be seen to be a periodic intensity variation related to the phase difference of the recombined waves. This phase difference results from the

Optical Methods in Engineering Metrology. Edited by D.C. Williams. Published in 1993 by Chapman & Hall, London. ISBN 0 412 39640 8

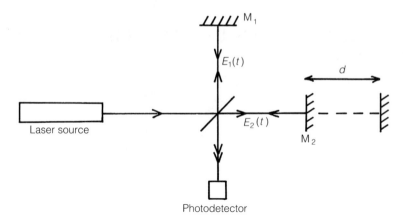

Fig. 5.1 Michelson interferometer arrangement.

different optical path lengths traversed in the two arms of the interferometer, and is zero for equal path lengths. If mirror M_2 is displaced by an amount d parallel to the measurement beam axis, the change in phase difference can be written

$$\Delta(\phi_2 - \phi_1) = (2\pi/\lambda)\, 2d \tag{5.2}$$

which results in intensity or fringe maxima according to the condition $4\pi d/\lambda = 2N\pi$. This reduces to the basic Michelson equation for displacement measurement

$$d = N\,(\lambda/2), \tag{5.3}$$

where N is the number of interference fringes recorded for an air wavelength λ and displacement d. For a 1 m displacement, $N \sim 3 \times 10^6$ for 633 nm laser light. Provided the wavelength is known to high accuracy, electronic counting techniques allow the determination of large displacements with single-fringe count resolution of $\sim 0.3\ \mu\text{m}$ ($\lambda/2$). In many applications, electronic fringe subdivision to $\lambda/100$ or better extends the precision of the device further.

5.1.2 Limitations to fringe formation

Before describing a typical single-wavelength fringe-counting interferometer, it is useful to consider various limitations to the quality of the interference fringes detected. The overriding limitation to fringe formation is the coherence length of the light source used. The coherence time τ_c of the source is defined as the time over which the source emits light without phase interruption or perturbation, and is inversely proportional to the source spectral bandwidth $\Delta\nu$. The coherence length is the

Table 5.1 Spectral bandwidths and coherence lengths for various sources

Source	λ (nm)	v (THz)	Δv	d_{max}
White light	400–600	700–450	250 THz	<1 μm
Multimode ion laser	515	482	10 GHz	1 cm
Mercury isotope lamp	546	550	300 MHz	30 cm
Single-mode diode laser	780	380	50 MHz	2 m
Single-mode He-Ne laser	633	473	1 MHz	100 m
Actively stabilized narrow linewidth He-Ne laser	633	473	50 kHz	2 km

corresponding greatest path difference d_{max} over which fringes can be formed and is given by $d_{max} = c\tau_c = c/\pi\Delta v$ where c is the speed of light. Table 5.1 demonstrates typical source bandwidths and coherence lengths for a variety of sources.

From Table 5.1, it can be seen that white light sources can only form fringes for very small path differences (~ 1 μm), and are not appropriate for the measurement of large displacements by fringe counting. White light fringes are, however, extremely useful fiducial reference indicators of zero path length difference in a variety of interferometry applications. In contrast, a typical single-mode He-Ne laser, where the light is emitted in a narrow frequency bandwidth determined by the laser cavity resonance (for example Siegman, 1986), will have a coherence length often well in excess of 100 m, and is thus well suited for length interferometry over displacements of a few tens of metres. The upper limit in this case is due to other constraints. Simple two-mode He-Ne lasers with typical 500 MHz mode separations are capable of forming fringes over similar path length differences, but in this case the fringe contrast periodically decreases to zero every half laser cavity length (~ 15 cm), resulting in fringe-counting difficulties for displacements in excess of this value. Finally, diode laser sources are found in increasing numbers of interferometry applications due to their small size, low cost, low power requirements and ease of electronic manipulation. However they do generally exhibit intrinsically broader single mode linewidths than a He-Ne laser, with corresponding coherence lengths of a few metres at most. Methods for improving upon this diode coherence limitation are discussed in section 5.7.1.

Given adequate source coherence properties, the contrast of the fringe amplitude relative to the overall detected intensity is dependent on the effectiveness of the overlap of the recombined beams $E_{1,2}(t)$, with contributions due to beam size differences, beam shear and non-parallelism (Rowley, 1972). The most significant contributor to degradation of effective interferometer operation over path differences greater than a few metres is the beam size variation with propagation. As explained in section 2.1.1, a collimated Gaussian beam diverges according to the relation

$$\omega^2 = \omega_0^2 [1 + (\lambda z / \pi \omega_0^2)^2], \tag{5.4}$$

where ω_0, ω are the minimum beam size at the waist and the size at distance z, respectively (Kogelnik and Li, 1966). Thus a 1 mm collimated beam diverges to 10 mm after 50 m, resulting in a fringe-contrast reduction of some 80% when recombined with the reference beam. To avoid this, interferometry over distances greater than a few metres needs expanded beams (e.g. greater than 5 mm), as do alignment applications (section 2.1.3).

Other contributions to fringe contrast reduction include beam shear, where the recombined beams are parallel but displaced, and non-parallelism, where subsidiary straight-line wedge fringes with a fringe spacing inversely proportional to the angle between the beams are formed. A difference in curvature of the interfering wavefronts also contributes according to equation (2.13), with the corresponding modification to equation (5.6) given by equation (2.14). Appropriate interferometer optical design and high-sensitivity angular adjustments should generally allow one to reach the optimum alignment situation of a 'fluffed-out' fringe, where the whole detector area senses a light field of uniform fringe intensity.

Assuming adequate coherence and contrast properties, the practical range limit in air for laser interferometry is ~50 m. This is caused by return beam wander due to atmospheric turbulence. Thermal gradients over these distances are very capable of shifting transiently the return beam off the detector, with resulting count interruption. This difficulty can of course be alleviated by housing the interferometer measurement arm within an evacuated chamber, or by use of other options outlined in section 2.3.3, but these generally are not feasible in simple arrangements.

5.1.3 Refractive index compensation

For an interferometer operating in air, the basic Michelson equation (5.3) is more completely written

$$N\lambda_{vac} = 2nd, \tag{5.5}$$

where λ_{vac} is the vacuum wavelength and n is the refractive index of air. This leads to a displacement given by

$$d = N (\lambda_{vac}/2) (1/n). \tag{5.6}$$

Most commercially available laser interferometers have means for automatic compensation of the laser wavelength for the effect of the air refractive index by monitoring of the environmental conditions. As explained in Section 2.3, the refractive index of air is itself a function of wavelength, as well as air temperature, air pressure, water vapour partial pressure (humidity), carbon dioxide partial pressure and any other significant gas constituent of the local atmospheric environment. A widely used equation (Edlén, 1966) for the refractive index of dry air n_s under standard conditions of temperature and pressure (15 °C, 1013.25 mbar) is

$$(n_s - 1).10^8 = 8342.13 + 2\,406\,030\,(130 - \sigma^2)^{-1} + 15\,997\,(38.9 - \sigma^2)^{-1}, \quad (5.7)$$

where $\sigma = 1/\lambda_{\text{vac}}$. The refractive index n_{Tp} at T °C and p mbar can then be computed from

$$(n_{Tp} - 1) = (n_s - 1) \times \frac{100p[1 + p(61.3 - T)10^{-8}]}{96\,095.4(1 + 0.003\,661T)} \quad (5.8)$$

Further small corrections need to be made for water vapour and carbon dioxide partial pressures. The value of the refractive index of air under typical conditions is ~1.00028.

It is helpful to gauge the magnitude of effect that changes in ambient conditions have on the refractive index and hence wavelength value and displacement measurement. An interferometer measurement error of 0.1 ppm can be induced by an air temperature change of 0.1°C, an air pressure change of 0.3 mbar, a humidity change of 10% and a carbon dioxide concentration level change of 600 ppm. As we shall see shortly, such magnitudes are fairly significant in the error budget for the absolute precision of a length measuring interferometer.

5.1.4 Single-wavelength fringe-counting interferometry

A simple form of a single-frequency fringe-counting displacement measuring interferometer is shown in Fig. 5.2. The heart of the system is the frequency-stabilized laser source which emits a highly stable vacuum wavelength of known value. Techniques for the frequency stabilization of He-Ne sources are considered in the next section, followed by a review of the methods for providing absolute calibration of the wavelength relative to the international definition of the metre.

As shown in Fig. 5.2, the interferometer comprises a fixed beam divider and reference arm, and a cube-corner retroreflector mounted on a moveable stage. The displacement of this retroreflector relative to the fixed part is monitored by counting the $\lambda/2$ fringes generated by the movement. A cube corner is generally used in preference to a plane mirror since the direction of the returned beam is insensitive to rotation about any axis (section 2.5.2). It also laterally displaces the return beam, preventing optical feedback into the laser source, which can seriously disturb correct operation of the stabilized laser. The direction of motion is determined by generating two sets of fringe signals which are in phase quadrature with each other. A simple method of producing this 90° phase shift is by use of a metallic film beam-divider (Peck and Obetz, 1953) and two photodetectors placed as shown in Fig. 5.2. The displacement is then computed from this bi-directional count and the wavelength value compensated for refractive index. Compensation can be effected either by use of air sensors to monitor temperature, pressure and humidity or by means of a subsidiary refractometer which interferometrically measures the magnitude of the

158 *Laser interferometry for precision engineering metrology*

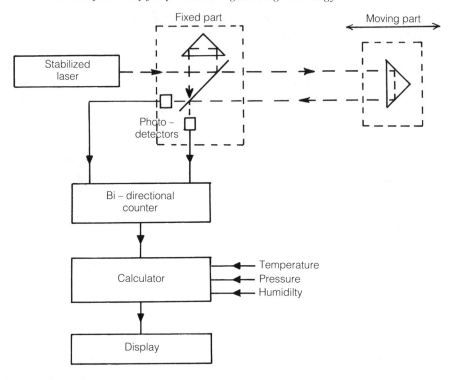

Fig. 5.2 A simple fringe-counting interferometer. (Courtesy of National Physical Laboratory.)

refractive index under measurement conditions (Terrien, 1965; Downs and Birch, 1983).

5.2 STABILIZED LASER WAVELENGTH SOURCES

5.2.1 International definition of the metre

The primary metrological unit of the laser interferometer is the frequency or vacuum wavelength value of the laser source radiation. The accuracy with which this is known determines the limiting capability of the measurement system, though other system and operator errors and uncertainties may prevent the achievement of this limit. Thus in any application requiring absolute measurement accuracy, the laser wavelength needs to be traceable to the internationally recognized standard of length: the metre. In 1983, the General Conference of Weights and Measures redefined the metre in terms of the length of the path travelled by light in vacuum in $1/c$ of a second, where c is the fixed value for the speed of light, 299 792 458 m s^{-1}. This definition was chosen, after considerable international deliberation, to be rather more conceptual than previous ones, and has the effect of rendering the metre dependent on the second. The major ramification of the generality of the definition is to allow the primary means of realizing it to change as laser technology evolves.

To provide traceability to this definition for a visible laser radiation, the frequency v of the radiation needs to be measured absolutely with respect to the caesium atomic clock microwave reference frequency which defines the second, and the wavelength derived from the relation $\lambda = c/v$. Whilst this remains a well-nigh impossible task for most users of laser interferometers, many national standards laboratories have developed certain highly stable reference lasers whose frequencies are traceable directly in this manner to an uncertainty of ±1 in 10^9. These can then provide traceability for the laser interferometer source radiation. This section presents a short review of common techniques for laser frequency stabilization for both interferometer sources and reference lasers, followed by discussion of methods of frequency calibration in terms of these references. The discussion concentrates on the single-frequency 633 nm red He-Ne laser but leads on to alternative wavelength He-Ne lasers operating at other visible wavelengths (e.g. 543 nm green) as these are of importance in multiple wavelength interferometers (Pugh and Jackson, 1986). A review of laser diode sources and techniques is delayed until section 5.7, which focuses on swept-wavelength interferometry.

5.2.2 He-Ne stabilization techniques

A typical unstabilized internal mirror He-Ne laser has two or more cavity modes separated by ~500 MHz, across a neon gain profile width of ~1.2 GHz. The frequencies of the modes are given by $mc/2l$ (for example Siegman, 1986), where l is the optical length of the cavity and m is the integer number of half-wavelengths within this length for a particular mode. In this basic form of laser, there is a relatively high uncertainty of average wavelength of ±1 ppm, due to thermal drift of the cavity length and of the corresponding cavity modes under the profile. A simple type of stabilization now in widespread use is the orthogonal polarization intensity balance scheme (Balhorn *et al.*, 1972; Bennett *et al.*, 1973). This makes use of the property of internal mirror He-Ne lasers where adjacent modes are orthogonally linearly polarized (Fig. 5.3(*a*)). The modes are separated using a polarizing beamsplitter and separate detectors monitor the intensity signal in each polarization. These signals are combined in a difference amplifier to provide an error signal, which is fed back to the laser via a heater wrapped around the laser tube. The tube length, and hence the laser frequency, are servoed to hold the intensity balance position. Laser frequency uncertainties of 10^{-7} are typical.

Another laser-stabilization technique used in various interferometer systems involves the application of a small magnetic field (~100 gauss) along the axis of a

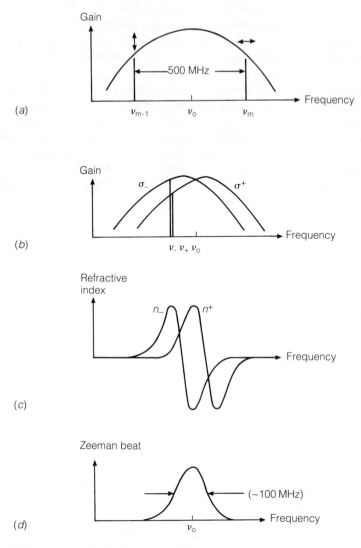

Fig. 5.3 He-Ne laser stabilization techniques.

short cavity internal-mirror laser tube. The magnetic field has two main effects. First there is a gross Zeeman splitting of the gain profiles for orthogonal $\sigma\pm$ polarizations (Fig. 5.3(b)). Thus for a single-cavity mode, there are two orthogonal polarization components to the mode in the region of profile overlap. Additionally there is a small shift (~1 MHz) between the orthogonal components, which arises because they see slightly different refractive indices and hence cavity lengths. The different frequencies are given by

$$v_-^+ = mc/2n_-^+ l. \qquad (5.9)$$

The intensities of the orthogonal polarizations are separated, detected and balanced to provide an equal intensity control point at the unshifted line centre of the transition v_o. The error signal so derived is then fed back onto the cavity length either by means of a heater or a piezoelectric transducer (PZT). This method is potentially slightly more accurate than the previous two-mode intensity-balance technique since the electronic balance point is the centre of the neon transition rather than a position on the slope of the gain profile which may exhibit some varying asymmetry with time.

A more accurate variant of this magnetic field approach is to derive the laser frequency control signal by detection of the ~1 MHz beat between the v_-^+ orthogonal components (Baer et al., 1980). Looking at Fig. 5.3(c) it can be seen that the beat $(v^+ - v_-)$ will vary over a small tuning range (~100 MHz) about a maximum at v_o line centre (Fig. 5.3(d)). The beat is detected by mixing polarizations on a fast detector, and slow cavitylength modulation techniques allow phase-sensitive detection of this maximum beat. A heater foil is used both for the cavity length modulation and error signal feedback (Rowley, 1990).

One of the most accurate and well-established techniques for providing long term stability is to reference the laser mode to saturated spectra in diatomic molecular iodine vapour (Hanes and Dahlstrom, 1969). Iodine has many thousands of fine-structure transitions across the visible spectrum due to vibrational and rotational substructure. If a short 10 cm iodine vapour cell is placed within a laser cavity with external mirrors and a Brewster-windowed 633 nm He-Ne plasma tube, one or more of these fine-structure 1 GHz wide Doppler-limited linear absorptions will modify the He-Ne gain profile. In addition, a series of weak iodine hyperfine saturated spectra (inverted Lamb-dips) can be observed over the profile tuning range. These are typically 5 MHz wide (1 in 10^8) and arise since the standing-wave cavity mode sees a narrow region of decreased absorption loss when it is coincident with the absorption natural linewidth of the hyperfine component centre frequency. This is a standard technique for the removal of Doppler spectral broadening due to the thermal motion of the absorbers. Both this and the use of a low-pressure absorption cell which does not need electrical excitation, provide the means for a long-term narrow spectral reference which is relatively perturbation free and very reproducible with time. Phase-sensitive detection and PZT transducers again allow saturated component signal recovery and cavity length feedback, providing a short-term frequency stability of $\sim 10^{-11}$ and reproducibility of 1 in 10^{10}. These iodine-stabilized He-Ne lasers are typically the current primary references in the realization of the metre, and have the ± 1 in 10^9 absolute uncertainty associated with their wavelength value when operated under internationally preferred conditions. Table 5.2 shows a comparison of the levels of reproducibility of the various types of stabilized laser.

Table 5.2 Stabilized He-Ne wavelength reproducibilities

Stabilization routine	Reproducibility ($\Delta v/v$)
Unstabilized	3×10^{-6}
Lamb dip stabilized	5×10^{-8}
Two-mode heater controlled	$10^{-7} - 10^{-8}$
Zeeman (intensity controlled)	$10^{-7} - 10^{-8}$
Zeeman (beat controlled)	$10^{-8} - 10^{-9}$
Iodine saturated absorption	10^{-10}

In recent years, the development of alternative wavelength He-Ne lasers operating, for example, in the green (543 nm), yellow (594 nm) and orange (612 nm) has had significant impact on the potential of multiple wavelength interferometry. These laser transitions have much lower gain than the 633 nm transition, and provide output powers in the range 200 μW to 2 mW. Techniques broadly similar to those for 633 nm stabilization have been developed for these alternative wavelengths, but with more attention being paid to problems of tube alignment and controlling increased random polarization flipping (Rowley and Gill, 1990). As a result a commercial multiple-wavelength interferometer incorporating such He-Ne devices is now available, predominantly for the calibration of gauge-block material standards. The technique of multiple wavelength interferometry is described in section 5.6.

5.2.3 Laser Wavelength Calibration and Traceability

At this juncture, it is appropriate to consider the methods used to provide traceability of the laser interferometer stabilized sources to the iodine-stabilized 633 nm reference lasers. The calibration of 633 nm lasers is by simple heterodyne measurement. The unknown and standard radiations are mixed on a beamsplitter and detected by a fast photodiode. Provided the frequency difference of the lasers is within the detection bandwidth, the r.f. beat frequency between them can readily be monitored over long periods with a frequency counter. For the calibration of other laser wavelengths away from 633 nm, the measurement is slightly more involved. Figure 5.4 shows the basic method. Both the unknown wavelength and radiation from a 633 nm transfer laser are directed through a 1 m evacuated Fabry–Perot interferometer. The etalon length is servo-controlled to a Fabry–Perot fringe maximum for the unknown radiation, and the transfer laser is frequency-servoed to another fringe maximum in the red. Prior knowledge of the etalon length and nominal wavelength values allow unambiguous determination of the fringeorder number ratio. Finally the red transfer laser is monitored relative to an iodine-stabilized laser by heterodyning. Accuracies of comparison of 1 in 10^{10} in both these techniques are possible.

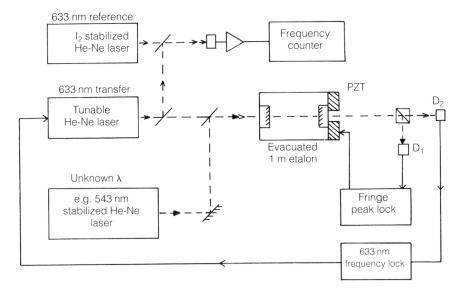

Fig. 5.4 Wavelength comparison. (Courtesy of National Physical Laboratory.)

5.3 HETERODYNE INTERFEROMETRY

One particular type of interferometer has been adopted by a number of interferometer manufacturers due to the advantages its conceptual design offers over simple d.c. fringe counting designs. The latter systems need to take careful account of intensity level changes both of the source and in the measurement arm as motion occurs, otherwise fringe-contrast changes and d.c. level shifts can cause fringe miscounting. The heterodyne interferometer is an a.c. device and thus avoids these problems.

A basic form of heterodyne interferometer is shown in Fig. 5.5. Typically, a Zeeman laser source emits the two closely spaced orthogonal polarization frequencies v_1, v_2 separated by ~ 1 MHz. A simple beamsplitter separates off part of the signal from both polarizations which are mixed on detector D_1 to provide a reference beat $v_1 - v_2$. The transmitted components are then split at a polarizing beamsplitter (PBS). Components v_1 and v_2 traverse the fixed reference arm and measurement arm, respectively, and are recombined at the polarizing beamsplitter and detected at D_2. In this case, however, the detected beat is only $v_1 - v_2$ when the measurement retroreflector is stationary. During motion of the cube corner, frequency v_2 is Doppler-shifted by Δv, the sign being dependent on the motion direction. The detected beat is thus given by $v_1 - v_2 \pm \Delta v$. The reference and Doppler-shifted beats are counted independently and subtracted to give Δv. This beat difference gives rise to a count rate dN/dt which is equal to $v(2s/c)$, where s is the velocity of the cube-corner. Integration of the count over time t leads to $N = 2d/\lambda$.

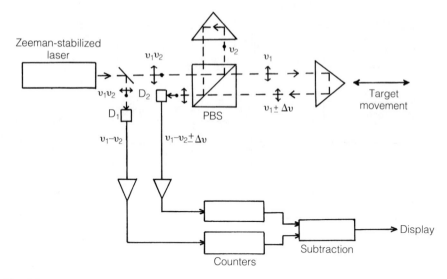

Fig. 5.5 A heterodyne interferometer.

With a typical reference beat of ~1 MHz, it is possible to monitor Δv values up to ± 1 MHz before introducing ambiguities due to the beat crossing through zero. This limits the target speed possible in this case to < 0.3 m s^{-1}, which could be a constraint in some applications. Systems using reference beats of a few megahertz have been developed. These might typically use external acousto-optic modulators to obtain the larger differences.

5.4 MACHINE TOOL CHARACTERIZATION

The description of single-frequency fringe-counting interferometers has so far concentrated on their application to a simple displacement measurement in one dimension. The interferometer systems now available, however, have a far greater capability for an almost complete characterization of machine-tool movements. The extent of the problem can be gauged by noting that there are some 21 different degrees of freedom for this characterization, with six degrees in any one direction of travel. Typical capabilities of these interferometers which help provide this characterization include displacement measurement in threeaxes simultaneously, velocity, angle (leading to pitch, yaw and flatness calibration) and straightness (leading to squareness and parallelism). Two of these options are discussed briefly.

5.4.1 Angle measurement

Angle measurement is achieved by placing a steering mirror above the polarizing beamsplitter in order to transmit the reference beam parallel to the measurement

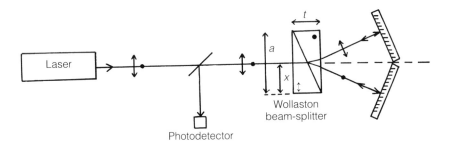

Fig. 5.6 A straightness interferometer.

beam but laterally displaced. The far retroreflector is a double cube-corner assembly with fixed separation a between the apices. As the machine stage to which this assembly is fixed tilts during travel, there is a displacement Δd of the reference arm relative to the measurement arm, and the angle of tilt θ can be computed from $\Delta d = a \sin\theta$.

5.4.2 Straightness Measurement

Straightness measurement requires a more significant change from the conventional displacement design, and makes use of a special Wollaston beamsplitter. The arrangement is shown in Fig. 5.6; like angle, the measurement is differential. The Wollaston prism comprises two birefringent wedges with orthogonal optical axes. One laser polarization experiences an ordinary refractive index n_o followed by an extraordinary one n_e and the other polarization encounters them the other way around. This has the effect of deviating the polarizations to separate fixed plane reflectors on either side of the axis of travel. The path difference experienced by the orthogonal polarizations on re-emergence from the prism after reflection is given by

$$\Delta d = \frac{t}{a}(a - 2x)(n_o - n_e), \tag{5.10}$$

where t and a are the prism thickness and width. Thus the path difference is proportional to x which is a measure of the beam sideways movement, or the lack of straightness. Such angle and straightness capabilities allow interferometric alternatives to the geometrical optical techniques for tilt and straightness measurement discussed in Chapters 2 and 3.

5.4.3 Plane mirror interferometers

Finally, there are a variety of plane mirror arrangements, where the measurement cube-corner is replaced by a plane mirror and quarter-wave plate combination. This is used either to increase the number of passes through the measurement arm and hence the system resolution, or to produce a common path arrangement for the

166   *Laser interferometry for precision engineering metrology*

reference and measurement arms, which can reduce the interferometer sensitivity to ambient changes (Baldwin and Siddall, 1984).

5.5   LASER INTERFEROMETER MEASUREMENT ERRORS

Concluding the discussion of single-wavelength fringe-counting interferometry, we examine the limiting accuracy possible with such systems, and additional errors which can further degrade these accuracies and which are due to operator error.

5.5.1   System specification

The basic specification typically claimed for a fringe-counting system (in which 0.32 μm or 0.16 μm corresponds to one $\lambda/2$ or $\lambda/4$ fringe) is 0.1 p.p.m. Here it is assumed that a correct wavelength compensation figure is supplied by the operator using error-free sensors for refractive index computation or an accurate refractometer. In most cases, however, use is made of the subsidiary automatic compensation unit which relies on input data from the various environmental temperature, pressure and humidity sensors. In these cases the limiting accuracy specification is downgraded significantly to a few parts per million, a fact which suitably indicates the potentially most serious causes of error. The overall error is thus the sum of various component errors arising from the stabilized laser wavelength and automatic air sensors. Errors may also arise as a result of fringe miscounting and errors in software used to compute the corrected displacement value.

5.5.2   Laser wavelength errors

There are a number of possible causes of laser error. A gross error can arise from parasitic laser operation at an additional wavelength such as 640 nm. In this case the extent to which the effective laser frequency is modified is not readily determinable, and can only be assessed through individual calibration of the particular laser tube. It is also possible for a malfunction of the laser-stabilization electronics to cause laser frequency drift up to ± 500 MHz and an associated length error of ± 1 p.p.m. This probably would also cause miscounting because of periodic loss of fringe contrast due to the existence of more than one mode at particular points on the laser tuning curve. Less significant laser frequency shifts occur over time as the laser tube gas pressure changes. For the Zeeman laser, for example, a shift of 0.01 p.p.m. per 1000 h is observed typically. Finally, changes by manufacturers to the isotopic constituents of the He-Ne mixture need to be taken into account in the laser frequency value used by the interferometer software.

5.5.3   Compensation errors

In general, the most serious errors arise from the automatic compensation sensors, with ± 0.1 p.p.m. overall length error arising from ± 0.1 °C temperature-sensor error,

±0.3 mbar pressure sensor error or ±10% humidity sensor error. Dependent on the type of sensor used, length errors of a few parts per million due to sensor inaccuracies are quite commonplace. In certain situations, complete failure of a sensor will result in default values being assumed by the interferometer software. Errors well in excess of 10 p.p.m. have been observed in some such circumstances.

5.5.4 Interferometer verification

In order to provide length traceability of commercial interferometers and to ensure that the high potential accuracy is achieved in practice, NPL provides a comprehensive calibration service to this end. First, the vacuum wavelength is calibrated by heterodyning against an iodine-stabilized He-Ne laser, with calibration uncertainty of 0.002 p.p.m. Automatic compensation sensors are then calibrated against working standards traceable to respective national standards of temperature, pressure and humidity. Finally, the overall displacement measurement facility is verified by comparison with an independently traceable NPL length interferometer, over distances of up to 30 m. This comparison can be achieved with an uncertainty of ±0.2 p.p.m. provided that the test device readings are sufficiently reproducible.

5.5.5 Operator errors

Whilst such considerations are necessary to verify the potential accuracy of the interferometer, system performance can be degraded seriously by the operator if attention is not paid to certain possible operator errors. The three most common operational errors arising are cosine error, Abbe or sine error and deadpath error.

Cosine error arises when the laser beam measurement direction is not parallel to the tracking direction of the stage upon which the measurement cube-corner is mounted. This can be a serious source of error for short displacements if care is not taken. For larger displacements, generally the need for good alignment to ensure adequate functioning of the counting system is sufficient to ensure that the cosine error is reduced below the intrinsic 0.1 p.p.m. accuracy limitation of the device. The fractional error is given by $\theta^2/2$ and amounts to ~150 p.p.m. for an angle θ of 1° between beam and track. In order to render this negligible, the angle needs to be reduced to $<0.01°$.

Abbe error arises when the observation axis is laterally displaced from the beam measurement axis. If the mechanical stage tilts during movement, then an Abbe offset error of magnitude $A\sin\theta$ will corrupt the interferometrically measured displacement, where θ is the tilt and A is the separation between observation and optical axes. For a tilt of 0.1° and 10 cm separation, this is ~150 μm, which is serious. Good metrological design always aims to minimize this separation.

Finally, deadpath error arises when the measurement cube-corner cannot be positioned close to the interferometer beamsplitter for the zero displacement position. In any subsequent displacement measurement there is then a deadpath common

168 *Laser interferometry for precision engineering metrology*

to both initial and final positions which suffers environmental changes that are not accounted for in the air wavelength compensation value.

5.6 MULTIPLE WAVELENGTH INTERFEROMETRY

In many cases of single-wavelength fringe-counting interferometry for displacement measurement, accuracies can be limited by motional and thermal effects associated with the cube-corner/workpiece movement, and by environmental effects. The techniques of multiple-wavelength interferometry for the precision length measurement of material standards and artefacts such as gauge blocks bypass such limitations. This is achieved by performing static interferometric measurements in semi-sealed environmental chambers, where atmospheric conditions can be monitored more closely. Whilst only really applicable to the measurement of small standards (such as gauge blocks and the smaller endbars), in general a better absolute accuracy is achieved, with absolute uncertainties at the 0.1 p.p.m. level and reproducibility to an order of magnitude better.

The technique of multi-wavelength interferometry has been used over several decades. However, only over the last few years, with the development of small and user-friendly stabilized lasers at wavelengths other than 633 nm, and the corresponding ability to use automatic fringe detection and fringe-field manipulation using TV displays, have methods for fast, automatic and efficient gauge block measurement become established firmly both in international standards laboratories and in some of the high-precision industrial metrology laboratories.

5.6.1 Method of excess fractions

The basic interferometer arrangement is a Twyman–Green variant of the Michelson interferometer shown in Fig. 5.7. The gauge to be measured is optically contacted or **wrung** to a flat base plate. Both the gauge end and the base plate are highly polished and reflecting. Stabilized laser light of suitable coherence is expanded, collimated and divided at the beamsplitter. The expanded beam in the measurement arm is reflected off the gauge top surface or base plate and recombined with the reference beam. A TV camera views the fringe field so formed, which clearly has different phase contributions due to the separation between gauge top and base plate. For ease of analysis the reference mirror is tilted slightly, so that straight-line wedge fringes are formed in the field of view, each successive fringe maximum representing a half-wavelength change in height relative to its nearest neighbour. The gauge block cross-section can thus be viewed with a set of wedge fringes discontinuous with those for the base plate (see Fig. 10.3(*a*)).

For any point on the gauge-block cross-section, the fractional fringe discontinuity ϵ between top and base can be measured. The length of the gauge is given by

$$2d = (M + \epsilon)\, \lambda_{\text{vac}}/n, \tag{5.11}$$

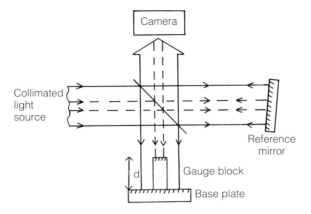

Fig. 5.7 A Twyman–Green interferometer arrangement.

where M is the unknown fringe order number corresponding to the path length between top and base. It is not possible to determine this integer with one wavelength only. However, this can be achieved with two or more known wavelengths and a preliminary less accurate knowledge of the gauge block length. Normally, the manufacturer's nominal value to a few micrometres is sufficient, in the case of a good gauge block. The measured set of fractions ϵ_i for wavelengths λ_i constitute an unambiguous set over a local range of several consecutive order numbers, but are liable to be repeated over larger ranges. Thus, provided the nominal value identifies this local range, the order integers can be determined precisely, leading to an improved value for the gauge length.

5.6.2 The NPL gauge-block interferometer

(a) Instrument design

A schematic of the automatic gauge-block interferometer developed at NPL is shown in Fig. 5.8. The laser sources are 633 nm and 543 nm stabilized He-Ne lasers with calibrated wavelength uncertainties of less than 0.01 p.p.m. (Pugh and Jackson, 1986). He-Ne lasers at other wavelengths such as 594 nm and 612 nm are also available. The laser light is coupled to the interferometer by means of a short multimode fibre link. The appropriate wavelength is selected by an acousto-optic shutter just prior to launching into the fibre optic coupler. Environmental conditions within the interferometer are monitored accurately by a platinum resistance thermometer, a vibrating cylinder pressure transducer and a dew-point hygrometer, and data is transmitted via an IEEE bus to the controlling computer. With such transducers, the refractive index correction to the wavelength can be monitored continuously and updated at the 0.1 p.p.m. level.

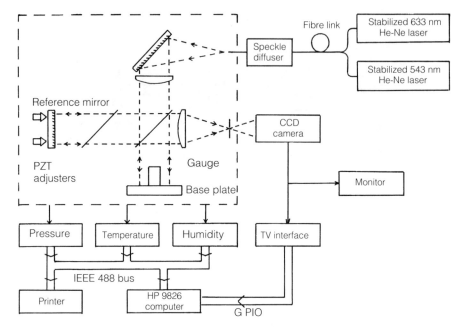

Fig. 5.8 NPL gauge–block interferometer. (Courtesy of National Physical Laboratory.)

(b) Fringe analysis

Automatic measurement of the fringe fraction ϵ_i at each wavelength λ_i is achieved by sampling electronically the CCD camera fringe display across the base and gauge along a selected line at right angles to the fringe contours and through the gauge centre. The base-plate fringe sequence is then interpolated to the centre of the gauge and subtracted from the fractional fringe value sampled at the centre of the top of the gauge. The fringe fraction can be measured to 0.01 fringe. A complete measurement at two wavelengths for one gauge typically takes two minutes. The base plate can accommodate up to 20 blocks at a time, and is rotated to bring each block in sequence into the camera field of view. In addition, gauge surface flatness and topography can be measured by varying the sampling line position along the gauge. Further detail is given in section 10.1.1. The device has proved to be a highly efficient instrument for routine gauge-block and end-bar measurement.

(c) Instrument refinements

One recent NPL refinement to the multiple-wavelength interferometer has been to add a white light interferometer to the instrument in order to increase its versatility.

Interferometry using laser diodes 171

This has the effect of dispensing with the need for prior knowledge of the nominal value of the gauge length. In this mode of operation the laser sources are replaced by a white light source, and the reference mirror is tracked so as to produce two sets of white light zero-path-length fringes for the gauge top surface and base plate, respectively. The displacement of the reference mirror is monitored in the normal way using a single-frequency fringe-counting commercial plane mirror interferometer.

5.7 MULTIPLE- AND SWEPT-WAVELENGTH INTERFEROMETRY USING LASER DIODES

So far multiple-wavelength interferometry has been discussed in terms of stabilized He-Ne laser sources. However, the recent emergence of laser diodes which are wavelength tunable over limited spectral ranges has resulted in what might be described as the rediscovery of multiple-wavelength interferometry, but for application to distance measurement. Such diodes have of course found applications in single-wavelength interferometry, generally over shorter lengths up to a few metres, limited by the coherence. In addition, the diode tunability characteristics have opened up the practical realization of techniques for swept-wavelength interferometry, which can be considered as the interferometric analogue to modulated infra-red (IR) beam electronic distance measurement.

This final section is devoted to an introduction to the techniques of multiple- and swept-wavelength interferometry using diode sources. This is preceded by a discussion of the properties of the laser diodes as interferometric sources of known wavelength.

5.7.1 Laser diode properties and stabilization

Manufacturing methods for semiconductor diode lasers have evolved considerably over the last 10 years, and have resulted in a wide range of commercially available diode laser products with varying spectral and optical properties and applications. Such applications include use as sources for optical storage and CD players, bar-code scanners and optical communications. Due to volume production for CD players, for example, continuous-wave milliwatt GaAlAs diodes with wavelengths near 780 nm retail at a few pounds. Diodes for communication purposes at 1.3 μm and 1.55 μm tend to be considerably more expensive, especially if spectral narrowing properties are introduced. Finally, it should be mentioned that with the recent emergence of high-power diodes with TEM_{oo} powers of 150 mW and diode arrays with continuous wave powers of 10 W, diodes are now competing with dye lasers and ion lasers as high-power pumps for other laser systems.

The present discussion centres on the low-power GaAlAs devices at the few milliwatts level operating in the near IR region of 650–850 nm. Further, the discussion is limited to index-guided diodes. Gain-guided diodes, where the cross-sectional gain region of the diode is governed predominantly by the transverse current flow through the p-n junction, tend to operate multimode and exhibit significant astigmatism. The index-guided diodes, however, have a gain cross-section that is guided by the surrounding cladding region of lower refractive index. The laser diode cavity is thus much better tailored, and the emitted light exhibits little astigmatism, and consists mainly of a single TEM_{oo} mode when the drive current is sufficiently above threshold.

The initial problem encountered with diodes is their elliptical and rapidly divergent output beam profile. This arises because the diode cross-sectional emitting area is typically of order less than 1 μm × a few micrometres. The smaller dimension reflects the width of the emitting junction, while the larger one is dictated by the index guiding. As a result the emitted beam is seriously diffracted by this aperturing, the more so across the narrow stripe width direction. Generally, the divergent beam can be collimated using a microscope objective located close to the diode emitting facet, though if there is serious astigmatism, a pair of cylindrical lenses may be used instead. The collimated beam is still elliptical, and one method for producing a circular beam is to use an anamorphic prism pair, whereby the diode beam diameter is expanded in the short axis to match that in the long axis.

The most significant laser diode property is its capability for tuning across small spectral ranges close to its nominal wavelength value. This nominal value is controlled at manufacture by the proportion of aluminium in the composition of the active layer. Wavelength selection of diodes to a few nanometres in the range 750–850 nm is possible. In addition, diodes have now been produced in the range 620–680 nm (and, more recently, down to 490 nm using II–VI compounds). Frequency tuning about these wavelengths is achieved simply by changing the diode operating temperature or the diode current.

A typical variation of diode frequency with temperature is shown in Fig. 5.9. This variation is often termed **staircase tuning** and has this characteristic shape due to two effects. The continuous tuning regions result from the change of diode cavity length, and hence the frequency of the emitted mode, with temperature. For short cavities of a few hundred micrometres and high material refractive index, cavity mode spacings are typically ~120 GHz. The tuning rate is about 30 GHz $°C^{-1}$ (0.06 nm $°C^{-1}$). Changes in temperature, however, also tune the emission band-gap, which determines the wavelength range, but at a much faster rate. As a result the laser eventually **mode-hop**s discontinuously to a new cavity mode. It can be seen from the distorted staircase of Fig. 5.9 that this hop is not necessarily to the next adjacent cavity mode. For a temperature change from 0°C to 55 °C, the overall frequency change is 6 THz (~13 nm), with a composite tuning rate of ~0.25 nm $°C^{-1}$. The diode can be temperature tuned in a controlled fashion by mounting it on a Peltier-cooled heat sink. If one is able to control the heat sink temperature to ~1 mK, the frequency is short-term stable to ~ 30 MHz or 0.1 p.p.m.

Typical diode operating currents are 50–100 mA. The variation of frequency with current is of order 3 GHz mA^{-1} (~0.006 nm mA^{-1}), mainly because of the change in diode local heating conditions with dissipated power. Current changes also produce output intensity changes, and so gross tuning normally is accomplished by tempera-

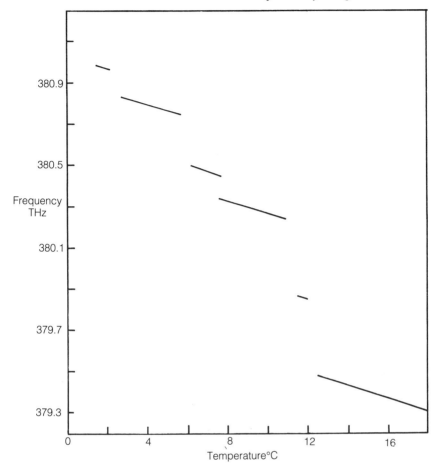

Fig. 5.9 Diode laser wavelength–temperature tuning curve. (Courtesy of National Physical Laboratory.)

ture means. However, the diode tuning response to current changes is very fast, and this allows very fast frequency modulation and feedback control in certain applications. For example, gigahertz modulation rates can readily be achieved.

Random mode hopping to non-adjacent cavity modes can result in certain spectral regions within the diode tuning range being inaccessible. Whilst not generally a problem for interferometric applications, this is a problem for developing diodes as frequency standards, where the mode is frequency stabilized to an atomic absorption. Very recently, techniques have emerged to regularize the mode hopping so that longer range continuous tuning and mode hopping to the adjacent cavity mode can properly be achieved. This can result in almost total spectral coverage across the range, rather than ~30%. The techniques generally involve application of a small amount of feedback to the diode from a thin glass plate placed ~50 μm from the diode front facet.

Another major property of single-mode TEM_{oo} laser diodes is their large linewidth compared to most other single-frequency lasers. This linewidth is theoretically dependent on the inverse square of the cavity mode spacing (Yariv, 1989), and results from internal noise generation within the laser due to random spontaneous emission into the cavity mode. This inverse square dependence becomes significant for the short cavity lengths of diode lasers. The resulting linewidths are typically ~ 30 MHz for a few milliwatts laser output. This limits coherence to a few metres, and as such is a significant drawback to interferometry and to stabilization of diode lasers to narrow Doppler-free spectroscopic features.

Closely related to the diode linewidth properties is its susceptibility to optical feedback. Uncontrolled feedback from optical surfaces can severely modify the lasing spectrum and some means of optical isolation is generally required. Controlled feedback, however, can have the extremely useful property of reducing the laser linewidth. Generally, this feedback is applied in one of two regimes: strong or weak feedback. For the strong-feedback case the diode front facet is antireflection-coated and a third external high-reflecting mirror added. This external cavity diode laser has a much larger cavity length of a few centimetres and hence a much reduced cavity mode spacing. This in turn leads to a much reduced theoretical laser linewidth of typically 20 kHz. For weak feedback $< 10^{-3}$ of the laser output is reflected back into the laser. If this feedback is from a resonant confocal optical cavity, for example, not only is the linewidth narrowed with a suitable cavity to values below 10 kHz, but also the laser passively locks to the cavity resonant frequency (Dahmani *et al.*, 1987; Laurent *et al.*, 1989; Barwood *et al.*, 1990). Tuning is then effected by tuning of the resonant cavity.

With narrowed diode linewidths it becomes very much easier to reference diodes to narrow spectroscopic features. Laser diodes at 780 nm and 795 nm can be tuned to coincidence with rubidium vapour absorptions, and diodes at 852 nm to caesium vapour absorptions. Techniques similar to those developed for stabilization of 633 nm He-Ne lasers have been applied to the stabilization of narrowed laser diodes to rubidium-saturated spectra. Short-term frequency stabilities below 10^{-11} have been achieved. The wavelengths of such narrowed rubidium-stabilized diodes have been measured absolutely by interferometric comparison with the 633 nm reference He-Ne laser, and are thus known to the 1 in 10^9 limiting accuracy of the red laser. These sources can thus serve as highly stable known wavelength standards for a variety of interferometric applications using diodes.

5.7.2 EDM equivalence

Before discussing the techniques of multiple- and swept-wavelength interferometry using tunable near-IR laser diodes, it is useful to make brief reference to the technique of electronic distance measurement (EDM). Here the basic method relies on the modulation of a carrier beam which is reflected from a target reflector. The phase of the returning modulation is measured relative to that of the outgoing reference signal. The integer value is of course unknown. In order to overcome this ambiguity, the modulation frequency is either stepped in a discrete manner to other frequencies, or swept continuously across a particular frequency interval. For the discrete stepping arrangement, a low modulation frequency can be chosen such that the target position is determined unambiguously from a phase measurement which is less than 2π (i.e. the target is within the half wavelength of the modulation). This provides an initial low-resolution value. The modulation frequency can then be increased by a factor of 10 for each successive phase measurement, allowing a progressive increase in measurement resolution, whilst continuing to retain unambiguity at each stage. The successful application of this type of technique has allowed the development of EDM instrumentation which is capable of measurement accuracies below 5 mm over ranges up to 1 km. The modern version of the Mekometer, which uses a microwave electro-optic frequency sweep applied to a He-Ne laser beam, is capable of submillimetre accuracy over a few kilometres.

These EDM techniques can be extended to multiple- and swept-wavelength interferometers with the difference that the phase of the IR or optical carrier itself is measured, either by interferometric or heterodyne methods. In these cases, provided the integer unambiguity can be maintained, then very high resolution is possible. The great advantage of this approach is that measurements can be made to a fixed retroreflector which does not have to be moved over the measurement range.

5.7.3 Multiple-wavelength diode theory and application

For a Michelson interferometer with path difference L between the two arms, the phase of the detected signal at any wavelength λ_i is given by $4\pi n_i L/\lambda_i$, where n_i is the refractive index. This phase value moves through 2π every $\lambda_i/2n_i$. If we measure the output phase at two different vacuum wavelengths λ_1, λ_2 the phase difference can be written

$$\Delta\phi = 4\pi n_i L/\Lambda, \qquad (5.12)$$

where Λ is the synthetic wavelength formed by λ_1 and λ_2 and given by

$$1/\Lambda = 1/\lambda_1 - 1/\lambda_2 \approx \Delta\lambda/\lambda^2, \qquad (5.13)$$

when $\lambda_1 \sim \lambda_2$ and hence $n_1 \sim n_2$. This phase difference varies over 2π every $\Lambda/2n_i$, i.e. much more slowly than in the single-wavelength case. By starting with appropriately closely separated diode wavelengths giving large synthetic wavelengths, target distances within this synthetic ambiguity length can be determined, but with relatively poor phase resolution. Moving to shorter synthetic wavelengths generated by other pairs of more widely spaced diode wavelengths allows a sequentially increasing resolution without loss of unambiguity. Finally, provided the ambiguity length of the shortest synthetic wavelength is within the single half-wavelength precision, the distance measurement can be switched to the highest resolution of the single-wavelength case. Many modern EDM instruments operate using this principle.

5.7.4 Swept-wavelength diode theory and application

The swept-wavelength analysis can be developed in a similar way to the multiple-wavelength case. Here, as the vacuum wavelength is swept from λ to $\lambda + \Delta\lambda$, the interferometer phase change for a Michelson arrangement is given by

$$\Delta\phi = 4\pi nL \, [1/\lambda - 1/(\lambda + \Delta\lambda)]$$
$$= 4\pi nL \, \Delta\lambda/\lambda^2$$
$$= 4\pi nL \, \Delta v/c \qquad (5.14)$$

leading to an expression for distance

$$L = Nc/2n\Delta v, \qquad (5.15)$$

where N is the number of fringes generated for the frequency sweep Δv. What is immediately evident from this expression is that standard fringe-counting methods can be applied for the measurement of large distances, and that the equivalent wavelength of $c/2\Delta v$ (i.e. one fringe) is only dependent on the frequency interval and not on the absolute frequency. Thus for simple fringe counting, basic resolution is $c/2\Delta v$ and this is optimized by the use of as large a sweep as possible. For a 10 GHz diode sweep, this corresponds to a single-fringe resolution of ~1.5 cm. Continuous sweeps up to 30 GHz are possible without mode-hopping. In order to obtain higher resolution some form of fringe interpolation or phase measurement is necessary, with the distance being expressed as

$$L = (N + \epsilon) \, c/2n\Delta v. \qquad (5.16)$$

Again, assuming a single-fringe count precision, the accuracy of the distance measurement is a summation of the frequency-sweep and phase-measurement uncertainties. Good phase-measuring instrumentation should be capable of 10^{-4} precision, and utilization of this level of precision would require sweep knowledge to a similar precision. For a 10 GHz sweep this corresponds to 1 MHz absolute uncertainty.

5.7.5 Future trends

This section has aimed to introduce some of the more versatile multiple-wavelength and swept-wavelength interferometric techniques that are possible with laser diodes. Whilst their low cost, size and tunability lead to their choice as preferred sources for these types of interferometry, it should also be stressed that there are significant developments in single-wavelength interferometry using fixed-wavelength diodes. Here, a great variety of interferometric techniques can clearly benefit from the use of diodes in preference to gas lasers. Whereas the He-Ne laser continues to provide a reasonably small, low-divergence source of nominally fixed wavelength, the continually improving low-cost mass-production techniques for near infra-red and visible laser diodes, and their low voltage drive and control capabilities, point towards the large-scale use of diodes in all types of future interferometric measurement.

REFERENCES

Baer T., Kowalski, F. V. and Hall, J. L. (1980) Frequency stabilization of a 0.633 μm He-Ne longitudinal Zeeman laser. *Appl. Opt.*, **19**, 3173.

Baldwin, D. R. and Siddall, G. J. (1984) A double pass attachment for the linear and plane mirror interferometer. *SPIE Proc.*, **480**, 78–83.

Balhorn, R., Kunzmann, H. and Lebowsky, E. (1972) Frequency stabilization of internal-mirror helium-neon lasers. *Appl. Opt.*, **11**, 742–4.

Barwood, G. P., Gill, P. and Rowley, W. R. C. (1990) An optically narrowed diode laser for Rb saturation spectroscopy. *J. Mod. Opt.*, **37**, 749–58.

Bennett, S. J., Ward, R. E. and Wilson, D. C. (1973) Comments on: frequency stabilization of internal mirror He-Ne lasers. *Appl. Opt.*, **12** (7), 1406.

Dahmani, B., Hollberg, L. and Drullinger, R. (1987) Frequency stabilization of semiconductor lasers by resonant optical feedback. *Opt. Lett.*, **12**, 876–8.

Downs, M. J. and Birch, K. P. (1983) Bi-directional fringe counting interference refractometer. *Precision Eng.*, **5**, 105–110.

Edlén, B. (1966) The refractive index of air. *Metrologia*, **2**, 71–80.

Hanes, G. and Dahlstrom, C. E. (1969) Iodine hyperfine structure observed in saturated absorption at 633 nm. *Appl. Phys. Lett.*, **14**, 362–4.

Kogelnik, H. and Li, T. (1966) Laser beams and resonators. *Appl. Opt.*, **5**, 1550–67.

Laurent, P. H., Clairon, A. and Breant, C. H. (1989) Frequency noise analysis of optically self-locked diode lasers. *IEEE J. Quant. Elec.*, **25**, 1131–42.

Peck, E. R. and Obetz, S. W. (1953) Wavelength or length measurements by reversible fringe counting. *J. Opt. Soc. Am.*, **43**, 505–9.

Pugh, D. J. and Jackson, K. (1986) Automatic gauge block measurement using multiple wavelength interferometry. *SPIE Proc.*, **656**, 244–50.

Rowley, W. R. C. (1972) Interferometric measurement of length and distance. *Alta Frequenza*, **XLI**, 887–96.

Rowley, W. R. C. (1990) The performance of a longitudinal Zeeman-stabilised He-Ne laser (633 nm) with thermal modulation and control. *Meas. Sci. Tech.*, **1**, 348–57.

Rowley, W. R. C. and Gill, P. (1990) Performance of internal-mirror frequency-stabilized He-Ne lasers emitting green, yellow or orange light. *Appl. Phys. B*, **51**, 421–6.

Siegman, A. E. (1986) *Lasers*, University Science Books, Ca 94941, Chapter 14.

Terrien, J. (1965) An air refractometer for interference length metrology. *Metrologia*, **1**, 80.

Yariv, A. (1989) *Quantum Electronics*, (3rd edn), Wiley, New York.

6
Laser vibrometry

N. A. HALLIWELL

6.1 LASER DOPPLER VELOCIMETRY

Laser doppler velocimetry (LDV) or laser doppler anemometry (LDA) is now a well-established technique which is used primarily for non-intrusive measurements in fluid flows. The first major conference concerning its development was held in Copenhagen in 1975 and there are currently several texts available which discuss the principle and practice in detail (see the list preceding the references).

Although the use of LDV for solid surface velocity measurement was recognized at an early stage, its development in this area received little attention compared with the effort in fluid mechanics. Fundamentally, LDV measures the velocity of a light-scattering particle which is seeded into the flow of interest and it is assumed that the particle follows the flow faithfully. Under normal operating conditions the velocity of a group of seeding particles in the measurement region is detected. Clearly, the number density of scattering particles is important since they essentially sample the flow and much of the early work in LDV was concerned with the validity or otherwise of statistical conclusions drawn from a particular measurement. An obvious practical problem to be overcome was the intermittence of seeding particles. An analysis system which claimed time-resolved measurement had to be able to cope with periods of signal dropout, where instantaneously either no particles were present in the measurement area or the vector addition of the light signals from particles combined to produce a very low amplitude.

For solid surface velocity measurement, light-scattering particles can be considered to be replaced by light-scattering surface elements. Then, theoretically, there is never any loss of signal due to absence of scatterers and it was probably this single reason which meant that this form of measurement was considered to be relatively straightforward compared with its fluid counterpart. However, when coherent light is scattered by a diffusely reflecting surface a laser speckle pattern is formed in space in front of the target (Dainty 1984) and the signal is produced by detecting the intensity of one or more speckles. It is this fact which distinguishes a solid surface velocity measurement and indeed it is the speckle pattern dynamics which limit the performance of the LDV system.

In what follows, a brief explanation of the physics of operation of an LDV system will be given before the particular problems which arise from solid surface measurements are addressed.

Optical Methods in Engineering Metrology. Edited by D.C. Williams. Published in 1993 by Chapman & Hall, London. ISBN 0 412 39640 8

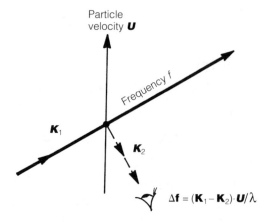

Fig. 6.1 The Doppler effect.

6.1.1 Principles of operation

The physical principle of all LDVs relies upon the detection of the Doppler frequency shift in coherent light which occurs when it is scattered from a moving object. Figure 6.1 is a schematic diagram of the effect when a particle moving with velocity U scatters light of wavelength λ in a direction K_2 from a laser beam of essentially single frequency travelling in a direction K_1. (K_1 and K_2 are unit vectors.) The change in light frequency Δf produced by the moving particle is given by

$$\Delta f = (K_2 - K_1) \cdot U/\lambda = K \cdot U/\lambda, \tag{6.1}$$

where $K = K_2 - K_1$.

A formal proof of this equation can be found in the text by Watrasiewicz and Rudd (1976). Figure 6.2 shows an optical geometry which is appropriate for a solid surface measurement where the laser beam is directed normal to the surface. In this geometry K_1 and K_2 are parallel, i.e. $K_1 = -K_2$, so that the Doppler frequency shift Δf_D

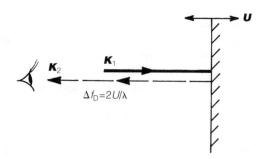

Fig. 6.2 Optical geometry for solid surface velocity measurement.

measured corresponds to the surface velocity component in the direction of the incident beam and is given by

$$\Delta f_D = 2U/\lambda \tag{6.2}$$

where $U = |U|$. In this way, by measuring and tracking the change in Δf_D it is possible to produce a time-resolved measurement of the solid surface velocity U.

The scattered light shown in Figs 6.1 and 6.2 has a frequency which is $\cong 10^{15}$ Hz and cannot be demodulated directly. The Doppler shift Δf_D is measured electronically by mixing the scattered light with a reference beam derived from the same coherent source on the surface of a photodetector. The latter responds to the *intensity* of the total light collected and this non-linear detection produces a heterodyne or beat in the current output whose frequency is equal to the difference in frequency between the two collected beams. This is shown schematically in Fig. 6.3, where a beamsplitter has been used to mix the two beams.

Unfortunately, the system as depicted in Fig. 6.3 is ambiguous in the measurement of the *direction* of motion. Figure 6.4 shows the frequency spectrum of the photodetector output and the Doppler signal which disappears twice per vibration period when the solid surface velocity is zero. When the Doppler signal is present it is not possible to distinguish whether the target surface is moving away from or towards the detector. Vibration engineers require amplitude and phase of the target surface motion and this is usually provided by frequency pre-shifting the reference beam.

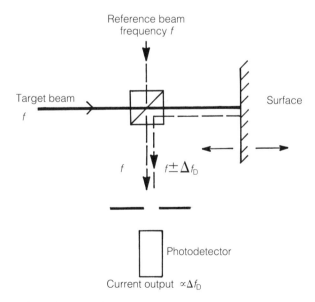

Fig. 6.3 Reference beam heterodyning.

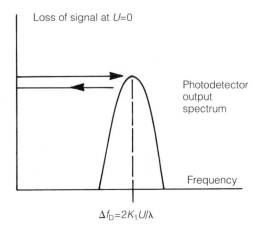

Fig. 6.4 Doppler signal ambiguity.

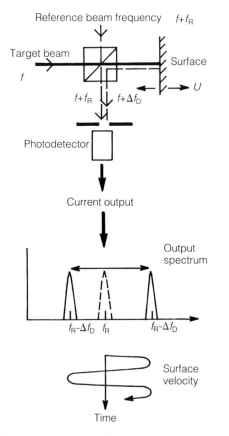

Fig. 6.5 Frequency shifting and time-resolved measurement.

Figure 6.5 shows an equivalent optical geometry where the reference beam has been frequency pre-shifted by an amount f_R. When considering the frequency spectrum of the photodetector output it is now clear that the frequency preshift provides a carrier frequency which the target surface velocity can frequency modulate. When the frequency modulation is tracked it is now possible to produce a time-resolved analogue measurement of the solid surface velocity in both amplitude and phase as shown schematically in the figure.

Laser doppler velocimeters for solid surface target use are often referred to as laser doppler vibrometers, or simply laser vibrometers. They all work on the physical principle described in this section and differ only in the choice of optical geometry and the type of frequency-shifting device used. Just as frequency shifting is paramount for solid surface vibration measurement, it is also included as a standard item in all commercially available LDV systems which are used for flow measurement. It is obviously necessary for measurements in highly oscillatory flows and in practice it is extremely useful to have a carrier frequency corresponding to zero motion for alignment and calibration purposes.

6.1.2 Frequency-shifting devices

The most commonly used form of frequency-shifting device is an acousto-optic transducer or Bragg cell, which is also described in section 11.4.3. The incident laser beam passes through a medium in which acoustic waves are travelling and the small-scale density variations diffract the beam into several orders. Quartz is a medium of choice and usually the first-order diffracted beam is used. In this way the frequency-shifted beam emerges at a slight angle to the incident beam, which requires compensation in some geometries.

Physical limitations restrict the frequency shift provided by a Bragg cell to tens of megahertz, which is rather high for immediate frequency-tracking demodulation. Consequently, two Bragg cells are often used which shift both target and reference beams by typically 40 Mhz and 41 Mhz, respectively. Subsequent heterodyning then provides a carrier frequency which is demodulated readily. An alternative scheme sometimes utilized is to pre-shift with one cell and to downbeat electronically the photodetector output prior to demodulation.

Figure 6.6 shows an optical geometry which can be used for solid surface velocity measurement and which incorporates two Bragg cells for frequency shifting. Their compactness, electronic control and freedom from mechanically moving parts make them popular as pre-shift devices in commercially available LDV systems for laboratory use. Their presence does, however, add to the expense of a system which then requires additional electronic and optical components.

Another means of frequency shifting is provided by rotating a diffraction grating disc through an incident laser beam. Just as the small-density variations in the Bragg cell diffract the beam, in the grating case the small, periodic thickness variations perform the same task. Advantages over the Bragg cell are the smaller shifts

184 *Laser vibrometry*

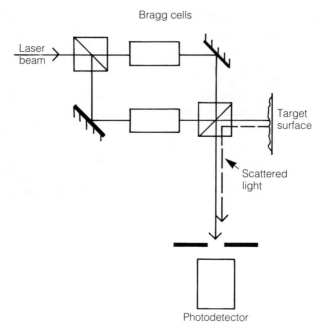

Fig. 6.6 Laser vibrometry: frequency shifting using Bragg cells.

obtained ($\cong 1$ MHz) and the easy and close control of the shifts through modification of the disc speed. Disadvantages are the mechanically moving parts and the inherent fragility of the disc itself which is expensive to manufacture. Extra optical components are again needed to control the cross-sectional areas of the diffracted orders. Readers are referred to the work of Oldengarm *et al.* (1973) for more details of this means of frequency shifting. A typical optical geometry for the measurement of solid surface vibration using a rotating diffraction grating as a frequency shifter is shown in Fig. 6.7.

Both the Bragg cell and rotating diffraction grating produce a reference beam which can be arranged to have a smooth Gaussian intensity distribution and uniform phase front. The latter point is particularly important if very low-level vibration measurement are intended and this point will be discussed later in the chapter.

A particularly simple and robust frequency shifter is provided by a rotating scattering disc (Halliwell, 1979) as shown in Fig. 6.8. The plane of the disc is at a slight angle to the incident laser beam so that the scattering elements passing through the beam have a velocity component in the same direction. Frequency-shifting magnitude is controlled simply by disc speed and radial position of the incident beam. In this way the reference beam takes the form of scattered light in a solid angle which is dictated by the aperture in front of the photodetector. The disc is covered with retroreflective tape or paint to ensure adequate intensity in direct backscatter and to obtain a self-aligning property which is not available with other

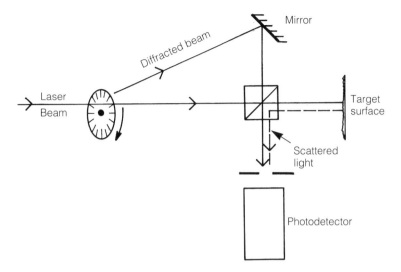

Fig. 6.7 Laser vibrometry: frequency shifting by diffraction grating.

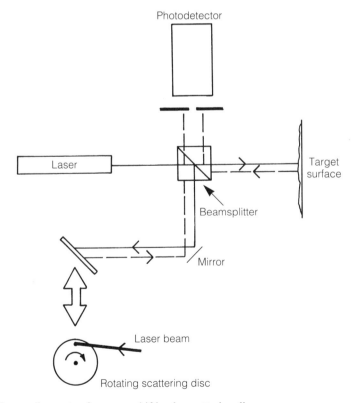

Fig. 6.8 Laser vibrometer: frequency shifting by scattering disc.

forms of frequency shifting, making this design particularly suited for a compact, robust and portable instrument. Frequency shifting in this way produces a noise spectrum for the vibrometer which is a periodogram and the implications of this will be discussed in the next section.

Other frequency shifting devices which have been utilized include Kerr cells (Drain and Moss, 1972), Pockels cells (Foord *et al.*, 1974) and piezoelectric elements (Baker *et al.*, 1990), but these are not common in commercially available instruments. It is also possible to produce a frequency shift by modulating the current drive to a solid state laser diode directly and employing a path length imbalance in the optical geometry (Laming *et al.*, 1986).

6.1.3 Doppler signal processing

A complete description of the various means of demodulating a Doppler signal is beyond the scope of this text and readers are referred to Chapters 6–9 of the text by Durst, Melling and Whitelaw (1981).

The choice of method is dictated by the characteristics of the Doppler signal itself, which are related directly to the particular measurement problem. In many fluid flows it is often necessary to seed the flow with scattering particles in order to detect sufficient intensities of Doppler shifted light. Clearly, for time-resolved measurements the ideal situation requires a continuous Doppler signal, but unfortunately in practice the signal is often intermittent due to changes in seeding particle density which occur naturally. There are various methods of Doppler signal processing available which employ burst-spectrum analysers, photon correlators, frequency counters and frequency trackers (Dantec Elektronik, Denmark, Publ. No. 9003E). Tracking and counting prefer nearly continuous Doppler signals, whilst burst analysers and photon counters can work with signals which are embedded in noise. The increased ability to deal with low signal-to-noise ratios and intermittent signals is usually indicative of the expense of the commercial processor, and the relationship is not linear!

In the case of solid surface vibration measurement Doppler signals are continuous, but they can be subject to periods of low amplitude due to speckle effects which will be addressed in a later section. Periods of low signal amplitude are analogous to dropouts in the fluid flow case.

Where it can reasonably be assured that a target surface is moving parallel to the incident laser beam, the Doppler signal takes the form of a frequency-modulated carrier of essentially constant amplitude. This form of signal can be demodulated with extreme accuracy and sensitivity, and measurements of vibration velocity as low as nanometres/second are possible. If the target motion induces spatial or temporal changes in the speckle pattern formed on the detector then performance is degraded; these effects are discussed in section 6.2.1.

The continuous nature of the Doppler signal from a solid surface measurement allows frequency-tracking demodulation and in what follows a typical scheme is explained (Wilmshurst, 1974). A voltage-controlled oscillator (VCO) is used to track the incoming Doppler signal and is controlled via a feedback loop. A mixer at the input stage produces an error signal between the Doppler and VCO frequencies which is bandpass filtered and weighted before being integrated and used to control the oscillator to drive the error to a minimum. The feedback loop has an associated slew rate which limits the frequency response to the processor.

With respect to the Doppler signal the tracker is really a low-pass filter which outputs the VCO voltage as a time-resolved voltage analogue of the changing frequency. The frequency range of interest for most vibration measurements, d.c. – 20 kHz, is well within the range of this form of frequency demodulation.

Some trackers incorporate sophisticated weighting networks which tailor the control of the VCO according to the signal-to-noise ratios of the incoming signal. A simple form of this network will hold the last value of Doppler frequency being tracked if the amplitude of the signal drops below a pre-set level. In this way the Doppler signal effectively drops out and careful consideration must be given to the statistics of what is essentially a sampled output, especially when high-frequency information of the order of a dropout period is required. Fortunately, for most practical applications this period, typically 0.2 μs, is negligible.

Some commercial vibrometer systems offer a compact optical geometry in some form of optical head which is remote from the signal processor. The latter can be expensive since it often carries the cost of the development work which was undertaken for fluid flow applications.

6.2 SOLID SURFACE VIBRATION MEASUREMENT

When used for a normal-to-surface velocity measurement, the laser vibrometer complements the accelerometer where use of the latter is precluded. This is typically in situations where target surfaces are hot, light or rotating. Accelerometers also prove troublesome if the vibration frequency of interest is in the region of contact resonances or indeed if the level of acceleration is very high, as in impact measurements.

Laser vibrometers also offer excellent time-saving when many measurements are needed. A typical example of this is the need for a vibration survey of a prototype automotive engine in a test cell installation. Use of laser vibrometry avoids the need for the laborious task of drilling and tapping holes for accelerometer mounts and therefore reduces test cell usage time significantly, thus offering high cost savings.

In what follows, the limitations of use of a laser vibrometer for normal-to-surface vibration velocity measurements are discussed. Performance of all commercially available systems is limited by the speckle pattern dynamics generated by the target surface and this is addressed in the next section. A direct consequence of this fact is that in some circumstances a degree of engineering judgement is needed before quantitative data can be accepted. This is a particular problem when laser vibrometers are used on rotating targets and the topic is discussed in section 6.2.2.

6.2.1 Laser speckle effects and theory of operation

It is the dynamics and characteristics of a laser speckle pattern which dictate the performance and sensitivity of a laser vibrometer. In order to understand this it is necessary to describe briefly how a speckle pattern is formed. For more in-depth reading concerning laser speckle patterns and their properties readers are referred to Dainty (1984).

When engineering target surfaces which are rough on the scale of the optical wavelength scatter coherent light, a speckle pattern forms in space in front of the target. This can be observed on a screen as a continuous random array of dark and bright spots which have a grainy, speckly appearance (hence the name **laser speckle pattern**). The phenomenon occurs because each scattering surface element within the laser spot acts like a point source of coherent light. At a point in space, individual scattered wavelets from these sources interfere constructively or destructively to produce the bright or dark speckle, respectively, which is observed on the screen. The speckle pattern is a continuous random distribution of light amplitude and phase although in practice we simply observe the corresponding intensity distribution.

An average speckle size can be calculated and attributed to the distribution observed on the screen (Françon, 1979). Within this correlation region the amplitude and phase of the light may be assumed to be constant. In laser vibrometry, the photodetector active area in the instrument is sampling a speckle pattern. One or more speckles may be sampled, depending upon the optical geometry used in a given measurement situation and also the collecting optics used (if any) for the scattered light. If the detector area collects several speckles, the current output is proportional to the instantaneous mean of the intensity distribution which has been sampled.

When a frequency-shifted reference beam is superposed on to the photodetector, a heterodyne beat occurs as described in section 6.1.1. For simplicity, consider what happens when the reference beam has a constant spatial phase distribution across the detector area and the target is stationary. This type of reference beam would be produced by a Bragg cell or rotating diffraction grating. It is useful to think of each speckle region as an interferometer which has a distinct value of intensity modulation and phase. In practice the detector output is a resultant addition of the outputs from all these separate interferometers corresponding to each speckle which is collected. In this way, for a stationary target the output from the detector is at the beat frequency with a constant amplitude and phase.

When the target moves normal-to-surface and parallel to the incident laser beam, the Doppler effect produces a uniform rate of change of phase across the detector area and a pure frequency modulation of the detector output occurs with negligible amplitude modulation. A word of caution is necessary here. Speckles form small, continuous cigar-shaped volumes in space and the detector plane effectively moves along the major axis of each volume when the target vibrates. Movement between speckle volumes would therefore cause amplitude modulation, but this effect usually is negligible.

If, however, the target tilts or moves other than normal-to-surface, the characteristics of the speckle pattern sampled by the detector will change spatially and/or temporally (Rothberg et al., 1989). This effect produces a phase modulation of the detector output, which is demodulated and is indistinguishable from the change in Doppler frequency associated with normal-to-surface movement. This is a major source of noise in laser vibrometers and has been referred to as pseudo-vibration.

When the scattering elements in the laser spot rotate locally due to surface tilt, the associated speckle pattern on the detector moves in sympathy and this is usually the dominant noise source. However, if a significant change in population of scattering elements occurs (perhaps due to in-plane target motion) then the speckle pattern will evolve or 'boil' in sympathy and produce noise. In practice, both these mechanisms contribute to the noise performance of the laser vibrometer and will be discussed further in the section devoted to practical considerations.

The light intensity incident *at a point* on the photodetector surface $I_p(t)$ can be written (Durst et al., 1981)

$$I_p(t) = I_R + I_T + 2\sqrt{I_R I_T} \cos[(\omega_R - \omega_T)t + \phi_R - \phi_T], \qquad (6.3)$$

where:

I_R, ϕ_R are the reference beam intensity and initial phase, respectively; I_T, ϕ_T are the target scattered light intensity and resultant phase, respectively; ω_R is the constant reference beam angular frequency pre-shift; and ω_T is the instantaneous Doppler angular frequency shift produced by target movement. Strictly, the phase change due to the Doppler effect at time t is $\int \omega_T \, dt$.

The above equation is similar in form to equation (5.1). It is derived after neglecting polarization effects and the very high-frequency modulation of I_p caused by the summed frequency term in the heterodyne process which is beyond the frequency response of the photodetector.

In practice the photodetector takes the average of $I_p(t)$ over its active area A and we must therefore consider the spatial distributions of I_R and I_T and their respective phases ϕ_R and ϕ_T. For convenience we can assume the reference beam intensity and phase is uniform in space and we need only consider the speckle pattern distribution. Accordingly, we can write the detector output current $i(t)$ as

$$i(t) = k I_R A + k \int_A I_T(a) \, da + 2k \int_A \sqrt{I_R I_T(a)} \cos[(\omega_R - \omega_T)t + \phi_R - \phi_T(a)] \, da, \qquad (6.4)$$

where A is the photodetector area; k is a photoelectric constant; $I_T(a)$ is the speckle intensity distribution; and $\phi_T(a)$ is the speckle phase distribution.

Examination of equation (6.4) confirms the earlier discussion that if the spatial distribution of the speckle pattern changes during a measurement period (i.e. $I_T(a)$, $\phi_T(a)$ are functions of time) then spurious information will be contributed when $i(t)$ is frequency demodulated. Unfortunately, in practice, this noise is usually linked to the vibration frequency of interest, introducing uncertainty into the measurement and necessitating the use of a degree of engineering judgement in interpretation of the results.

6.2.2 Measurements on rotating targets

A full treatment of this topic is beyond the scope of this text and readers are referred to the papers by Rothberg et al. (1989, 1990). Only a brief discussion, which is aimed at pointing out the pitfalls in measurement interpretations, will be given here.

When a laser vibrometer is pointed at a rotating target, the spatial and temporal characteristics of the speckle pattern sampled by the photodetector change with time. With reference to equation (6.4), the terms $I_T(a)$, $\phi_T(a)$ describing the speckle pattern intensity and phase distribution across the detector surface become periodic at the rotation frequency and are demodulated as pseudo-vibration. Consequently, when the target surface is not vibrating but simply rotating about a fixed axis in space, the laser vibrometer will produce a noise floor which has a spectrum typical of a pseudo-random signal. This consists of a fundamental at the rotation frequency and higher-order harmonics of similar magnitude. It is produced because the changing speckle pattern produces random phase and amplitude noise which repeats exactly in sympathy with the rotating target.

Unfortunately, in practice, the vibration frequencies of interest are usually direct integer multiples of the rotation frequency and consequently the distribution of this speckle noise is a worst possible case. It is at this stage that a degree of engineering judgement is required before deciding whether the vibration levels measured in a given bandwidth are not those associated with the pseudo-vibration of speckle noise.

In attempting to quantify this form of speckle noise, it is useful to consider the design of a laser vibrometer mentioned in section 6.1.2 which utilizes the rotating scattering disc as a frequency-shifting device (Pickering et al., 1986). The noise spectrum of this design is, as anticipated, a periodogram at a 5 Hz fundamental disc frequency as shown in Fig. 6.9. This is comparable with the noise spectrum which would be produced by use of a laser vibrometer with frequency shifting by Bragg cell or diffraction grating when the latter was pointed at a 5 Hz rotating target and emphasizes the point made in the last section that noise floors of laser vibrometers are target-surface specific.

The vibrometer of Pickering et al. (1986) is fully portable and in Fig. 6.9 the effects of holding the instrument in the hand whilst pointing it at a stationary target are demonstrated. The spurious data below 30 Hz (predicted by Griffin (1990)) is evident and it can be seen that operator body movement has the effect of averaging out the speckle noise since the strictly periodic nature of the noise is destroyed partially by such movement. It is clearly preferable when working with this design of vibrometer to operate between the periodic noise peaks and for a given measurement situation this can be achieved by changing the rotational speed of the disc.

Typical heights of the speckle noise peaks are 10^{-4} m/s and it is reasonable to assume that levels above this, *at the same frequency*, can be accepted as valid vibration data. It is in this interpretation that a degree of engineering judgement is required (Rothberg et al., 1990a, b).

For measurements on rotating surfaces (or on any which are moving other than strictly parallel to the incident laser beam), frequency shifting by scattering disc

Solid surface vibration measurement 191

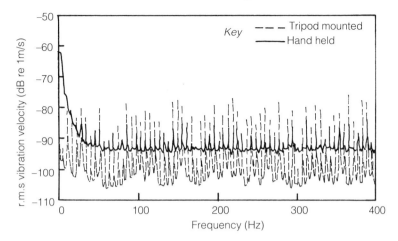

Fig. 6.9 Laser vibrometer noise floor (0 – 400 Hz): effect of hand–held use.

offers the advantage to the operator of more knowledge of the inherent speckle noise distribution and levels above which vibration data are clearly valid. Further to this, the complex addition of two moving speckle patterns from target and reference arms has the effect of reducing the speckle noise peaks at the target rotational frequency. This is not true of designs which frequency shift by Bragg cell or rotating diffraction grating where the majority of speckle noise energy remains at the fundamental rotational speed and higher-order harmonics and thus contributes a higher effective noise level at these frequencies.

6.2.3 Choice of laser vibrometer: practical considerations

There are now several commercially available laser vibrometers which differ in cost by factors of as much as three and four times. Commercial instrument makers aim their product for as wide an application range as possible and each design will have advantages and disadvantages compared with others. Choice of a laser vibrometer is problem specific. The user should decide what the major use of the instrument would be and choose the design offering the most advantages within the budget limit. Industry is not willing to pay a very high price for a technology which may spend the majority of its life on a shelf waiting for occasional use.

Commercially available laser vibrometers now claim dynamic ranges of better than 120 dB. Typical noise floors for vibration velocity measurements are 10^{-6} m/s, whilst frequency responses extend up to 100 kHz. Target-to-instrument working distances extend from a fraction of a metre to hundreds of metres.

Users need to be wary of the very low noise floors and sensitivities quoted by instrument manufacturers in the literature. These have usually been measured on a stationary target and in practice, as discussed in the last section, the true noise floor is peculiar to the speckle pattern behaviour caused by target surface dynamics during the actual measurement.

When a full vibration survey of a target surface is required it is worthwhile considering a scanning laser vibrometer system. Commercial systems can scan the laser beam across a specific area of target surface and produce vibration velocity maps, although these tend to be very expensive compared with the single-point instruments.

Laser technology in the engineering workplace is still new and if its use is to become commonplace then safety and user friendliness are prime concerns. With current codes of practice for safety this means that the output beam should be class II, which ensures safety for momentary accidental viewing of the beam and corresponds to 1 mW or less from a helium-neon laser. For a user-friendly instrument the operator should be able to point the laser beam at the target from distances of typically one metre and immediately obtain an output with the minimum of setting-up time.

Laser vibrometers are available which will operate off an untreated diffusely reflecting surface. Their output beam power is often higher than class II and they may require some form of optical adjustment from a fixed position, thus being tripod-mounted instruments. For a great many cases it is possible to attach a small piece of retroreflective tape or dab a spot of retroreflective paint on the target surface. In these cases it possible to avoid the need for any beam transmission adjustment and still work with a class II laser source. In this situation there is the further advantage of insensitivity to gross solid body movement of the target. For a focused beam instrument, if the target moves outside the waist region (see section 2.1.1) loss of the Doppler signal will occur. This is also true, of course, if the instrument itself undergoes gross solid body movement relative to the target. Such designs, therefore, preclude hand-held use. For vibrometer designs which can be hand-held, Griffin (1990) has shown that spurious vibration information due to human body movement is negligible above 30 Hz.

It should be remembered that a laser vibrometer measures the rate of change of optical path length between the instrument and the target. This normally is interpreted as the Doppler effect produced by the target movement, but if during the course of a measurement the instrument moves so as to change the path length then this too will be construed as vibration. An accelerometer placed on the front of the vibrometer can be used to check the level of this source of noise over the frequency range of interest. When the instrument is tripod mounted it is usually negligible.

In the initial set-up procedure when a laser vibrometer is pointed at the target surface, a very low Doppler signal level is often obtained and the signal processor will usually flag this condition. This is important for designs which focus the laser beam on to the target, since in these cases only one speckle is sampled by the detector and the most probable intensity level for a speckle selected at random is low (Dainty, 1984). Other designs which sample more than one speckle are less prone to this, but it is still not unusual to find that the addition of intensities due to the

Solid surface vibration measurement   193

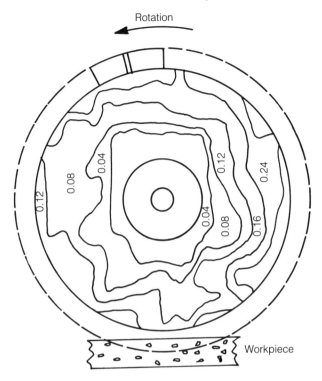

Fig. 6.10 Sawblade vibration whilst cutting: contours show surface r.m.s. velocity in metres per second.

interference of each speckle with the reference beam results in a low signal level. The cure is to move the laser spot slightly until the detector samples a speckle distribution with a high resultant amplitude before beginning the measurement, thus ensuring a high amplitude Doppler beat frequency. As long as it can reasonably be assured that the target is moving parallel to the incident laser beam during the measurement, the noise floor recorded for a stationary target in this situation will be valid (Rothberg *et al.*, 1989).

Figure 6.10 shows an example of a use of a laser vibrometer to measure the r.m.s. vibration velocity contours of a diamond-impregnated circular saw blade whilst cutting granite. This result was used in producing a design of a retrofit noise control pad for the saw blade (Shaw, 1982). An example of the use of a laser vibrometer to measure the vibration of a very light surface is shown in Fig. 6.11. This is a tweeter from an audio system demonstrating a rocking mode at 3 kHz. The contours shown are of vibration velocity and the result was obtained by phase referencing the results to a specific point, pointwise scanning the laser beam and producing the results on a waterfall plot. The author wishes to acknowledge communication of this result by P. Fryer (Wharfedale Ltd).

194 *Laser vibrometry*

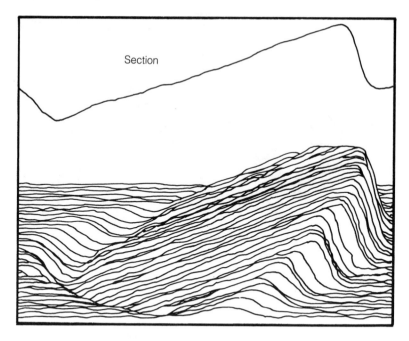

Fig. 6.11 A loudspeaker rocking mode at 3 kHz (after Fryer).

6.3 TORSIONAL VIBRATION MEASUREMENT

When designing rotating machinery components an engineer must be careful to suppress torsional oscillations, since incorrect or insufficient control may lead to fatigue failure, rapid bearing wear, gear hammer or fan belt slippage and produce associated excessive noise problems. Torsional oscillations are a particular problem in engine crankshaft design, where tuned torsional dampers are commonly used to maintain oscillations at an acceptable level over the working speed range of the engine.

To date, torsional transducers include optical, seismic and mechanical torsiographs, strain gauges and slotted discs. The latter system has found common use in the automotive industry and consists of a slotted disc fixed to the end of the crankshaft. A proximity transducer monitors the slot passing frequency which is then demodulated to provide a voltage analogue of the crankshaft speed and hence torsional oscillations, but within a limited frequency range. Strain gauges and associated telemetry or slip-ring systems are notoriously difficult to fix, calibrate and use successfully. In short, the measurement of torsional oscillations provides difficult problems for contacting transducer technology and, of course, necessitates machinery downtime and special arrangements being made for fitting, calibration, etc. Very often the cost of this machinery stoppage will preclude a measurement being attempted, even though the vibration engineer has concluded that it is vital if a design improvement is to be made.

There is therefore a real need for a torsional vibrometer which is user friendly and can provide data immediately in on-site situations. It was not until the advent of laser technology that a solution was found and in what follows a laser torsional vibrometer is described which allows the engineer to point low-powered laser beams at a rotating target component and obtain torsional vibration information. The system is insensitive to solid body oscillation of the component or operator and will operate on a target irrespective of its cross-section.

6.3.1 Cross-beam torsional vibrometer

The cross- or dual-beam laser Doppler velocimeter (Durst *et al.*, 1981) provided a means of measuring tangential surface velocities without contact and this is shown schematically in Fig. 6.12. With this optical geometry the backscattered light produces a Doppler beat frequency Δf_D in the photodetector output which is given by

$$\Delta f_D = \frac{2\mu U}{\lambda} \sin\frac{\theta}{2}, \tag{6.5}$$

where U is the tangential surface velocity at the point of intersection of the beams, $\mu \cong 1$ is the refractive index of air, λ is the wavelength of the laser light and θ is the included angle between the incident laser beams. Frequency tracking the Doppler signal produces a time-resolved voltage analogue of U, the fluctuating part of which is the torsional vibration velocity.

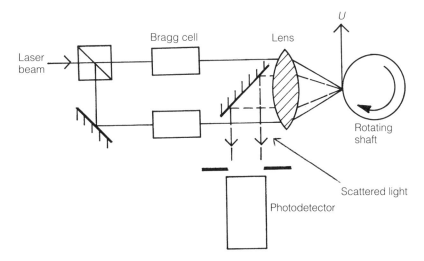

Fig. 6.12 Cross-beam laser velocimeter.

196 *Laser vibrometry*

It should be noted that the Bragg cells shown in the figure are not essential in this application since, given a sufficiently steady speed of the rotating component, it is not necessary to frequency shift the laser beams in this application, as we have a unidirectional surface velocity. They do, however, provide a convenient means of testing the Doppler signal-to-noise ratio during set-up if the target is not rotating.

A laser torsional vibrometer based on this cross-beam geometry was constructed (Halliwell *et al.*, 1983) without the Bragg cells and this is shown schematically in Fig. 6.13. This design was used to measure torsional oscillations of the crankshaft of a six-cylinder in-line diesel engine. A comparison of the results obtained with those using a slotted disc system us mentioned in the introduction is shown in Fig. 6.14. A further laser measurement taken with the engine operating under no-load conditions confirms the expected overall reduction in vibration level. Broad agreement with the slotted disc system is achieved except at very low levels, which can be attributed to the limited frequency response of the disc system used.

With reference to Fig. 6.13, the cross-beam laser velocimeter suffers two major disadvantages in practice. First, the intersection region of the laser beams, where the target surface must remain at all times, is typically less than 1 mm in length. Consequently the target must have a circular cross-section and the instrument must be tripod mounted at a fixed distance *l*. Clearly, gross solid body movement of the target or instrument will prevent a measurement being taken. Second, because the

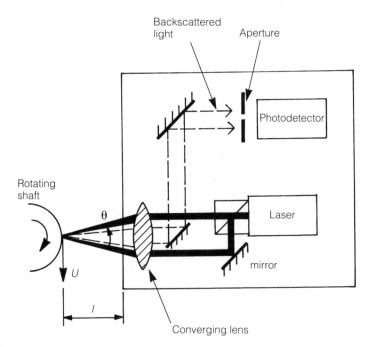

Fig. 6.13 Cross-beam velocimeter instrument.

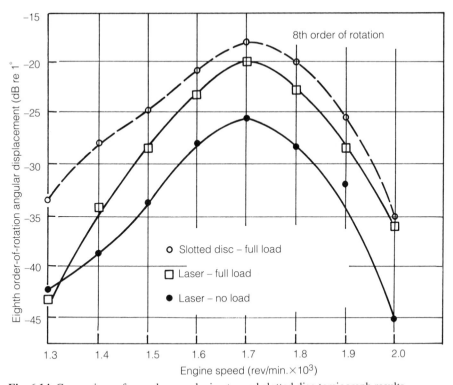

Fig. 6.14 Comparison of cross-beam velocimeter and slotted disc torsiograph results.

target has a solid body oscillation in practice, the component of oscillation in the direction of the tangential surface velocity contaminates the data, so that with this cross-beam geometry torsional oscillations cannot be distinguished from solid body movement. A final practical point is that the mean Doppler frequency is dictated by the angle of the beam intersection θ, which is fixed. If the instrument is to be used over a wide speed range (100–10 000 rev/min), then the associated electronic processing must have a very wide bandwidth which adds expense and complexity to the final design.

These points prevent the cross-beam geometry providing a reliable and user-friendly instrument, and laser technology for torsional vibration measurement did not develop for over a decade until the invention of a new geometry which will be described in the next section.

6.3.2 The laser torsional vibrometer

The optical geometry for the laser torsional vibrometer is shown schematically in Fig. 6.15. A full theoretical analysis of this light-scattering geometry is beyond the scope of this text and only a brief description of the physical principles will be given

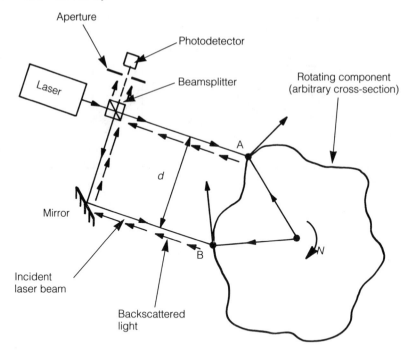

Fig. 6.15 Optical geometry of the laser torsional vibrometer.

here. Readers are referred to the papers by Halliwell *et al.* (1984, 1985) for a full derivation.

With reference to Fig. 6.15, the beam from a low-powered (typically 2 mW) helium–neon laser is divided by a beamsplitter and aligned by a mirror so that two parallel beams are incident on the rotating target at the points A and B. Light backscattered from these points is collected by the same components and is incident on the surface of a photodetector where a heteterodyne process occurs. The difference or beat frequency f_D in hertz is given by

$$f_D = \frac{4\pi N}{\lambda} d \sin \alpha, \tag{6.6}$$

where N is the target revolutions per second, d is the incident beam separation, λ is the laser light wavelength and α is the angle between the plane defined by the incident light beams and the rotational axis of the component. In this way, frequency demodulation of the photodetector signal provides a time-resolved voltage analogue of the speed of rotation of the target component, the a.c. part of which is the torsional vibration. The frequency response of the instrument is dictated by that of the frequency tracker used and the usual region of practical interest (up to 1 kHz) is well within this range.

The optical geometry used makes the instrument insensitive to solid body oscillation of the target or operator. The incident laser beams are parallel and therefore any solid body motion produces an identical Doppler frequency shift in the light scattered from the points A and B. The subsequent heterodyning of this light at the photodetector, which produces a current output proportional to any frequency difference, is therefore insensitive to this form of motion.

With reference to Fig. 6.15, the frequency shift in the backscattered light from A or B is not dependent on their radial distance from the rotational axis and is therefore independent of the cross-sectional shape of the component. When the light is heterodyned, the beat frequency is dependent upon the beam separation d. This fact allows the instrument to work successfully on targets of arbitrary cross-section such as gear wheels and there is no restriction to components of circular cross-section as with the cross-beam geometry discussed in the last section.

In order to assess the accuracy of the new geometry, results were compared with those obtained from the cross-beam laser velocimeter under closely controlled conditions. The target consisted of a rotating disc driven by a brushless d.c. motor upon which was superimposed an a.c. voltage at specific frequencies in order to induce torsional oscillation. Figure 6.16 compares the vibration energy levels measured by the two instruments at the fundamental torsional frequency. Agreement to within 0.5 dB is demonstrated over the range 30–130 Hz achieved in this test. Figure 6.17

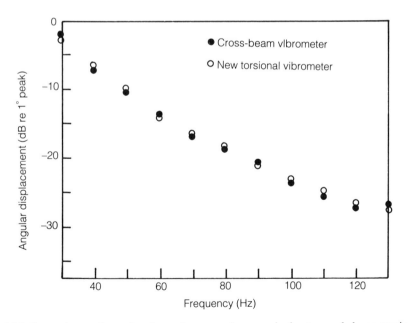

Fig. 6.16 Comparison of results from the cross-beam velocimeter and laser torsional vibrometer.

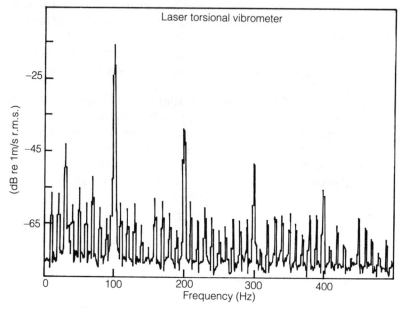

(a)

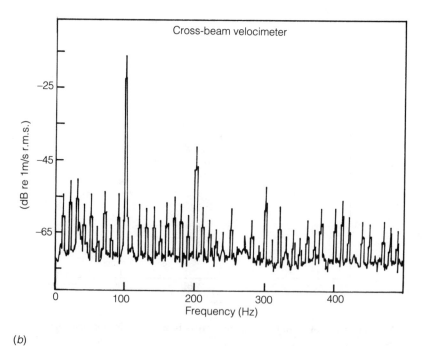

(b)

Fig. 6.17 Comparison of torsional vibration spectra: (*a*) laser torsional vibrometer, and (*b*) cross-beam velocimeter.

compares two torsional vibration spectra with a fundamental frequency of 100 Hz which were obtained in this way, and excellent agreement is observed.

6.3.3 Practical considerations

In practice, the highest levels of torsional vibration will be found on the crankshafts of large diesel engines or reciprocating compressors. These may be as high as a few degrees of peak displacement, but in frequency terms these levels still represent a relatively small modulation of the mean beat frequency resulting from the target speed. Consequently, measuring high levels does not present a problem and it is simply necessary to ensure that they are within the frequency response range of the Doppler signal processor.

It is the range of possible rotational speeds which represents much more of a practical problem for signal processors. At a beam separation d of 1 cm for beams perpendicular to the rotational axis ($\alpha = \pi/2$), speed variation of 500 to 15 000 rev/min produces a beat frequency range of 1.5 to 45 MHz! The theory for the instrument is developed assuming the incident laser beams are infinitely thin and therefore, by an order-of-magnitude argument, a minimum beam separation of 1 cm is required. The lower speed range can therefore be extended by widening the beam separation in order to match the input bandwidth of the processor. At the higher rotational speeds, using the instrument so that the incident laser beam plane is at an angle to the rotational axis will reduce the mean beat frequency.

In practice, it is defining the lowest torsional vibration levels that can be measured accurately which is difficult. This is again dictated by laser speckle pattern dynamics, which were discussed in section 6.2.1. If the target has no torsional vibration whatsoever, the frequency spectrum produced by the instrument will take the form of a periodogram at the fundamental rotational frequency. An example of this is shown in Fig. 6.18, where the instrument was used to measure the speed fluctuation of a d.c. motor from a tape-recorder drive. The displacement spectrum shown clearly exhibits the periodic nature of the speckle noise which masks the intended measurement completely.

Fortunately, the insensitivity of the instrument to solid body oscillation means that the speckle periodicity can be destroyed by moving the laser beam plane during the course of the measurement. Holding the instrument by hand and moving the beam from side to side means that the speckle pattern incident on the photodetector does not exactly repeat. Figure 6.19 shows an equivalent measurement taken on the d.c. motor when the incident laser beam plane was moved from side to side at a very low frequency of about 1 Hz. The figure shows the average noise spectrum of the instrument and the 'wow' of the motor can now be distinguished. For most mechanical engineering applications, a level of –60 dB re 1° peak means that torsional vibrations have ceased to be a problem and speckle periodicity is therefore not troublesome.

When a low-powered helium-neon laser is used as a light source so that the outgoing beams have about 1 mW power, the target distance for operation is up to several metres when retroreflective tape or paint can be used. Again, there is a trade-

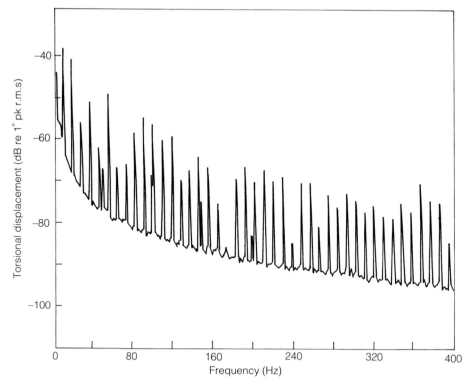

Fig. 6.18 Torsional vibration spectrum: speckle pattern periodicity.

off between target distance, target surface treatment, collection and/or focusing optics and ease of use, as discussed in section 6.2.3. Choice of a particular instrument should therefore be problem specific.

Calibration of the laser torsional vibrometer should be carried out *in situ*. It is not necessary to attempt to measure the beam separation d or angle of inclination of the laser beam plane α to the rotational axis. In most circumstances it is possible to rotate the target at several known speeds so that an *in-situ* calibration of mean d.c. volts per rev/min is possible. In practice, it is often more convenient to mount the instrument on a tripod so that the laser beams are incident on the end face of the rotating target at an angle to the rotational axis. It is worthwhile to remember that if the instrument is hand-held in this situation the spectra produced will only be contaminated below a frequency of 30 Hz (Griffin, 1990).

If it is not possible to rotate the target at known speeds, then a small disc can be used which is driven at known speeds and interposed in the laser beams in place of the target. It is only necessary to monitor the d.c. voltage from the instrument at three known speeds to produce a calibration factor giving negligible error. It is usual to output the instrument signal straight to a real-time analyser which will integrate the output directly to produce vibration spectra in terms of torsional vibration

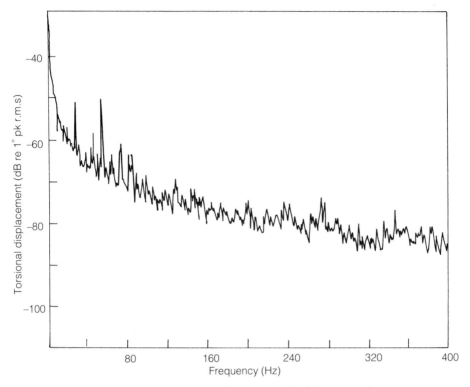

Fig. 6.19 Torsional vibration spectrum: speckle pattern periodicity removed.

displacement re degrees peak. The alternative, of course, is to tape record the analogue output on site for later analysis, but it must be remembered that the d.c. level is important if an *in-situ* calibration has taken place.

6.4 FIBRE OPTIC VIBRATION SENSORS

Optical fibres were developed primarily for use in the communications industry and their use in sensing applications quickly followed on from this. Fibre sensors are treated in detail in Chapter 11. They may be divided into two types which are termed **extrinsic** and **intrinsic**. Intrinsic sensors are so called because the measurand is arranged to affect a property of the light as it is transmitted inside the fibre. The largest range of these sensors is interferometric where the measurand affects the phase of coherent light being transmitted. This is usually achieved through straining the fibre in order to change its refractive index. The phase change is measured by interference with light from the same source which has passed through an undisturbed reference fibre. This form of detection offers extreme sensitivity which, although advantageous in the optical laboratory, has proved to be a problem for practical instrumentation where environmental noise can often swamp the intended measurement.

204 *Laser vibrometry*

Extrinsic fibre sensors employ the fibre only as a light guide and the measurand affects the light outside the fibre. When optical fibres are incorporated into existing open-air-path optical sensors it is often possible to produce a much more compact and versatile optical geometry. They also offer successful operation without the need for direct line-of-sight access to the target. A further advantage of fibre optic sensing is one of passive operation, where it is necessary to sense a physical parameter without the presence of electrical voltages at the target end, for example in coal mines or on oil platforms.

Vibration sensing to a very high sensitivity can be achieved in the optical laboratory, but particular problems arise when on-site instrumentation is considered. The high sensitivity of interferometric intrinsic sensors makes them highly susceptible to environmental effects such as noise, temperature and pressure changes. A consequence of this has been that, to date, commercially available fibre optic vibration sensors have been extrinsic and based on existing laser Doppler velocimeters with the incorporation of fibre optics. There are still problems in achieving this successfully for on-site instrumentation and these will be discussed in the following sections.

6.4.1 Intrinsic vibration sensors

Intrinsic vibration sensors have largely been confined to the optical laboratory. There are several classical optical geometries available such as Fabry–Perot, Michelson or Mach–Zehnder, but the basic physical principle remains the same. The vibration is made to strain the fibre and modulate the phase of the transmitted light. The induced phase change is then detected interferometrically by mixing this light with a reference beam taken from the same source on the surface of a photodetector. The current output from the detector is then modulated by the cosine of the phase difference between the two. Electronic processing of this signal then measures the phase difference to high accuracy. For the practicalities of light launching and transmission through fibres, readers are referred to the book by Marcuse (1981) for details.

The source of coherent light for an intrinsic vibration sensor is usually a helium-neon laser or single-mode semiconductor laser. Output powers are typically 1–5 mW and 1–40 mW, respectively, at wavelengths of 633 nm and 780–850 nm. The coherence length of a helium-neon laser is typically 0.3 m compared with several metres for the semiconductor lasers used. Photodetectors commonly used are either PIN or avalanche photodiodes.

Several fibre optic accelerometers have been suggested by Tveten *et al.* (1980) and Kersey *et al.* (1982), and a schematic diagram of the latter is shown in Fig. 6.20. The sensor geometry chosen is that of a Mach–Zehnder interferometer where the sensing arm consists of a length of monomode fibre. The vibration acceleration is arranged to strain the fibre through squeezing of a compliant cylinder which supports the mass M shown in the figure. The fibre is tension wrapped around the cylinder. Subsequent to this the fibre is tension wrapped around a piezoelectric element (PZT) which is driven electronically in a breathing mode so that it too strains the fibre. In this way

Fibre optic vibration sensors 205

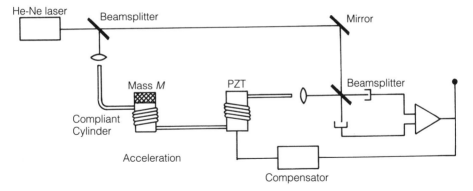

Fig. 6.20 High-sensitivity fibre optical accelerometer (after Kersey *et al.*, 1982).

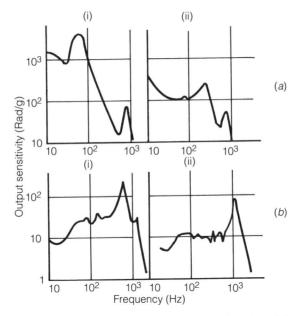

Fig. 6.21 Variation in accelerometer sensitivity (rad/g) as a function of frequency: (*a*) solid rubber cylinder with load mass (i) 660 g, (ii) 25 g; and (*b*) twin-walled perspex cylinder, with load mass (i) 260 g, (ii) 60 g. (After Kersey *et al.*, 1982)

the PZT element is driven to null out the phase changes induced by the vibration, and the output of the PZT is conveniently a representation of the vibration acceleration due to the linearity of the device.

Figure 6.21 shows the sensitivity of the fibre accelerometer as measured by Kersey *et al.* (1982). The single-degree-of-freedom system employed meant that it was possible to tune the accelerometer by appropriate choice of the mass *M* and

206 *Laser vibrometry*

compliant cylinder material. Jackson *et al.* (1980) have demonstrated that Mach–Zehnder fibre interferometers can detect phase changes of less than 10^{-6} rad over a frequency range 10 to 10^4 Hz. The results shown in Fig. 6.21 therefore show that it is possible to measure accelerations as low as 10^{-7} g over this frequency range. This compares well with the figure of 10^{-4} g for a commercial piezoelectric accelerometer, although the size of the device is limited by the diameter of the fibre coil (> 15 mm), compliant cylinder and mass M.

This fibre accelerometer provides a typical example of the levels of sensitivities which can be achieved with intrinsic vibration sensors. It is clear that this type of vibrometer will have a somewhat specialized application, and in the next section extrinsic sensors for more general use are introduced.

6.4.2 Extrinsic vibration sensors

Extrinsic vibration sensing has been based on incorporating fibre optics into laser Doppler velocimeter systems. This immediately offers the advantages of access to the target where no conventional line of sight exists and remote sensing of the vibration. The first fibre-optic laser Doppler velocimeter was reported by Dyott (1978), although this instrument did not incorporate the frequency shifting which is necessary in laser vibrometry, as explained in section 6.1.1.

Nokes *et al.* (1978) incorporated a Bragg cell for frequency shifting in a design of a fibre optic velocimeter, and a schematic diagram of this optical geometry is shown in Fig. 6.22. Practical points to be considered when a length of fibre is used to deliver the light to the target in this way are as follows.

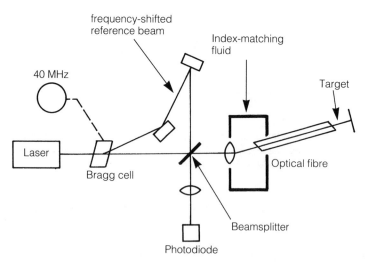

Fig. 6.22 Fibre-optic velocimeter incorporating a Bragg cell (after Nokes *et al.*, 1978).

Fibre optic vibration sensors 207

1. Launching the laser light into the fibre must be achieved without any back reflection which would otherwise produce a persistent carrier frequency from the photodetector through interference with the frequency-shifted reference beam. Nokes *et al.* (1978) used index-matching fluid and angled, polished fibre ends.
2. Light transmission out of the fibre to the target must be achieved without back reflection, for the same reason as given in 1. This is not easily achieved when collimation of the laser beam is required by a further optical component such as a selfoc lens.
3. Remote detection through a long length of fibre is only possible if the path length of the reference beam is also lengthened so that the path difference between the two arms of the interferometer is less than the coherence length of the laser source used.
4. Any environmental disturbance to the fibre which affects the phase of the light transmitted to the target will appear as spurious noise in the detector output. The high sensitivity of intrinsic sensing discussed in section 6.4.1 which was advantageous is now a major disadvantage when a length of fibre is used in this way. In practice, noise, vibration, electromagnetic effects and heat all contribute to the output of this form of sensor.

It is clear that the above points make the design of a practical fibre vibrometer for on-site engineering use particularly difficult to achieve, because conventional frequency-shifting devices are not readily compatible with the use of fibre optics. The point made in 4 will always be true whenever the reference and target light do not traverse the same optical path within the instrument.

6.4.3 A practical all-fibre laser vibrometer

Laming *et al.* (1986) have suggested a design of fibre optic vibrometer which avoids the problems inherent with the use of conventional frequency shifters. A schematic diagram of the optical geometry is shown in Fig. 6.23. Coherent light from a semiconductor laser is launched into one port of a four-port fibre coupler, which serves to both deliver light to the target and collect scattered light which is returned through the coupler to a photodetector. One arm of the coupler is redundant and is index matched to prevent any back-reflection from this port. The reference beam of the interferometer is provided by a back-reflection of the target beam at the interface between the fibre and the selfoc lens which is used to collimate the beam. In this way the reference light and scattered light traverse identical optical paths inside the fibre and so produce an interferometer which is insensitive to environmental effects on the fibre, since the heterodyne process at the photodetector only responds to phase differences in the incident light.

Frequency shifting is achieved by sinusoidally modulating the output light frequency of the laser diode through control of the drive current to the device. With reference to Fig. 6.23, scattered light returning from the target has travelled an extra

208 Laser vibrometry

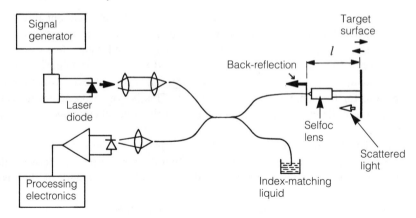

Fig. 6.23 Fibre optic vibrometer (after Laming *et al.*, 1986).

optical path length $2l$ before it mixes with the reference beam at the photodetector. Since the light frequency is a function of time, this produces an inherent time-varying frequency shift between the reference beam and scattered light which is a function of the fibre-end-to-target distance. The actual form of the photodetector output is a Bessel function expansion at the laser modulation frequency and higher-order harmonics. Laming *et al.* (1986) show how this can be demodulated successfully to produce a voltage analogue of the change in fibre-end-to-target distance l per unit time, i.e. the target vibrational velocity.

This fibre optic vibrometer, being an all-fibre design with integral frequency shifting where the reference and signal arms are identical, effectively provides a solution to all the practical problems outlined in points 1–4 made earlier. As such it provides an ideal geometry from which construction of a practical fibre vibrometer for engineering on-site use can be considered. There are still, however, some practical considerations which will limit its use and these are considered in the next section.

Figure 6.24 compares results taken with the fibre vibrometer of Laming *et al.* (1986) and with a commercial piezoelectric accelerometer. The target used was a vibrating surface driven by an electrodynamic shaker at 494 Hz. The two measurements are in excellent agreement; the low-frequency peaks produced in the fibre vibrometer spectrum are the result of mains feedthrough in the electronic processing and can be ignored. The noise floor of the vibrometer for this measurement is $\simeq -105$ dB re 1 m/s r.m.s. This is limited by the electronic processing used and there is potential for a further reduction. Figure 6.25 shows the dynamic range of the fibre vibrometer, which is only limited in its frequency response to 20 kHz by the frequency tracker used in the testing.

6.4.4 The fibre optic vibrometer: practical considerations

When used for solid surface vibration measurement, the fibre optic vibrometer is limited by speckle effects as described in section 6.2.2. This means that the noise

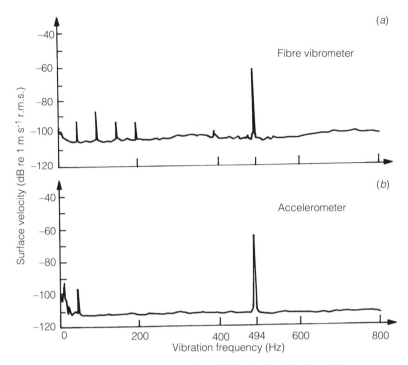

Fig. 6.24 Spectral plot of test surface velocity obtained using (*a*) fibre vibrometer compared with that given by (*b*) an accelerometer (after Laming *et al.*, 1986).

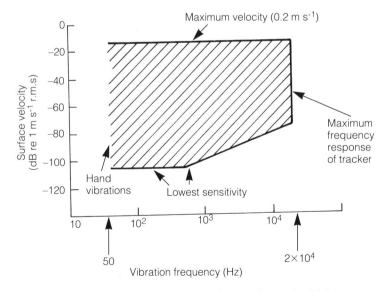

Fig. 6.25 Dynamic range of the fibre vibrometer (after Laming *et al.*, 1986).

floor of the instrument is again target surface specific and will change if the target surface tilts, rotates or moves in a direction other than parallel to the incident laser beam. With reference to Fig. 6.23, this is because the selfoc lens which captures the scattered light still samples a speckle pattern. The pattern will change in sympathy with the target surface dynamics to produce amplitude and phase variations at the photodetector which are not directly related to the true normal-to-surface motion.

A very important practical point concerns vibration isolation of the fibre end when remote sensing is used. It must be remembered that, like the open-air-path laser vibrometer, the fibre optic vibrometer measures relative rate of change of path length. The rate of change of path length between the fibre end and target is what is actually measured. Consequently, if the fibre end moves relative to the target then this will be interpreted as vibration velocity. This would clearly be of concern during remote sensing of vibration, so the operator needs to take care that the fibre end is isolated from the intended measurement. Holding the fibre end by hand during a measurement would introduce speckle noise at all frequencies and produce the highest levels below 30 Hz (Griffin, 1990).

REFERENCES

1 Laser Doppler velocimetry texts

Drain, L. E. (1980) *The Laser Doppler Technique*, Wiley.
Durst, F., Melling, A and Whitelaw, J. H. (1981) *Principles and Practice of Laser Doppler Anemometry*, 2nd edn, Academic Press, London.
Proc. lst Int. Symp. Laser Doppler Anemometry (1975). Tech. Univ. of Denmark, Lyngby, Copenhagen.
Watrasiewicz, B. M., and Rudd, M. J. (1976) *Laser Doppler Measurements*, Butterworths, London.

2 Other references

Baker, J. R., Laming, R. I., Wilmshurst, T. H. and Halliwell, N. A. (1990) A new high sensitivity laser vibrometer. *Optics and Laser Technology*, 22 (4), 241 – 4.
Dainty, J. C. (ed.) (1984). *Laser Speckle and Related Phenomena*, Springer Verlag.
Drain, L. E. and Moss, B. C. (1972). The frequency shifting of laser light by electrooptic techniques. *Opto-Electronics* 4, 429.
Dyott, R. B. (1978). The fibre-optic Doppler anemometer. *Microwaves, Optics and Acoustics*, 2, 13 – 16.
Foord, R. *et al.* (1974). A solid state electro-optic phase modulator for laser Doppler anemometry. *Report of Physics Group*, Royal Radar Establishment, Malvern, UK.
Françon, M. (1979). *Laser Speckle and Applications in Optics*, Academic Press.
Griffin, M. J. (1990). *Handbook of Human Vibration*, Academic Press.
Halliwell, N. A. (1979). Laser Doppler measurement of vibrating surfaces: a portable instrument. *Journal of Sound and Vibration*, 62, 312 – 15.
Halliwell, N. A. and Eastwood, P. G. (1985) The laser torsional vibrometer. *Journal of Sound and Vibration*, 101 (3), 446.
Halliwell, N. A., Pickering, C. J. D. and Eastwood, P. G. (1984) The laser torsional vibrometer: A new instrument. *Journal of Sound and Vibration*, 93, (4), 588.
Halliwell, N. A., Pullen, H. L. and Baker, J. (1983). Diesel engine health: laser diagnostics. *Transactions of the Society of Automotive Engineers*, USA.

Jackson, D. A., Dandridge, A. and Sheem, S. K. (1980) Measurement of small phase shifts using a single mode optical fibre interferometer. *Optics Letters*, **5**, 139 – 41.

Kersey, A. D., Jackson, D. A. and Corke, M. (1982) High sensitivity fibre optic accelerometer. *Electronics Letters*, **18**, 559.

Laming, R. I., Gold, M. P., Payne, D. N. and Halliwell, N. A. (1986) Fibre optic vibration probe. *Electronics Letters*, **9**, 378 – 80.

Marcuse, D. M. (1981) *Principles of Optical Fibre Measurements*, Academic Press.

Nokes, M. A., Hill, B. C. and Barelli, A. E. (1978) Fibre optic heterodyne interferometer for vibration measurements in biological systems. *Review of Scientific Instruments*, **49**, 722 – 8.

Oldengarm, *et al.* (1973) Laser Doppler velocimeter with optical frequency shifting. *Optics and Laser Technology*, **5**, 249 – 52.

Pickering, C. J. D., Halliwell, N. A. and Wilmshurst, T. H. (1986) The laser vibrometer: a portable instrument. *Journal of Sound and Vibration*, **107,** (3), 471 – 85.

Rothberg, S. J., Baker, J. R. and Halliwell, N. A. (1989) Laser vibrometry: pseudo vibrations. *Journal of Sound and Vibration*, **135**, (3), 516 – 22.

Rothberg, S. J., Baker, J. R. and Halliwell, N. A. (1990a) On laser vibrometry of rotating targets: effect of in-plane and torsional vibration. *Journal of Laser Applications*, **2**(1), 29 – 36.

Rothberg, S. J., Baker, J. R. and Halliwell, N. A. (1990b) Practical laser vibrometry for rotating targets. *Proc. Int. Conf. Vib. Problems in Eng.*, Wuhan-Chongquing, China, 831 – 7.

Shaw, S. (1982) Noise control of stone cutting circular saws. MSc Thesis, ISVR, Univ. of Southampton.

Tveten, A. B., Dandridge, A., Davies, C. M. and Giallorenzi, T. G. (1980) Fibre optical accelerometer. *Electronics Letters*, **16**, 854 – 5.

Wilmshurst, T. H. (1974) An autodyne frequency tracker for use in laser Doppler anemometry. *Journal of Physics E: Scientific Instruments*, **7**, 924.

7
Industrial application of holographic interferometry

R. J. PARKER

7.1 HOLOGRAPHY

Holography has provided the modern engineer with one of the most revealing measurement and visualization tools of the twentieth century. This chapter leads the reader through the basic principles of holography to the practicalities of holography in an industrial environment. The applications of holographic interferometry to non-destructive testing, vibration analysis, flow visualization and dimensional measurement are all described with examples of industrial use.

The principles of holography were laid down in the 1940s and 1950s by Gabor (1948, 1949), who subsequently was awarded the Nobel prize for his work in **wavefront reconstruction**. It was he who introduced the word **hologram** into modern usage (derived from Greek words for *whole* and *writing*). This pioneering work was well ahead of its time. It was not until 1962 that technology caught up with it and holography of solid objects became a reality.

With the invention of the laser, researchers were given the high-intensity source of polarized, coherent, monochromatic light which they needed to form a hologram. The other significant development was the introduction of an off-axis reference beam (Leith and Upatnieks 1964), made possible by the high coherence of the laser source. This allowed clear three-dimensional holograms to be reconstructed without the confusion of multiple overlapping images.

The succeeding 25 years have seen many significant developments in the technology surrounding holography. Most significant amongst these has been the development of reliable commercial lasers for both continuous and pulsed holography. In parallel with laser development, the development and refinement of photographic materials for recording holograms has helped to turn holography into a practical engineering tool (Gates, 1986).

When it first appeared, holography, like the laser, was described as 'a solution in search of a problem'. There are now many applications of holography ranging from

Optical Methods in Engineering Metrology. Edited by D.C. Williams. Published in 1993 by Chapman & Hall, London. ISBN 0 412 39640 8

214 *Industrial application of holographic interferometry*

engineering measurement to credit-card security, and from scanning laser beams to selling breakfast cereals.

An important part of the theme of the Essex University workshop series, on which this book is based, was the communication of the authors' practical experience of the techniques being presented. This chapter discusses the equipment and materials required to set up a practical holographic laboratory as well as real examples of holographic interferometry in action.

7.1.1 Recording the whole picture

The hologram is most commonly thought of as a three-dimensional photograph. It is instructive to consider the essential differences between holography and photography in order to gain a better understanding of this novel medium.

When we observe a three-dimensional object, it is the scattered light from the object which contains all the information which we use in stereo-perception. When a conventional photograph is formed, the light field is imaged by a lens on to the film. The developed photograph shows only where the object is bright and where it is dark from the viewpoint of the camera. All of the three-dimensional information has been lost and we are left with a flat two-dimensional image which provides the same information irrespective of the angle we view it from. It is possible to fool the brain to some extent by taking two photographs from different viewpoints and providing each of these as a separate image for the observer's two eyes. This stereo-photography still retains the fixed viewpoints of the original camera, but fills in some of the perspective information required by our brains.

In the case of a hologram we require to record and replay the scattered light field exactly as it comes from the object. Light is a wave phenomenon, and hence the wavefield scattered from the object is characterized by amplitude and phase. Photographic materials are an example of what is known as a **square-law detector**, i.e. they record only the intensity of the light that falls on each point and all of the phase information is lost. It is this phase information which is vital for the 'whole writing', since it contains information on the direction in which the light is travelling as well as its brightness. To form a hologram the phase information must be encoded on the photographic material. This is achieved by interference, which allows the phase information to be converted to intensity in the recording medium.

7.1.2 Two-beam interference and holography

If two beams of light from the same coherent source cross at some angle 2θ (Fig. 7.1) an interference pattern will result which fills the region of overlap of the beams. The fringes may now be thought of as planes of interference. The planes will bisect the angle 2θ. A detector with a very small aperture moved through this region along the x-axis would see a sinusoidal variation in intensity. The spacing of the

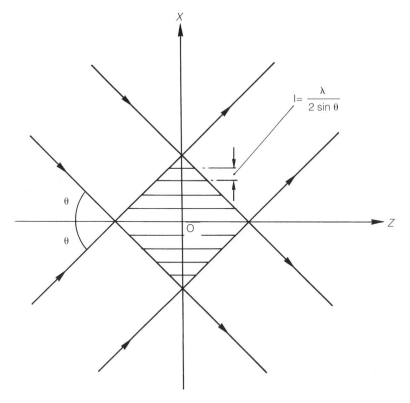

Fig. 7.1 Principles of two-beam interference. Two plane wavefronts intersect at angle 2θ.

interference planes l is controlled by the wavelength λ of the light and the angle θ according to

$$l = \frac{\lambda}{2 \sin \theta}. \tag{7.1}$$

If a photographic plate is placed in the plane defined by the x- and y-axes, then, when developed, it will be covered in parallel interference fringes with the sinusoidal profile seen above. This plate would in fact be a simple hologram!

If the plate is placed back in just one of the two beams above, part of the light will pass straight through, unaltered, but part will be diffracted by the microscopic pattern on the plate. This diffracted beam will emerge at exactly the same angle as the second beam. In this way a wavefront has been reconstructed. If one of the beams had a more complicated form (e.g. it was scattered from a diffuse object), the diffracted, reconstructed wavefront would adopt this complicated form when the plate was illuminated only with the simple plane wave. The interference pattern

216 *Industrial application of holographic interferometry*

recorded on the hologram plate would now look very complicated, but the fringes would still have an average spacing determined by the mean angle between the two beams, as in equation (7.1). To an observer looking into the diffracted beam from the plate, it would appear that the original laser-lit object was still there.

7.1.3 Recording transmission holograms

In the last section it was shown how a simple two-beam interference system can form the basis for recording a hologram. This type of hologram, in which the diffracted beam travels through the hologram and emerges as a replica of the original, second beam, is called a transmission hologram.

A practical system for recording a **transmission hologram** of a solid object is shown in Fig. 7.2. Some similarities with the simple two-beam system above and even with the Michelson interferometer can be seen. The light from the laser is split into two parts by the beamsplitter. One part (the **object beam**) is expanded and used to illuminate the object. The light scattered from the object travels in all directions, but some of it falls on to the photographic plate. It is this scattered light that we wish to record. The second part of the laser beam is expanded and shines directly on to the photographic plate. This beam with a simple geometrical wavefront is called the **reference beam**. At the plate the two beams interfere to form a microscopic interference pattern. It is not possible for the photographic material to record the phase of the scattered light, but by interfering the scattered light with the reference beam, the phase has been encoded in the intensity pattern recorded by the plate.

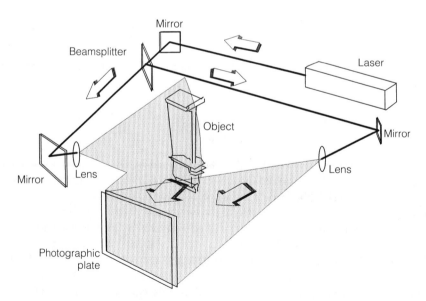

Fig. 7.2 Recording a transmission hologram.

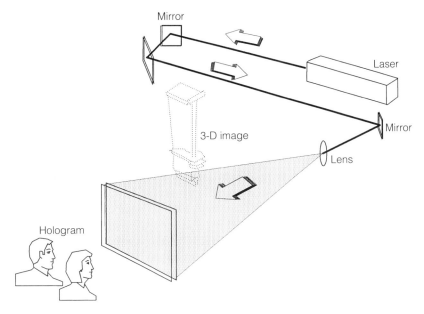

Fig. 7.3 Reconstructing a transmission hologram.

To reconstruct the holographic image (Fig. 7.3), the plate is processed chemically and replaced in a laser beam similar to the original reference beam (now called the **reconstruction beam**). As in the simple two-beam case, part of the beam is diffracted by the pattern recorded on the plate to form a replica of the light originally scattered from the object. An observer looking in through the 'window' formed by the hologram plate sees a realistic image of the laser-lit object. This image, behind the hologram plate, is a virtual image.

7.1.4 Viewing in white light

If a transmission hologram as discussed so far was illuminated with a white light source, the result would be a blurred, multi-spectral image because of the dispersion of the light by the hologram. The one exception to this is when the image is located on, or very close to the hologram plate in an **image-plane hologram** (Brandt, 1969). The dispersion does not then blur the image, but the object and image must have little or no depth to them. Since it is not possible to record much three-dimensional information in this form it is of little interest for general display holography.

A variant on the thin transmission hologram which can display three-dimensional information in white light is called the **rainbow hologram** (Benton, 1969, 1980). The credit-card security holograms are of this form. They preserve only parallax information in the horizontal plane; the vertical plane is sacrificed in order to control the dispersion of the white light.

218 *Industrial application of holographic interferometry*

There is a form of hologram which can display full three-dimensional images in white-light illumination. This is called the **volume reflection hologram** (Syms, 1990). In order to record reflection holograms the reference and object waves approach the recording plate from opposite sides. The simplest way to achieve this was proposed by Denisyuk (1962). In this process, the object is placed immediately behind the hologram plate and illuminated through the plate. The light passing through the plate is both the reference beam and the illuminating beam. Light scattered from the object towards the plate interferes with the incoming light. The planes of interference created are now running almost parallel to the recording plate, and are thus recorded through the 'volume' of the emulsion.

When the developed hologram is illuminated by a white-light source, the planes of developed silver within the hologram act like a crystal lattice, and select a narrow band of wavelengths which satisfy the Bragg (1912) condition (Longhurst, 1957, pp. 276–8) for constructive interference. The remainder of the light is either transmitted, or absorbed in the hologram. The hologram acts like a selective mirror, reflecting only a narrow band of wavelengths. It also shapes the reflected wavefronts into the form of the original scattered light so that a monochromatic, three-dimensional image is observed. The image is fully three dimensional, having both vertical and horizontal parallax.

There are two disadvantages to the Denisyuk approach:

1. The reference and object beam intensities cannot be controlled independently, except by suitable coating of the object to increase diffuse reflection; and
2. A long coherence length is required, since the beams are mismatched by twice the distance from the plate to the object.

Despite this, it remains a very attractive technique because of its simplicity. It is also very tolerant of ambient vibration, because of the close proximity of the plate and object.

It is possible to record volume reflection holograms by a conventional two beam technique, but with the reference beam approaching the plate from the opposite side to the object beam. It is also possible to copy a normal transmission hologram into volume form. These techniques are not considered further here, as transmission holograms have generally been used for the majority of industrial applications of holography to date. Those interested will find further information on these and other types of hologram in the literature (Hariharan, 1984; Syms, 1990).

7.2 EQUIPMENT FOR HOLOGRAPHY

7.2.1 Lasers

In the laboratory small continuous-wave (cw) lasers can be used to form holograms. Lasers commonly used range in output power from He-Ne lasers ($\lambda = 632.8$ nm) giving typically 1 to 25 mW to argon ion lasers (λ = 514.5 nm and 488.0 nm main lines) providing several watts. Diode lasers are, however, rapidly increasing in power, coherence and beam quality (section 5.7.1), but the majority have wavelengths beyond the red end of the visible spectrum. They are used currently in TV holography, or electronic speckle pattern interferometry (ESPI) systems (Chapter 8). As wavelengths decrease to coincide with the sensitivity of both the human eye and holographic recording materials, they will find many applications in holography.

The helium-neon lasers typically require exposures of between one-tenth of a second and several seconds. Because of the extremely high spatial frequencies present in the hologram, any movement of the components forming the hologram, or of the object itself, during the exposure will reduce the efficiency of (or even destroy) the hologram. When using these small lasers, vibration-isolated tables are essential to remove any ambient vibration. These lasers also have fairly short coherence lengths, 10 cm to 30 cm, which restricts the size of object which may be recorded.

The argon ion lasers permit exposures as short as 1 ms. This is still normally too long to remove the effects of ambient vibration. Once again a vibration-isolated table must be used. These lasers are capable of laser action at several spectral lines. The particular line or wavelength required is normally chosen using a dispersive element in the laser cavity. Argon ion lasers can be fitted with etalons to increase their coherence length by selecting one narrow line from those which the cavity is capable of supporting within a given spectral line. When fitted with such an etalon the coherence length may be several metres.

In order to work away from the laboratory, large pulsed lasers are required. A variety of pulsed lasers is available (Koechner, 1976), but to date only the ruby laser has found widespread use in holography because of its high energy and good spatial and temporal coherence properties.

Ruby lasers (λ = 694.3 nm) have high output energy (up to 10 J) and good longitudinal and transverse mode structure. The major disadvantage currently is their low repetition rate (maximum 1 Hz, but typically once in 30 s for high-energy versions). The short pulses (20–50 ns) are sufficient to freeze all but the most violent of ambient vibrations and movements. It is possible to generate rapid double exposures by double-pulsing the laser. Time intervals from 0.5 µs to 1 ms can be used to study dynamic events by holographic interferometry.

Pulsed lasers do not always remove the need for vibration isolation. Where a long time interval is required between the two exposures, it is important that the object or optical system should not move between exposures, or else error fringes will appear in the final hologram.

7.2.2 Laser safety

Even when using relatively small He-Ne lasers care must be taken to avoid exposure to the unexpanded laser beam. A code of safe working practice will need to be agreed with local safety inspectors within the context of national laser safety procedures (in the UK BS7192, 1989).

Most argon ion lasers and all pulsed lasers come into safety category IV. For this category, it is necessary to provide all personnel in the area with suitable laser goggles to block out the laser beam. Interlocks must be provided on all doors into the laser area so that the laser is disabled if a door is opened. These precautions should again be backed up by a suitable code of safe working practice. In open sites it is not practical to have interlocked doors. In this situation physical barriers are used to prevent accidental exposure of others. Where this is not possible, it is necessary to clear personnel from within the NOHD (nominal ocular hazard distance). This can be a large distance if pulsed lasers are used with unexpanded beams. For example, it was calculated for a recent job that the beam from a one joule pulsed ruby laser, 1 cm in diameter at the exit, would have to travel 5 km before it would have diverged sufficiently to be safe to view!

7.2.3 Recording materials

The recording materials generally available for holography are listed below.

1. Silver halide
   High-resolution (up to 2000 l/mm)
   Exposure 1–30 μJ cm^{-2}
   Chemical development
2. Thermoplastic
   Band-limited resolution (300–1000 l/mm)
   Exposure 0.5–100 μJ/cm^{-2}
   Heat-developed *in situ*
3. Photopolymers
   High-resolution (up to 2000 l/mm)
   Exposure >1 mJ/cm^{-2}
   Chemical development
4. Photoresist
   High-resolution (up to 4000 l/mm)
   Exposure >10 mJ/cm^{-2}
   Chemical development
5. Non-linear crystals (BSO, BTO)
   High-resolution (limited by lattice)
   Exposure 100 mW to 100 W/cm^{-2}
   Instant

Only the first two materials (silver halide and thermoplastic) have found widespread use in practical holography for engineering applications. The others are not considered further here. A complete description of all these materials can be found in Smith (1977).

(a) Silver halide materials

Conventionally, very fine-grained silver halide emulsions coated on to glass plates have been used for holography. These can be produced in large formats and can be sensitized for various laser wavelengths. They are capable of recording the very fine interference fringes which form the hologram (up to 2000 lines/mm). Because of the fine grain structure, they are photographically very slow, requiring typically 3–100 $\mu J/cm^2$ exposure. The major manufacturers are Agfa-Gevaert, Fuji, Ilford and Kodak.

The Agfa plates come in two wavelength sensitivities for green (argon ion) and red (He-Ne and ruby) lasers. They are also available in two speeds requiring typically 3 and 30 $\mu J/cm^2$ exposures. The Ilford materials have only recently become available. Like the Agfa materials they are available in red and blue/green sensitivities. They are photographically similar in speed to the slowest of the Agfa materials.

The processing of silver halide holograms is by fairly conventional photochemistry. The author has used Agfa plates, Kodak D19 developer and Amfix (not SuperAmfix) fixer. A typical development cycle would be:

Develop	2–3 min in stock D19 (at 25 °C)
Stop	30 s in 2% acetic acid solution (optional)
Fix	1 min in Amfix (dilution 1+3)
Wash	20 min in running water
Rinse	water + Photoflo or similar wetting agent
Dry	upright without heating

The author has found that it is only necessary to process the plates to give an optical density of 0.5 to 1.0 to obtain bright images provided care is taken to ensure that the majority of light falling on the hologram plate is actively participating in the formation of the hologram, i.e. ambient light is kept to a minimum and coherence conditions are maximized. Traditionally optical densities of 2 or 3 have been recommended.

This type of processing produces a simple amplitude hologram, with all the exposed silver halide converted to silver and all the unexposed silver halide fixed out. The maximum diffraction efficiency which can be achieved from such a hologram is 6%. The diffraction efficiency is defined as the percentage of the light used to reconstruct the hologram which emerges in the required image beam.

A vast number of processes are published (and still more are secret!) for producing phase holograms by bleaching. This process involves converting the silver in the developed hologram back into a transparent silver salt, or removing it altogether. This can greatly improve the efficiency of the reconstruction process, since much less of the light is absorbed by the hologram. In theory 33% diffraction efficiency can be achieved for these bleached, thin holograms. Several bleach formulae are given by Hariharan (1984) or Saxby (1988).

The author in general avoids bleaches for engineering holography for two reasons:

1. The bleaches almost invariably increase the scattered light noise from the holograms. Thus, while the signal is increased, the signal-to-noise ratio may actually fall.
2. Most bleach processes require the plate to be initially developed to a higher density. This in turn requires a longer exposure or higher output from the laser.

It is also possible to obtain silver halide emulsions coated on to film in a variety of formats. This greatly facilitates engineering applications, since several exposures may be recorded before processing is possible, or convenient. One requirement when using film is that it should be held flat, otherwise the hologram will appear distorted when reconstructed. Vacuum backs or glass plates are commonly employed to hold the film flat. It is possible to modify standard cameras so that different formats of holographic film can be used; examples include a 35 mm Nikon camera (Fig. 7.4) and a 70 mm Hasselblad (Fig. 7.5). The author has also used a reconnaissance camera cassette for 127 mm wide film. The first two required additional ports to be made for the reference beam to enter. In all cases a data-recording system is incorporated so that data can be written on to the film for future reference.

The major disadvantage of all silver halide materials is the need for chemical processing, which is time consuming and difficult to perform *in situ*. In the author's

Fig. 7.4 A modified Nikon camera for use in holography. The viewfinder has been removed. (Courtesy of Rolls-Royce plc.)

Equipment for holography 223

Fig. 7.5 Modified metrological Hasselblad camera for holography. The side port allows reference beam access. (Courtesy of Rolls-Royce plc.)

experience this has been one of the major barriers to the transfer of holographic techniques from research to test or production environments.

(b) Thermoplastic materials

While many other recording materials have been applied to holography (Smith, 1977) nearly all are less sensitive than silver halide. The major exception is photo-thermoplastic recording materials, often referred to simply as thermoplastics (Parker, 1990). These record phase holograms by a heat-developed xerographic process. In the past decade a number of commercial holographic cameras employing this technique have become available.

The recording process is illustrated schematically in Fig. 7.6. The film or plate consists of a thin layer of thermoplastic resin on top of a layer of organic photoconductor. This is on a plastic or glass substrate. A transparent metal electrode is either coated on to the substrate, or mounted directly behind the substrate (for thin films). The recording process is as follows

1. Charge the thermoplastic layer evenly.
2. Expose to the holographic light field. This modulates the charge distribution on the surface by acting on the photoconductor layer.
3. Recharge. This enhances the charge variation. Some systems miss out this step.
4. Heat. The plastic layer melts and takes up a corrugated shape defined by the charge pattern.

224 *Industrial application of holographic interferometry*

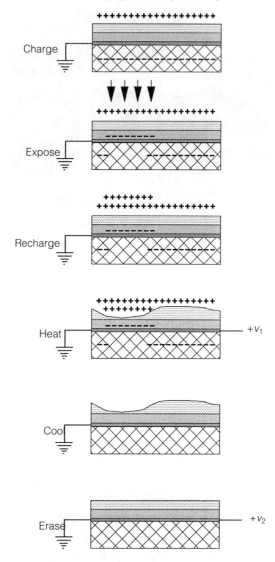

Fig. 7.6 The thermoplastic recording process. Some systems omit the second charging stage.

5. Fix by cooling. This creates a 'permanent' deformed surface which is a phase hologram.

This whole cycle can take as little as 10 seconds. Since the hologram formed is a phase hologram, formed only by the variation in thickness of the thermoplastic layer, the diffraction efficiency can be very high (up to 30%). Thermoplastics therefore provide bright, instant holograms.

Equipment for holography 225

Fig. 7.7 Thermoplastic holocamera produced by Rottenkolber Gmbh. This uses a reel of thermoplastic film.

The Rottenkolber thermoplastic camera is shown in Fig. 7.7. This uses a roll of thermoplastic film (produced by Kalle-Hoechst). After each shot the film is moved on and the holograms can be retained for future reference. Several of the other commercial systems employ this process. Another commercially available system produced by Newport (Fig. 7.8) employs a reusable glass plate. After each hologram

226 *Industrial application of holographic interferometry*

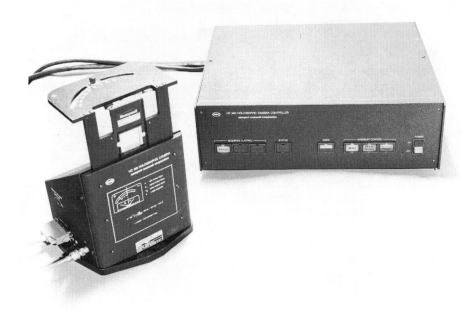

Fig. 7.8 The Newport research thermoplastic recording system uses a reusable thermoplastic plate.

has been reconstructed, the plate is heated to erase the image. The next hologram is recorded on the same plate. While this provides no permanent holographic record, for many NDT and inspection applications this is not necessary.

Thermoplastic recording combines push-button electronic technology with almost instant access to the holographic image. This technology has helped greatly the transfer of holographic techniques to non-research areas.

7.2.4 Other holographic equipment

In addition to the laser and the recording materials there are several other items of holographic equipment. If a pulsed-laser holocamera is purchased, all of the necessary optics will be built into a single unit ready to point and fire.

For cw holography a stable vibration-isolated table is required, because of the relatively long exposures. Depending on the environment, it is possible to use car

Equipment for holography   227

innertubes and a massive iron table to give a stable working platform. More sophisticated air-supported tables are available, and are recommended for most industrial working environments.

The optical components must be fixed rigidly to the table. This can be achieved by using magnetic pucks or bases, or else by bolting components directly to the table. Bench rods are now available with internal damping to give an even higher level of vibration immunity. It is important not to transmit vibration to the table having isolated it.

Figure 7.9 shows a typical holographic recording table. In this example it is equipped with a thermoplastic recording camera. Note that a large, uncluttered area in the centre of the table has been left free to allow for rapid mounting of the test objects.

All of the optical surfaces must be kept clean. It is necessary that all mirrors are front surface coated to avoid multiple reflections. When using high-power lasers, special dielectric optical coatings will be required to prevent damage from the laser. These can be very expensive.

Fig. 7.9 A cw holography system for vibration analysis. The thermoplastic camera is positioned at the top of the picture. The argon laser is mounted on the table off the right-hand side of the picture. (Courtesy of Rolls-Royce plc.)

7.3 MEASUREMENT WITH HOLOGRAMS

Having recorded a detailed three-dimensional image of the object, it is possible to make measurements directly from the image (Gates, 1986). Provided that the reconstruction wave has the same geometrical form and wavelength as the original reference wave, the image will be reproduced faithfully, without distortion.

If the image lies close to the hologram, it can be measured directly using conventional rulers or callipers and holding these inside the holographic image. This crude but effective technique allows measurements to be made on, for example, holographically stored dental records (Keating *et al.*, 1984).

It is possible to make more precise measurements by moving a small light pointer, for example the end of a fibre-optic, around inside the holographic image. If the pointer is moved under computer control on a three-coordinate traverse, accurate measurements of the positions of features can be made. Such a system has been developed by Rolls-Royce for the measurement of shock-wave positions in three-dimensional flow holograms (Parker, 1987).

To make even more accurate and detailed measurements from a hologram, it is possible to use the very high resolution of the hologram to full effect by projecting the 'real' image and examining this directly with a TV camera. The real image is produced by illuminating the hologram from behind with a reversed (conjugate) version of the original reference wave. The real, projected image cannot be observed directly, but can be seen on a screen placed into the object space, or by placing a CCD detector in the same region. Such a system has been developed for inspecting radioactive fuel elements (Tozer and Webster, 1980; Tozer *et al.*, 1985). A similar system of projected images can be used to make very high resolution measurements on small particles, droplets and sprays (Trolinger, 1975). With such a system it is possible to image and measure droplets down to 5 μm in diameter.

7.3.1 Holographic interferometry

Having captured the phase distribution of the wave field scattered from the object, it is possible to use this phase information as the basis for very precise differential measurements. If the phase of the field changes, and a double-exposure hologram is recorded before and after this change, the reconstructed hologram will produce not one, but two wave fields. These fields will themselves interfere. The new, macroscopic interference pattern will contour the change in phase which has occurred between the two holographic recordings. This is the basis of holographic interferometry (Powell and Stetson, 1965).

Holographic interferometry involves the interference of two or more complicated wave fields from holograms. In its simplest form the two images are recorded sequentially on the same holographic plate, each image representing a different state of the object under investigation (Fig. 7.10). The interference pattern observed in the reconstructed hologram contours the change in the object (Fig. 7.11). An example of

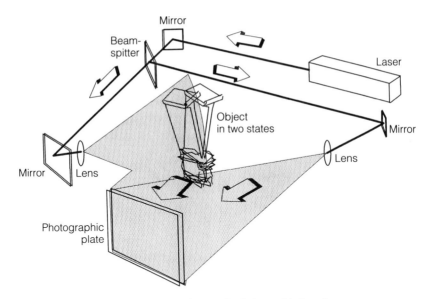

Fig. 7.10 Recording a double-exposure hologram for holographic interferometry.

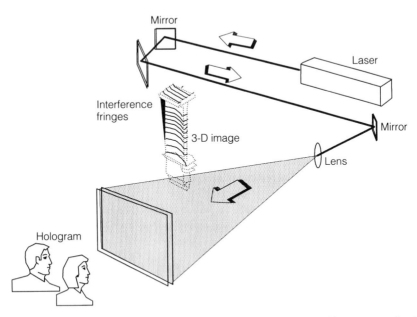

Fig. 7.11 Reconstructing a holographic interferogram. Changes in the object are perceived as interference fringes superimposed on the three-dimensional image.

230 *Industrial application of holographic interferometry*

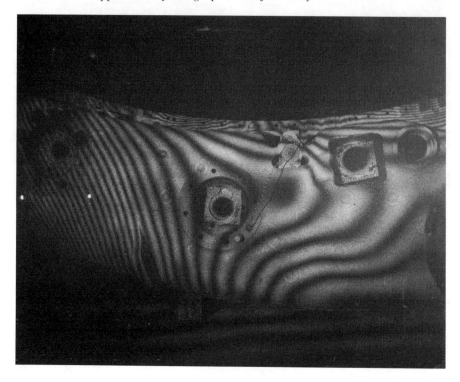

Fig. 7.12 Double-exposure hologram showing interference fringes due to deformation. The object was a gearbox housing and deformation was produced by local heating between holographic exposures. (Courtesy of Rolls-Royce plc.)

such a holographic interferogram is shown in Fig. 7.12. This shows the way in which the interference fringes form a contour map on the image of the object, in this case a large gearbox.

It is also possible to compare the holographic wave field with the actual field from the object. This is done by recording a hologram of the unstressed object, developing it and then observing the laser-lit object with the holographic image superimposed. This provides a 'live' interference pattern which allows changes to be observed in real time. The method is convenient for inspection as the holographic pattern may be viewed by a video system and the operator/inspector can observe changes in real time.

Conventionally, with holograms recorded on silver halide material, the hologram has to be repositioned with sub-micrometre accuracy after development so that the live interference fringes can be seen. Any small amount of mispositioning of the hologram would result in erroneous fringes, not associated with the object deformation. If the mispositioning error is too large, the hologram and the object wave fields are decorrelated and no fringes can be observed.

Measurement with holograms

It is possible to develop the hologram *in situ* using a **liquid gate** around the photographic plate (van Deelen and Nisenson 1969). The development chemicals are circulated through this gate. Once the film has been exposed it is necessary to keep it submerged in water while the real-time fringe pattern is studied. One commercially available system uses this process (Laser Technology Inc.).

More recently, the availability of thermoplastic recording systems (section 7.2.3(b)) has greatly facilitated real-time holographic interferometry. Since the hologram is instantly developed *in situ*, it requires no repositioning and the live interference pattern can be studied immediately.

7.3.2 Fringe pattern interpretation and analysis

The fringes seen in double-exposure holographic interferometry can be shown (Vest, 1979) to have an intensity I at any point (x, y) in the field of view described by

$$I(x, y) = 2|a(x, y)|^2 [1 + \cos \Delta\phi(x, y)], \tag{7.2}$$

where $2|a(x, y)|^2$ represents the intensity at the point due to the individual images of the object and this is modulated by the cosinusoidal fringes. The expression assumes that the two reconstructions are of equal intensity so that the fringes exhibit full modulation. The phase term $\Delta\phi$ represents the change in the phase of the optical wave field from the object between the two exposures. The phase is given by

$$\phi = 2\pi \frac{n\,d}{\lambda} \tag{7.3}$$

and

$$N = \frac{\Delta\phi}{2\pi}, \tag{7.4}$$

where n is the refractive index of the medium through which the light passes, d is the physical path length from the laser to the object and thence to the plate and λ is the wavelength of the laser light used to record the hologram. The fringe order (contour number) N increases in integer steps for each 2π change in phase. The applications of holographic interferometry arise from variation of one of the parameters n, d or λ. By contouring the difference in one of these three parameters, holographic interferometry can be used for solid object deformation measurement, flow visualization and shape measurement.

In double-exposure holographic interferometry of diffuse objects the change in form between exposures can be investigated. This change in form is most often produced by the application of some external stress between exposures. Any change in shape of the object will change the physical path length d from the laser to the hologram plate via the object. It can be seen from equations (7.3) and (7.4) that the interference fringes seen in the hologram will contour the change in shape of the object. The fringe order at any point (x, y) is given by

$$N(x, y) = \frac{n}{\lambda}[d_1(x, y) - d_2(x, y)]. \tag{7.5}$$

This simplistic approach shows the basis of the interference pattern forming process. In any real system the value $d_1 - d_2$ will vary for every point on the object and every viewing position on the hologram.

7.3.3 Resolving components of displacement

For collinear viewing and illumination, the interference pattern in the hologram contours the component of displacement along the line of sight. For non-collinear viewing and illumination, the interference pattern contours that component of displacement along the **sensitivity vector**. This bisects the illumination direction and the viewing direction at the particular point on the object. The component of the displacement l_i of a point on the object is related to the fringe order observed at that point by

$$l_i = \frac{N\lambda}{2n \cos(\alpha_i/2)}, \tag{7.6}$$

where α_i is the angle between the viewing and illumination directions for that point on the object.

The hologram can be made sensitive to different components of the displacement by suitable choice of viewing and illumination directions. This latter property allows the full three-dimensional nature of the object distortion to be investigated. A minimum of three independent interference patterns will be required fully to resolve the components of motion for all points on the object (Shibayama and Uchiyama, 1971). For greater accuracy and for more redundancy in the measurement process, more views may be used.

For small objects close to the hologram plate, it is possible to obtain a number of views through the hologram, each of which allows a different component of motion to be resolved (Ennos et al., 1985). For larger objects, separate holograms are required to record each component of the distortion, as the range of viewing angles (and therefore sensitivity vectors) available through a single hologram plate is insufficient to accurately resolve the full, three-dimensional motion. A useful alternative is to keep the hologram position fixed and vary the illumination direction (Hung et al., 1973). In this way the interference patterns for different components are mapped on to the same, two-dimensional view of the object. This greatly facilitates the analysis, since it is not necessary to allow for perspective changes between individual images.

The preceding treatment of holographic interferometry allows the reader to grasp the principles behind the fringe-forming process and to analyse holograms. A useful and more rigorous vector approach is used by Jones and Wykes (1983, pp. 79–96). Those requiring even more detail and rigour will find it in the works of Schumann et al. (Schumann and Dubas, 1979; Schumann et al., 1985).

7.3.4 Automatic analysis

The subject of fringe-pattern analysis is covered elsewhere in this book (section 8.5 and Chapter 10). The description given here is less detailed and is intended only to show that it is possible to use computer image analysis to extract data automatically from holographic interference patterns.

Using the simple equations given in the previous sections it is possible to analyse holographic interference patterns. For many years this had been performed manually using photographs of the interference patterns. However, with advances in computer image processing, it became practical to automate the process of extracting engineering data from holograms. A recent review of the subject provides useful background information (Creath, 1990).

The analysis of holograms can be divided into two basic approaches. There are amplitude techniques (section 10.1) which work on a single image of the holographic interference pattern, detecting, much as a human observer would, the light and dark interference fringes; and there are phase techniques (section 10.2), which usually require some manipulation of the phase of the interference pattern and may require more than one image to be taken.

The basic approach for the amplitude techniques is to determine by one means or another the peaks and troughs (or the mid-height points) of the sinusoidal intensity distribution in the interference patterns. There are many different ways of doing this which include fringe following (Trolinger, 1985), skeletonizing (Yatagai et al., 1982) and peak-picking from scans through the image (Robinson, 1983).

Figure 7.13 shows a screen display of a single fan blade hologram undergoing analysis (Robinson and Williams, 1986). The computer has found each of the bright fringes and marked them. It has also determined correctly the turning point in the displacement map. The display shows a plot of the centreline displacement.

Time-average holograms (section 7.5.1) have also been analysed using similar techniques. The fringe patterns can be interpreted automatically using computer image analysis techniques (Robinson and Williams, 1986). Figure 7.14 shows the screen display during the operation of the analysis program. The computer has identified correctly the bright interference fringes (marked by crosses) and has also determined automatically the nodal line position (identified by the brightest fringe). The program then goes on automatically to ascribe fringe order numbers to each fringe, from which it can derive the vibration amplitude at each point on the blade.

The phase techniques imply direct evaluation of the phase $\Delta\phi$ of the interference pattern (equation (7.2)). From the phase, the displacement can readily be calculated. One approach is heterodyne analysis (Dändliker et al., 1973). A **double-exposure hologram** is recorded with a different reference beam for each exposure. The two holographic images are reconstructed using two beams of slightly different optical frequency. The interference pattern is then modulated continuously at the beat frequency. The relative phase in the electrical signal produced by a discrete detector

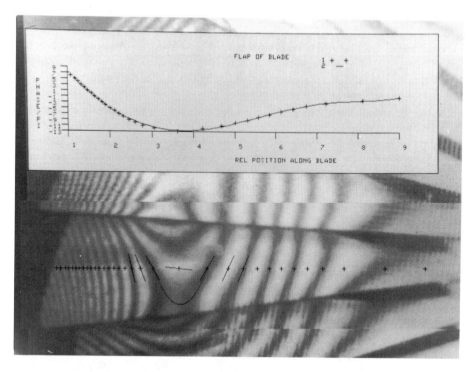

Fig. 7.13 Computer display during automatic analysis of double-pulsed vibration hologram. Flap is a measure of the out-of-plane motion of the blade. (Courtesy of Rolls-Royce plc.)

Fig. 7.14 Computer display during automatic analysis of time-averaged vibration hologram. The computer identified correctly the bright fringes and the nodal lines (brightest fringes). (Courtesy of Rolls-Royce plc.)

at one point in the composite image and a second at a reference point is a measure of the phase in the interference pattern. (More detail is given in section 10.2.1.)

In the phase-stepped or **quasi-heterodyne** (Hariharan *et al.* 1982) approach, three individual holographic interferograms are produced in which the phase of the interference pattern is changed by a known amount of one-third of a cycle between each exposure. In this way it is possible to create a set of simultaneous equations whose variables are the background intensity function A, the fringe-modulation function B and the unknown phase $\Delta\phi$. These are

$$I_1 = A + B \cos(\Delta\phi) \tag{7.7}$$
$$I_2 = A + B \cos(\Delta\phi + 2\pi/3) \tag{7.8}$$
$$I_3 = A + B \cos(\Delta\phi - 2\pi/3) \tag{7.9}$$

yielding

$$\phi = \tan^{-1} \frac{\sqrt{3}\,(I_3 - I_2)}{(2I_1 - I_2 - I_3)}. \tag{7.10}$$

In this way the phase (and hence displacement) at each point on the object is related only to three readings of intensity I_1, I_2 and I_3 taken from the individual interferograms. This is relatively easy to achieve and to automate. The technique is extremely robust, allowing automatic detection of regions of poor data and absence of fringes (e.g. outside the edges of the object) by calculation of the degree of modulation. This removes the need to provide details of the object geometry for the analysis system and allows a single analysis programme to cope with a wide variety of different interferograms. The claims in the literature are for accuracies of 1/200 fringe (Hariharan, 1985), or greater. However, other sources of error must be considered in any practical system (McKee and Parker, 1989) which reduces the likely accuracy considerably, but still leaves these phase-stepping techniques with obvious advantages over their more conventional predecessors. Other algorithms involving more phase steps are given in section 8.5.3 (*b*) and a more general analysis can be found in section 10.2.2.

Another interesting approach to the extraction of the phase data from the image is to manipulate the data from a single image by computing the spatial Fourier transform (Green *et al.*, 1988). This approach has many of the advantages of the phase-stepping technique, but without the need for multiple images which may not be practical in many dynamic situations. The major disadvantage of this method derives from the need to introduce a 'fringe carrier' by purposefully introducing tilt to one of the wavefronts, which ultimately limits the dynamic range of the process. The approach is discussed further in section 10.2.3.

7.4 APPLICATIONS OF HOLOGRAPHIC INTERFEROMETRY

The applications of holographic interferometry derive from the change between holographic exposures of one of the three variables in equation (7.3). Solid body

236 *Industrial application of holographic interferometry*

deformations under static or dynamic load produce changes in the optical path length d which can be seen in the hologram. Changes in refractive index n of the medium through which the light passes also give rise to interference patterns which can be used to visualize flow fields. Finally, there are a number of techniques which can render as interference patterns contours of physical shape. This is achieved by changing the laser wavelength λ, the illumination direction, or the refractive index of the medium surrounding an object.

The remaining sections of this chapter consider in detail these methods of holographic interferometry and illustrate, through selected case studies, the diversity of application of these powerful techniques.

7.4.1 Non-destructive testing

It has already been shown that holographic interferometry can reveal as a pattern of light and dark fringes the change in the object scene captured in the hologram. This effect may be used in a variety of ways to investigate changes in form of the object under investigation. The manner in which an object deforms when loaded, stressed, heated, pressurized, etc., can be studied in detail. Because of the great sensitivity of the hologram, which is capable of detecting movements of one micrometre, or less, the loading required to study deformation does not cause any permanent damage to the object. A secondary effect is that deformation of the object may also render visible in the hologram an internal mechanical fault. This provides a method of inspection, many examples of which are to be found in the literature, e.g. Erf (1974), Vest (1979) and Abramson (1981).

Some of the industrial applications of holographic NDT known to the author include testing of:

1. Composite helicopter rotor blades (France);
2. Helicopter body parts (Germany);
3. Tyres (Germany, USA (Brown, 1974), UK);
4. Adhesion of sprayed coatings (USA);
5. Integrity of honeycomb panels (USA, Japan);
6. Composite wing tips for damage (UK);
7. Composite materials for delamination (USA, France (de Smet, 1985)); and
8. Plastic beer crates! (Japan).

These are all routine shop-floor applications.

There is also a vast amount of NDT holography carried out to aid the design process and as proof of design concept. These studies have included:

1. Thermal deformation of disc drives;
2. Thermal deformation of circuit boards around 'hot' components;
3. Deformation of pipe joints;
4. Deformation of welds under stress/internal pressure;
5. Deformation of bones under body weight;

Applications of holographic interferometry 237

6. Deformation of artificial limb joints; and
7. Crack propagation in materials.

In Japanese industry in particular, with its emphasis on quality and 'right first time', holography has found many applications (Suzuki and Yuzo, 1983).

The lists above comprise a very small cross-section intended to show the diversity of the holographic technique. There follow case studies providing more detailed examples of the NDT use of holography.

(a) Case Study 1   Bearing housing distortion

In this case the problem arose from wear in a bearing. The bearing housing mounted in its gear box was placed in the holographic system and a thermoplastic recording camera was used to allow the real-time deformation of the housing to be studied. An argon ion laser was used to record the holograms.

The first exposure was made with the nuts holding the bearing in place tightened to just below their normal torque. The hologram was then developed and it was immediately possible to see the resulting image superimposed on the laser-lit object. If there had been no change, the scene would have appeared bright and free from interference fringes.

The nuts holding the bearing housing in place were then tightened up to the normal torque while observing the live interference pattern. The pattern seen is shown in Fig. 7.15. It would be possible to make accurate measurements from this

Fig. 7.15 Photograph of real-time interference pattern observed during bolt tightening. It shows the asymmetric deformation of a small bearing housing. (Courtesy of Rolls-Royce plc.)

photograph of the deformation of the entire structure. In this case, however, a more qualitative approach sufficed to reveal the cause of the problem. It was immediately apparent from the hologram that the deformation of the housing was asymmetrical; the housing was twisting around the line joining the two location bolt-holes. It was this twist that was the cause of the original problem.

Using the holographic data, the mounting features on the body of the gear box were redesigned and subsequent testing showed that this had removed the twist component from the distortion.

(b) Case Study 2 Honeycomb panel inspection

A testing technique was required for assessing the quality of brazing inside a titanium honeycomb panel. The panel consisted of two outer metal skins separated by and bonded to an internal honeycomb structure of the same metal. This type of construction is becoming increasingly common in both metal and plastic materials, since it provides very high stiffness and strength but low weight.

The holographic system was the same as the one for the bolt-tightening tests above. Again, an argon laser was used, but this time frozen-fringe holograms were produced on holographic film.

It was necessary to find a way of producing a differential displacement between the metal skin and the internal honeycomb. The one finally adopted was to vibrate the panel, but stressing by vacuum or local heating was also considered. At low frequencies the panel vibrates in modes similar to those of a solid plate, which shows little information about its internal structure. At higher frequencies the individual cells of the honeycomb act as small diaphragms and their outer skins vibrate. Holograms were taken to establish the frequencies at which this occurred. It was noted, however, that at any one frequency not all of the cells were vibrating, because of minute differences from cell to cell. To overcome this the frequency of vibration was 'chirped' (increased linearly over a small range) during the exposure of the hologram.

The result can be seen in Fig. 7.16. Over the majority of the panel individual cells have turned dark and are bounded by bright lines where there is no movement over the internal honeycomb web. However, in some areas it can be seen that there is movement over the web, indicating missing or defective braze.

This forms the basis for an inspection technique which, with modern computer hardware and software, could easily be fully automated. A similar technique could also be used to check for damage on panels in service.

(c) Case Study 3 Thin wall detection

This case study was an exercise to develop an optical method for examining thin wall sections in turbine blades with internal cooling passages, as an alternative to the ultrasonic techniques normally used to measure wall thickness.

Applications of holographic interferometry 239

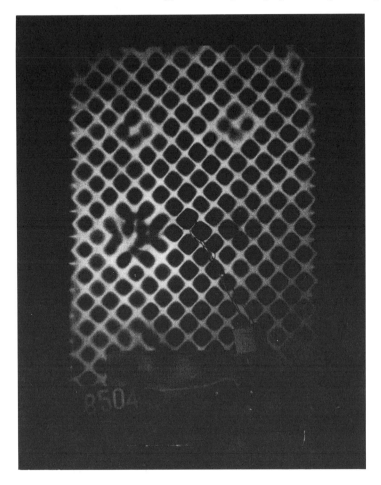

Fig. 7.16 Defects observed in the internal structure of a titanium honeycomb panel. (Courtesy of Rolls-Royce plc.)

It was decided to try a holographic technique. By applying a small internal pressure to the appropriate cooling passage and blocking the other end, the wall of the blade could be made to deform outwards. The amount of deformation for a given pressure would be an indication of the thickness of the blade. A thermoplastic camera was used so that live deformation could be observed.

An initial test piece was made using a square diaphragm of the blade material of a representative thickness. This was pressurized at various internal pressures and the results are shown in Fig. 7.17. They show the expected increase in deformation and hence fringe number with internal pressure. It should be noted that a large number of interference fringes can be resolved.

240 *Industrial application of holographic interferometry*

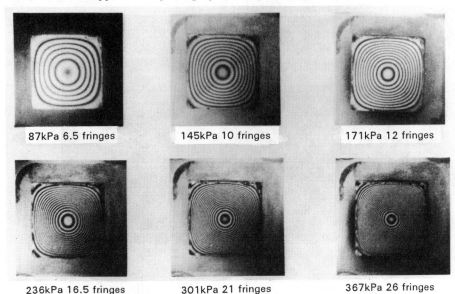

Fig. 7.17 A thin metal diaphragm is subject to steadily increasing pressure. The deformation is observed by real-time holographic interferometry. (Courtesy of Rolls-Royce plc.)

The experiment was then repeated with real blades. The blade was held in a jig which connected an air line to one end of the large trailing-edge cooling passage and blocked off the other end. A mirror was included in the jig so that both faces of the blade aerofoil surface could be inspected simultaneously. Figure 7.18 shows a typical result, in which the interference fringes contouring the deformation can be seen. At the left hand end, a significant increase in the number and density of fringes indicates a thin spot on the blade.

Unlike the previous example where the holographic image could be used directly as a pass/fail indicator, this technique requires calibration. It is necessary to know what thickness of blade wall corresponds to what number of fringes at a given pressure. In this case two approaches to calibration were used. The first was a destructive method which involved the sectioning of a number of test blades to measure the actual wall thicknesses. A finite-element model of the blade was also constructed to predict the wall movement for a given pressure. There was insufficient confidence in the computer model to use this as the sole method of calibration. However, the measured blade data and the holograms can be used to verify the computer prediction, so that for future similar blade types the computer model could be used with reasonable confidence. Once again, it is apparent that for a bulk-testing application, computer image analysis could be used to turn this into a fully automatic process.

(d) Case Study 4 Pneumatic tyre testing

One of the first successful applications of holography to be adopted for routine use in industry was the testing of pneumatic tyres. Tyres are composed of multiple layers of

Applications of holographic interferometry 241

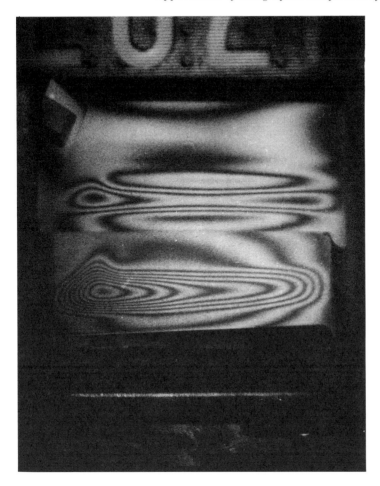

Fig. 7.18 A thin spot in a turbine blade wall is detected by holographic interferometry. Both surfaces of the blade are viewed simultaneously by placing a mirror below the blade. (Courtesy of Rolls-Royce plc.)

fabric or metal cable and rubber. One of the common manufacturing defects to occur is the inclusion of a void or air bubble between successive layers in the tyre, which forms a weak spot.

Early work by Brown (1974) showed that holography could be used to detect these voids. The tyre is put in a vacuum chamber and the external pressure is lowered; the area of the tyre above the void then bulges outwards as the void tries to expand. A double-exposure hologram recorded at two different external pressures shows an interference pattern which contours these bulges. Each disbond is marked by a bull's-eye pattern of closed fringes, which will be clearly visible against the general distortion pattern on the tyre.

242 *Industrial application of holographic interferometry*

Fig. 7.19 Aircraft tyre testing was one of the first routine industrial inspections to be performed holographically. The tyre is loaded into a vacuum chamber for testing. (Courtesy of Rottenkolber GmbH.)

A recent improvement on the technique has been effected by the use of thermoplastic cameras. These allow the distortion to be observed in real time on a video screen, which greatly speeds up the testing process. A system produced by Rottenkolber GmbH in Germany (Schorner and Rottenkolber, 1983; Rottenkolber, 1985) is shown in Fig. 7.19. The large vacuum chamber can be seen, as well as the control unit which houses the video monitor. A typical result, photographed from the video screen, is shown in Fig. 7.20. The fringe patterns contouring several defects are apparent.

Many large tyre companies have adopted this testing technique; it is of particular value in the testing of aircraft tyres.

7.5 VIBRATION ANALYSIS BY HOLOGRAPHY

Normally vibration is anathema to the holographer, causing the destruction of the holographic image. As was seen in section 7.2.4, vibration isolation is used to ensure that there is no movement of the object or optical equipment during the exposure. There are, however, ways of using holograms to study the vibration modes of objects.

7.5.1 Time-averaged holograms

The ability of the hologram to record the mode shape of a vibrating object was discovered very early in the history of holography (Powell and Stetson, 1965).

When using small continuous-wave lasers, the exposure of the hologram is usually long compared to the vibration period of the object, so that it vibrates through many

Vibration analysis by holography 243

Fig. 7.20 Tyre test results. The operator sees internal disbonds and delaminations as bull's-eye patterns of interference fringes. (Courtesy of Rottenkolber GmbH.)

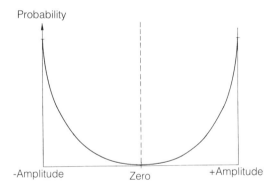

Fig. 7.21 For a sinusoidally vibrating object, any point on the object has a probability of being found at a particular displacement. The greatest probability is that it will be stationary at one of the extremes of the displacement.

cycles during the recording of the hologram. The hologram then records a **time-average** of the incident light field. If the vibration is random, or non-repetitive, the microscopic interference pattern which forms the hologram will become blurred and no hologram will be formed. However, if the object's vibration is sinusoidal and repetitive, the phase field at the hologram will be stationary at each extreme of the object's travel (Fig. 7.21). There will be some points on the object where the extremes of travel produce pathlengths which differ by whole numbers of wavelengths. These points will appear bright in the hologram. Conversely, those points which produce pathlengths differing by an odd number of half wavelengths will appear dark in the hologram. The image of the vibrating object will thus be covered with bright and dark fringes contouring the vibration amplitude. As the amplitude increases, any point on the object will spend smaller proportions of its time close to the stationary end points; the brightness of the bright fringes is thus observed to decrease with amplitude. The nodal points on the object which are always stationary will produce the brightest fringes.

This phenomenon is discussed by Vest (1979) who shows that the fringe pattern is described by a zero-order Bessel function of the first kind J_0 (Fig 7.22) according to the formula

$$I(x,y) = a^2(x,y) J_0^2 \left(\frac{\pi n \Delta d}{\lambda} \right). \tag{7.11}$$

The symbols have the same meaning as in equation (7.2). It is important to remember that the term Δd is the path-length change and not the amplitude of vibration. Δd is equal to twice the component of the peak-to-peak vibration amplitude along the line of sight for collinear viewing and illumination.

Coupling holography with time-resolved single-point measurements by accelerometer or strain gauge provides a complete description of the vibration: the hologram provides a full-field 'snapshot' of the vibration at discrete frequencies; the accelerometer provides a time history of the vibration at discrete points. For some objects, however, the weight of the strain gauge or accelerometer is sufficient to modify the vibration behaviour. In these cases holography provides one of the few non-contact techniques available.

As seen in Chapter 6, laser Doppler probes are now available which perform time-resolved measurements at discrete points on the object without physical contact. They can be scanned to form images of the vibrating surface. These are compared with holographic techniques in Hancox et al. (1989).

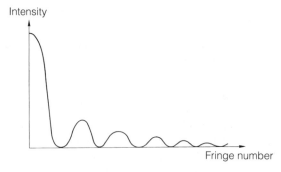

Fig. 7.22 The intensity of interference fringes seen in time-averaged holography of vibrating objects is governed by a zero-order Bessel function of the first kind. Intensity of the bright fringes decreases with increasing amplitude of vibration.

(a) Case Study 5 Turbine blade vibration

For small objects, it is reasonable to take the objects to the holographic test system and mount them on a vibration-isolated table. In this case continuous-wave (cw) lasers may be used. Figure 7.23 shows a typical time-averaged hologram of a turbine blade. The nodal lines and contour fringes described by equation (7.11) can be seen clearly. The Bessel function envelope of the interference fringe intensity produces bright fringes which decrease in brightness as the amplitude increases. The brightest line is always the line which has zero component of vibration along the sensitivity

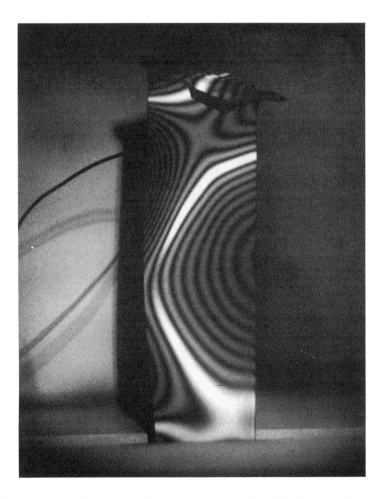

Fig. 7.23 Time-averaged hologram of a vibrating turbine blade. The brightest fringes are the stationary nodal lines. (Courtesy of Rolls-Royce plc.)

246 *Industrial application of holographic interferometry*

vector. For simple objects, this will normally be the nodal line. For more complicated objects and vibrations, however, this may not always be so.

The object must be driven in a pure single mode of vibration, using single-frequency sinusoidal excitation, in order to produce the standing-wave pattern recorded by the hologram. If the object was vibrating randomly, or at more than one frequency simultaneously, the result would be that either no hologram was recorded or else a very complicated and undecipherable pattern would be seen.

Figure 7.24 shows a series of time-averaged holograms of the same vibrating blade. The resonant frequency at which the blade was excited was increased from left to right along each row. At each frequency the modal pattern can be seen clearly. These maps of mode shape are useful as an indication of likely modes of failure of particular blades, should they experience high vibration at these resonant frequencies in service. In the early days of holography it was only employed as a diagnostic tool after a problem had occurred. Now, however, holograms are taken routinely of new component designs. These allow potential problems to be spotted and corrected, and also provide a useful database to assist the designer.

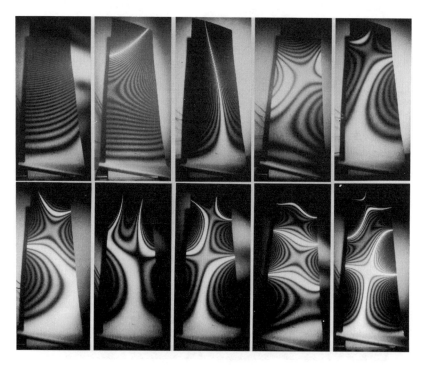

Fig. 7.24 Sequence of time-averaged holograms showing response of a compressor blade at different resonant frequencies. Frequency increases left to right along each row. (Courtesy of Rolls-Royce plc.)

Vibration analysis by holography 247

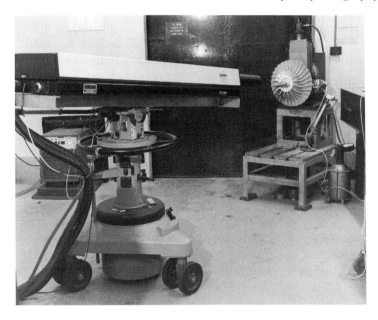

Fig. 7.25 A 10 J pulsed ruby holocamera (left) is used to take double pulsed holograms of an aero-engine component during vibration testing. (Courtesy of Rolls-Royce plc.)

7.5.2 Large-object or large-amplitude analysis

For objects and assemblies which are too large or unwieldy to mount on a vibration-isolated table, or where large-amplitude (>50 μm) vibrations are expected which cannot be resolved by time-average methods, pulsed lasers are used (Hockley *et al.*, 1978). A typical arrangement is shown in Fig. 7.25. The freestanding assembly is on the right of the picture; the ruby laser is on the left. The laser box contains all the optical components required for holographic interferometry as well as the laser itself. It is thus a fully portable and self-contained holocamera which can be taken to any object under test.

A double-exposure hologram is taken using the pulsed laser. The interference pattern then contours the change in shape between the two exposures. The two laser pulses are normally timed to occur during a single vibration cycle and to be spaced equally about the zero-amplitude position. If the pulse interval is made short compared to the period of vibration it is possible to capture any part of the vibration cycle and to produce a hologram with a resolvable number of contour fringes, even when the amplitude of vibration is very large. If the pulse interval is very much shorter than the vibration period, the hologram can be thought of as an instantaneous map of velocity in which the fringes map $\Delta d/\Delta t$.

248   *Industrial application of holographic interferometry*

(a)   Case Study 6   Fan-blade high-amplitude vibration

It is sometimes necessary to study the vibration behaviour of an object at realistic levels of vibration. If the vibration amplitude increases beyond 50 μm in a time-averaged hologram, there will be too many fringes to resolve covering the object.

Figure 7.26 shows a double-pulsed hologram of a vibrating aero-engine fan blade, approximately 0.4 m long. It was excited using a loudspeaker behind the tip. The tip amplitude was high (5 to 10 mm) but, by using a pulsed ruby laser and a short pulse separation, the tip movement observed during the holographic exposure was reduced to just 3.5 μm.

Figure 7.26 demonstrates one disadvantage of the pulse approach compared to time average. In the time-averaged holograms the nodal lines are clearly visible as the brightest fringes. In pulsed holograms, however, all the fringes have equal brightness and contrast. The pulsed laser can, however, be used to locate the nodal line using an **open-lase** technique. If the Pockels cell inside the pulsed ruby laser is switched off, the laser will produce a longer pulse lasting up to one microsecond. When looked at in detail, this pulse can be seen to be an irregular train of many short pulses. Figure 7.27 shows a hologram recorded in this way, in which the nodal lines are once again clearly visible. This can be used in conjunction with Fig. 7.26 to determine the vibration amplitude at all points on the object.

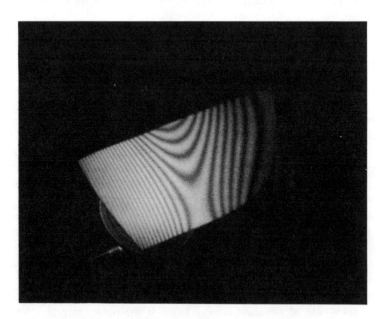

Fig. 7.26 Rapid double-pulsed hologram of high-amplitude vibration. The hologram contours just 13 µm of tip movement out of the total of 10 mm.

Vibration analysis by holography 249

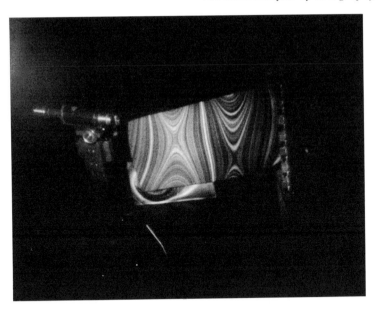

Fig. 7.27 'Open lase' hologram of the vibration of a large fan blade. A rapid succession of pulses from the laser produces a quasi-time-average hologram. (Courtesy of Rolls-Royce plc.)

(b)   Case Study 7   Fan asembly

In the previous sections individual objects were tested using holography. However, it is also necessary to study the way in which objects interact in their vibrational behaviour when they are assembled together.

Figure 7.28 shows a hologram of a 0.86 m fan assembly recorded using a pulsed ruby laser. The fan was vibrated using an electromagnetic shaker to drive it in a single 2F3D mode. This notation implies that the vibration mode is in the second family, i.e. there is one turning point in the radial distribution of displacement and it is the third diametral member of that family: the visible pattern repeats six times around the fan, but adjacent groups of blades are moving in antiphase, so that the vibration pattern repeats three times. The fringe orders contour the mode shape in equal increments from the stationary centre of the fan.

The fringes provide information about a single component of the object motion along the sensitivity vector. It is possible to extract information about all components of the object's movement by taking holograms from several different positions. Figure 7.29 shows a hologram taken from an off-axis position; the fan and vibration mode are the same as in Fig. 7.28. The data extracted from the holograms are used to compare with and verify finite-element computer models of these vibrating assemblies. Figure 7.30 shows a computer-generated contour map of the mode shape of a similar fan.

250 *Industrial application of holographic interferometry*

Fig. 7.28 Double-pulsed hologram of blades assembled into a fan. The structure exhibits a coupled blade and disc mode of vibration. (Courtesy of Rolls-Royce plc.)

7.5.3 Holography of rotating objects

The use of pulsed holography for routine vibration analysis of non-rotating vibrating objects was described in the previous section. The rapid double pulse from the ruby laser freezes the object, such as a fan, at two points in the vibration cycle. The reconstructed hologram produces a fringe pattern contouring the displacement during this short time interval, revealing the mode-shape of the vibration. If the object is rotating, it will have moved between the first pulse and the second. If it rotates too far between the two laser pulses, the two holographic images will become decorrelated and no interference fringes can be observed. If the vibration amplitude

Fig. 7.29 Holograms taken from several angles allow different components of displacement to be resolved. Same mode of vibration as Fig. 7.28. (Courtesy of Rolls-Royce plc.)

is large, so that the pulse interval can be kept very short, the fringes will be visible clearly. As the amplitude decreases, the pulse interval must be increased until a point is reached where significant rotation of the object has occurred between the two pulses. The limits to this regime were considered by Storey (1986).

A solution is to introduce a rotating optical element into the holographic system so that the image of the rotating object appears stationary (Stetson, 1978; McBain *et al.*, 1979; Fagan *et al.*, 1981). One arrangement uses a rotating mirror Abbe prism as described by Storey (1983, 1984). Another, commercially available, system (Fagan *et al.*, 1981) uses a reflective Porro prism as the rotating element. The rotating prism must rotate at exactly half the speed of the rotating component (see section 2.5.2)

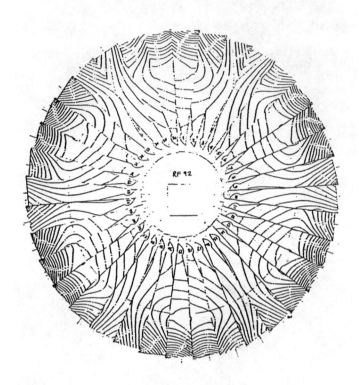

Fig. 7.30 Computer prediction of fan mode shape. Holograms can be used to validate computer (FE) models of component vibration behaviour.

and must be set accurately to have an axis of rotation collinear with that of the rotating object. It is also necessary to have accurate timing of the laser firing in order to choose the correct point in the vibration cycle and the rotation of the component.

(a) Case Study 8   Rotating fan vibration

Derotated holography has been applied to studies of automobile tyres (Tiziani 1983), automotive cooling fans and aero-engine fans. Rolls-Royce has used holography to study vibrating fans in aerodynamic test facilities (Storey 1983; Parker and Jones, 1988b).

Flow visualization 253

A mirror Abbe prism derotator was used for this study. Its speed was electronically matched to that of the fan by phase-locked circuitry. It is important that the rotational axes of the derotator and the fan are aligned to a high degree of accuracy (less than 1 mm in 6 m).

The fan was painted with a retro reflective paint (3M 7210 Silver Reflective Liquid) to increase the amount of laser light returned to the optical system. Figure 7.31 shows a 0.56 m (22 in) clappered fan vibrating predominantly in a 2F2D mode while rotating at 4125 rev/min.

7.6 FLOW VISUALIZATION

Modern engineering design confronts the engineer with increasingly complicated flow fields. Holography has provided a powerful new weapon in the arsenal of techniques for flow visualization (Bryanston-Cross, 1986a,b; Parker, 1990), allowing interferometric techniques to be used in various flow regimes and engineering environments.

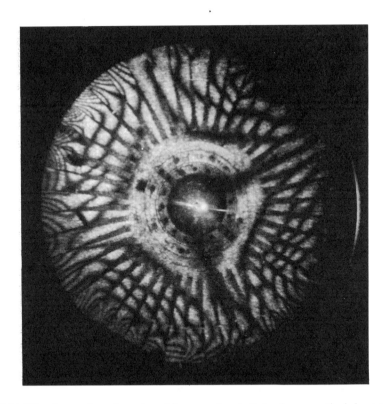

Fig. 7.31 Vibration modes of rotating objects can be studied using an optical derotator. Fan rotating at 4125 rev/min. (Courtesy of Rolls-Royce plc.)

254 Industrial application of holographic interferometry

Holography is one of several optical techniques which can be applied to visualization of fluid flows; others commonly used are the shadowgraph (Dvorak, 1880), the schlieren method (Toepler, 1864) and the Mach–Zehnder interferometer (Zehnder, 1891; Mach, 1892). A useful description of these and other techniques can be found in Merzkirch (1974). Optical techniques possess the major advantage that they are non-contacting and non-intrusive. They do not, therefore, perturb the flow field which is being measured.

Interferometric flow visualization techniques, such as Mach–Zehnder interferometry, have a major shortcoming in that they visualize all refractive index features along the line of sight. Some of these features may arise from the flow field, but others come from imperfections in the windows around the flow field and the other optical components in the system. Holography, as will be demonstrated, possesses the ability to remove the effects of these extraneous phenomena (Tanner, 1966), revealing only the refractive index field of the flow under observation.

Holographic flow visualization is performed using holographic interferometry. In this case it is the change in refractive index and hence density of the flowing medium between the two exposures which produces the interference pattern, rather than the physical changes of path length involved in mechanical distortion of objects. Most commonly, the first holographic exposure is made without the flow field and the second with the flow present. The interference pattern then contours the total change in density of the gas between ambient temperature and pressure and the running conditions.

Transonic and supersonic flow fields exhibit densities significantly different from ambient. Holographic techniques are, however, not only applicable to such flow fields, but can be applied to much slower flows provided that a change in refractive index is produced, e.g. by heat input or by a change in the fluid species between the two exposures. It is important to remember that the interference fringes are not stream-lines, but are contours of change in density.

7.6.1 Fringe interpretation

As well as visualization of flows, the engineer can also extract numerical data from the hologram. In order to do this it is necessary to understand what the optical fringe patterns mean in terms of the thermodynamic properties of the flow.

In a double-exposure hologram, the phase change $\Delta \phi$ at a point (x,y) in the image is an integrated effect along each optical ray r through the flow field, being given by

$$\Delta \phi(x,y) = \frac{2\pi}{\lambda} \int [n_1(r) - n_2(r)]\, dr, \qquad (7.12)$$

where λ is the wavelength of the laser light and n is the refractive index of the fluid. The subscripts 1 and 2 refer to the two exposures. As seen in section 2.3.2 for air, the refractive index of a gas is related to its density by the relationship

$$n = C\rho + 1. \qquad (7.13)$$

C is known as the Gladstone–Dale constant ($C = 2.25 \times 10^{-4}$ for air with $\lambda = 694$ nm (ruby laser) or 2.27×10^{-4} for 515 nm (argon laser)). The phase change $\Delta\phi$ is related to the fringe order or contour number N in the image by the sinusoidal nature of the interference pattern, as given by equation (7.2). Thus we have

$$\Delta\phi = 2\pi N. \tag{7.14}$$

For simple two-dimensional flow fields, the integral along each ray can be replaced by the optical path length L. Substituting from equations (7.12) and (7.13), equation (7.14) becomes

$$N(x,y) = \frac{C}{\lambda} L \Delta\rho(x,y). \tag{7.15}$$

The density change $\Delta\rho$ at each point in the flow field maps directly to each point (x,y) in the hologram image. In the normal double-exposure hologram, the image can be seen to be a contour map of the absolute change in density between ambient and flow conditions. In the rapid double-exposure it is a map of the local rate of change of density $\Delta\rho/\Delta t$, where Δt is the pulse separation.

In the case of isobaric or isothermal density changes, the perfect gas laws may be applied to obtain temperature or pressure, respectively, from the fringe pattern (Brownell et al., 1989). In the special case of an isentropic flow, with $P/\rho^\gamma =$ constant, where P is the pressure and γ is the ratio of the specific heats, the Mach number M may be derived directly from the holographic fringe pattern (Bryanston-Cross et al., 1981) as

$$M = \sqrt{\left\{ \frac{2}{\gamma-1} \left[\left(1 - \frac{N\lambda}{\rho_0 LC}\right)^{-(\gamma-1)} - 1 \right] \right\}}, \tag{7.16}$$

where ρ_0 is the density at the stagnation point.

7.6.2 Two-dimensional flows

In the laboratory, it is possible to use small continuous-wave (cw) lasers to record holograms. The laser, equipment and flow subject must all be mounted on a vibration-isolated table. For less steady flows an argon ion laser giving several watts of laser light may be used. This allows exposure times of the order of a millisecond and the system is then much more tolerant of unsteadiness in the flow field.

Figure 7.32 shows schematically a laboratory system for recording the flow in a simple model. It has all the essential components of the basic recording apparatus discussed earlier. In this case the model is two-dimensional, having a constant section along the line of sight. It is thus the property of the hologram to record phase which is important, rather than the ability to record a three-dimensional image. An

256 *Industrial application of holographic interferometry*

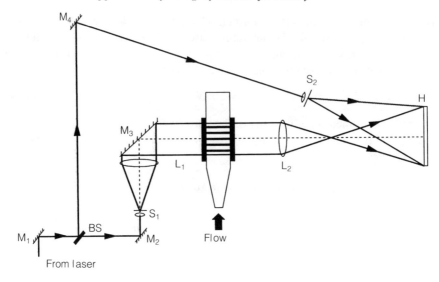

Fig. 7.32 Holographic flow–visualization in two-dimensional models using image-plane holography. M are mirrors, L are lenses, S are spatial filters, BS is a beamsplitter, and H is the hologram.

image-plane hologram is employed (Brandt, 1969), the test section being magnified by a lens and imaged on to a holographic plate. Such holograms can be reconstructed with a broad-band source instead of a laser (Fig. 7.33). The holograms display colour dispersion, but the image remains sharp since it is localized at the hologram plate. One important feature of this method of reconstruction is the removal of the coherent speckle noise produced by laser illumination (Parker and Brownell, 1985).

To study simple flow fields in wind tunnels, it is possible to apply very similar holographic techniques to those described above. The major difference is that it is very difficult to achieve vibrational stability in this environment to the required tolerances for holography with cw lasers. In addition, the flow itself is rarely stable enough over the relatively long periods of time required for exposure with small lasers. For this reason pulsed lasers are employed. The short pulses freeze any temporal flow disturbances, allowing holograms of highly turbulent flows and transient flow phenomena to be recorded. A pulsed laser is essential when studying the flow in piston tunnels, shock tubes or blow-down facilities, the desired flow conditions in such facilities being achieved for only a fraction of a second (Bryanston-Cross *et al.*, 1981).

Figure 7.34 provides an example of an image-plane hologram recorded using a pulsed ruby laser. It shows the transonic flow in a turbine blade cascade. The interference fringes once again form an isodensity contour map and a number of flow features can be seen in the hologram.

Flow visualization 257

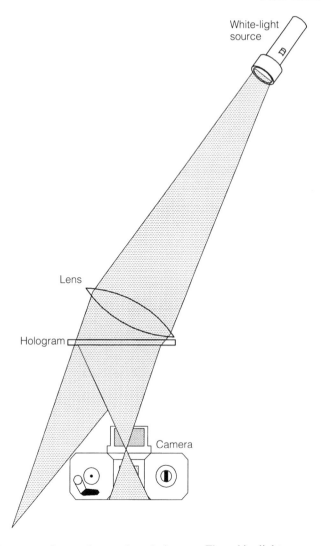

Fig. 7.33 Reconstructing an image-plane hologram. The white-light source results in dispersion, but not blurring. Speckle noise is greatly reduced.

Holograms recorded using rapid double pulses do not show the full flow field, as above, but produce interference fringes only where a change in density has occurred during the short time interval. Double-pulse holography thus provides a tool for studying the temporal variations of the flow field. Figure 7.35 shows a hologram recorded with $100\,\mu s$ between the two exposures in which only regions of turbulence and instability are visualized.

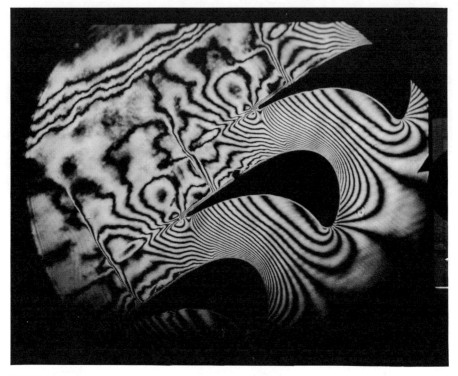

Fig. 7.34 Flow around two-dimensional turbine blade profiles in a transonic cascade. The double-exposure hologram records the total difference between flow and non-flow conditions. Fringes are isodensity contours. (Courtesy of Rolls-Royce plc.)

(a)   Case Study 9   Labyrinth seal flows

The apparatus described above with a 2 W argon ion laser was used to record the flow inside a life-sized model of a labyrinth seal (Brownell *et al.*, 1989). The labyrinth seal is designed to effect a gas seal between rotating and stationary parts without physical contact. It achieves this by forcing the gas to flow through a series of constrictions and expansions which induce losses in the flow. The seals are very small, and the high spatial resolution achievable with holography is used to good effect to visualize the flow through gaps as small as 250 μm.

Figure 7.36 shows the interference patterns obtained for a 207 kPa pressure drop across the seal. The first exposure of the hologram was made at ambient temperature and pressure and the second with the pressure applied across the seal. A pair of contrarotating vortices is produced within each cell of the seal, shown by the closed-loop interference fringes which form iso-density contours. The density information can only be interpreted in terms of pressure and temperature if the nature of the thermodynamic processes involved is known.

Flow visualization 259

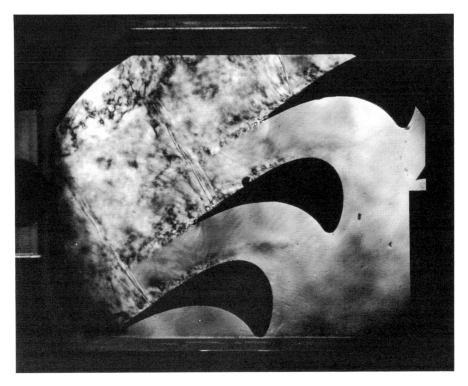

Fig. 7.35 A rapid double-exposure hologram (100 µs) reveals unsteady flow features. Shocks, wakes, and separations are all delineated by interference fringes. (Courtesy of Rolls-Royce plc.)

7.6.3. Three-dimensional Flows

Three-dimensional flow fields may also be studied by holographic flow visualization, using the same apparatus and techniques as used for two-dimensional flows. The hologram, for any particular viewpoint, will present a density map integrated along the line of sight. This may still be used for visualization purposes. If, however, quantitative information is to be derived to find the density at a particular point, then a tomographic approach must be employed using several viewing positions to build up a complete three-dimensional database (Craig, 1983). If the flow field exhibits some symmetry, such as radial symmetry, the process can be simplified and an Abel transform may be used (Vest, 1974).

7.6.4 Real-time flow visualization

One important variant on the standard technique described above is the live viewing of holographically formed interference patterns. A single-exposure hologram is taken

260 *Industrial application of holographic interferometry*

Fig. 7.36 Flow through a life-sized model (5 mm x 15 mm) of a labyrinth seal is contoured in a double-exposure, image-plane hologram. Pressure difference across the seal is 207 kPa. (Courtesy of Rolls-Royce plc.)

of the flow model at ambient temperature and pressure. This is developed and replaced exactly in the recording apparatus. The live wavefront transmitted through the flow field now interferes with the stored wavefront reconstructed by the hologram. When the flow is applied, it can be observed in real time in the interference pattern.

A photograph of a live fringe pattern recorded using photothermoplastic material is shown in Fig. 7.37 (Parker, 1978). It shows the thermally induced flow field inside a gas-filled light bulb. This technique has been applied successfully to study the flow instabilities inside photometric standard lamps (Parker and Gates, 1979). Once the live fringe pattern is produced, it may be recorded by still or cine photography or by video. It is also possible to use high-speed cine techniques to observe rapidly varying or transient flow features.

Various methods have been proposed which allow manipulation of the fringe pattern after recording of the hologram. These are summarized by Trolinger *et al.* (1983).

Flow visualisation 261

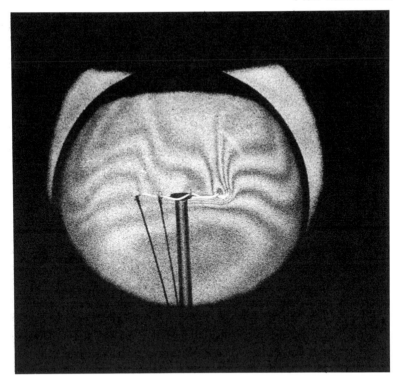

Fig. 7.37 Gas flow inside a gas-filled light bulb is observed in real time using thermoplastic recording materials. (Courtesy of Rolls-Royce plc.)

7.6.5 Rotating flows

Many of the flow fields of interest to engineers are in rotating machinery. Techniques have been developed which allow shock waves and other flow features to be visualized in three dimensions in rotating aero-engine fans (Wuerker *et al.*, 1974; Moore *et al.*, 1981; Parker and Jones, 1985: Bryanston-Cross, 1986b).

(a) Case Study 10 Flows in aero-engine fans

The optical system used to visualize flow in aero-engine fans is shown schematically in Fig. 7.38. A double-pulsed ruby laser freezes the rotation of the fan in each exposure. The object beam is injected through a small window upstream of the fan and illuminates the far side of the compressor casing, which has been painted white. A hologram is recorded through a large window over the blade tips. Some of these features can be seen in Fig. 7.39, which is a photograph of the equipment in a test facility.

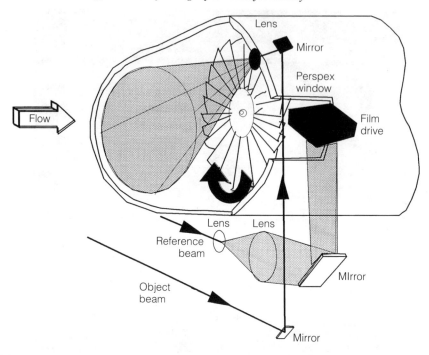

Fig. 7.38 Diagram of the holographic installation for rotating-flow visualization.

The rotating blades are seen in the hologram image in three-dimensional silhouette against the bright background. Rotation of the blades and flow field occurs between the two exposures. The pulse separation is kept short (1 – 3 μs) so that the movement of the blade tips will also be small (0.5 – 1.5 mm). The interference pattern in the hologram will contour the change in density during this short time interval, as seen from the hologram's static frame of reference. This change in density is effected by the rotation of strong density gradients through the field of view. The resulting interference fringe pattern produces a three-dimensional image of the shock wave or other flow feature.

Figure 7.40 shows a holographic image of a fan which was rotating at 7000 rev/min. A normal shock can be seen. The flow in the photograph is from left to right and the rotation of the fan is downwards.

7.7 HOLOGRAPHIC CONTOURING

A hologram records a three-dimensional image of an object. It is possible to make measurements, with great accuracy, of the distance between features on the object by measuring the image. This has been put to good use in a **hologrammetry** system for measuring radioactive fuel rods (Tozer and Webster, 1980). As with conventional

Holographic contouring 263

Fig. 7.39 Holographic equipment installed on a compressor rig. The film drive (top) has a view through a large perspex window over the rotating blades. (Courtesy of Rolls-Royce plc.)

photogrammetry (Chapter 4), this approach works well only if there are distinct features, markers or other depth clues within the scene. On many engineering components, the surface is a continuous curved form with no distinct features, which does not lend itself to measurement by this feature extraction approach.

To measure a surface shape by conventional contact methods with great accuracy is time-consuming and uses very expensive coordinate measuring machines. The techniques often rely on contact between touch probes and the surface, although some optical triangulation heads are now available (section 2.5.1). On very delicate materials such as wax patterns or some soft plastics, there is the additional danger that a contact technique might scratch or deform the surface under test.

There is clearly a need for a full-field, rapid contouring technique which would provide accurate coordinate data at all points on the surface. There are a number of ways (Varner, 1974; Hariharan, 1984) of recording holographic interferograms so that the interference fringes produced are absolute contours of surface height above some reference surface.

264 *Industrial application of holographic interferometry*

Fig. 7.40 In rotating-flow visualization, the blades are seen in 3-D silhouette against a bright background. The shock wave is seen in the double-exposure (3μs separation) hologram as a smoky surface. (Courtesy of Rolls-Royce plc.)

7.7.1 Two-angled illumination

In one technique, a double-exposure hologram is formed of an object illuminated first with one source and then with a second displaced laterally by a small amount. This is most easily achieved by reflecting the illuminating beam from a small mirror on to the object. By tilting the mirror between exposures the angle of the illumination beam is changed and the source appears to translate.

When the hologram is reconstructed, a pattern of interference fringes is seen. They represent the intersection of a family of hyperbolic shells with the object surface.

Holographic contouring 265

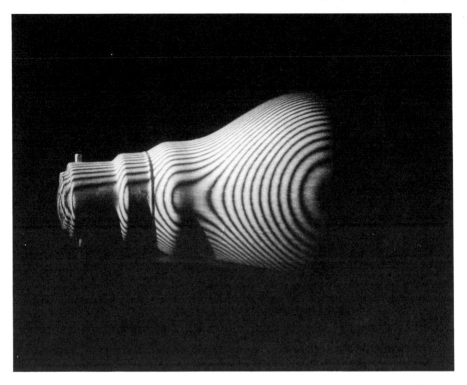

Fig. 7.41 A light bulb contoured by two-angled illumination holographic contouring. (Courtesy of Rolls-Royce plc.)

These will approximate to plane surfaces only if the object is small and a long way from the illuminating source, or if the illuminating beam is well collimated.

Figure 7.41 shows an example of a pearl light bulb contoured using this technique. The contour interval is approximately 1 mm. To obtain contouring planes normal to the line of sight, the object must be illuminated from the side. The consequent problem of shadowing can clearly be seen: large areas of the object cannot be measured even with this relatively simple geometry.

To understand why this technique works, suppose that the two illuminating beams are present at the same time and well collimated. Where these two beams cross they form planes of interference. Any object placed into this light field will appear to be cut by the planes of interference into a number of slices, forming the contours. Even though the object is illuminated sequentially with each of the beams, and no contours are directly observable, both beams appear to be present simultaneously when the hologram is reconstructed. Thus the object is seen as if illuminated by the two crossed beams. From this it can also be seen that for collimated beams the fringe or contour spacing Δh is given by

266 *Industrial application of holographic interferometry*

$$\Delta h = \frac{\lambda}{2 \sin(\theta/2)}, \qquad (7.17)$$

where θ is the angle between the two beams and λ is the wavelength of the laser light. The reference planes for the contours will bisect the two beams.

The two-angled technique would in fact seem to offer little advantage over a projected fringe or moiré technique (section 9.1.2) (Takasaki, 1970), unless the three-dimensional nature of the holographic image was felt to be important.

7.7.2 Two-refractive-index techniques

In these techniques, the object is mounted inside a cell with a plane-glass window (Tsurata *et al.*, 1967; Varner, 1974). By changing the refractive index of the fluid around the object between the two exposures of the hologram, interference fringes are produced which contour the distance from the object to the window. The interpretation of the fringes is much simpler than in the previous technique, since a physical reference plane is provided by the cell window. The process is very similar to the use of holography for flow visualization. In visualization of two-dimensional flow fields the fringes contour a variable change in refractive index over a constant path length (Zelenka and Varner, 1969). In contouring, a variable path length is contoured for a constant change in refractive index.

For two-refractive-index contouring, equations (7.13), (7.14) and (7.15) show that the fringe orders are given by

$$N = \frac{d}{\lambda}(n_1 - n_2), \qquad (7.18)$$

where n_1 and n_2 are the two refractive indices of the fluid surrounding the object during each of the two exposures. The contour interval is simply the change in surface height Δh which corresponds to an increment of one in the fringe order, that is

$$\Delta h = \frac{\lambda}{n_1 - n_2}. \qquad (7.19)$$

By suitable choice of the fluids, it is possible to obtain a wide range of contour intervals. In an extreme case, such as water in exposure one and air in exposure two, the contour interval would be approximately 1 μm. These fringes could only be resolved on near planar objects set parallel to the windows. With various liquids, contour intervals between 50 μm and 500 μm can be produced.

(a) Case Study 11   Contouring using gas pressure change

It is not particularly convenient to use liquids for routine measurement applications on more complicated objects. Instead, different gases can be used. The author has

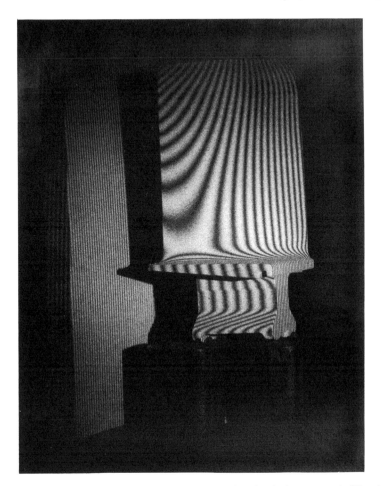

Fig. 7.42 A compressor blade is contoured by a two-refractive-index approach. The change in refractive index is achieved by changing gas pressure in the chamber surrounding the object between exposures.(Courtesy of Rolls-Royce plc.)

developed a system using air and varying the air pressure (Parker and Jones, 1988 a). By measuring the pressure accurately, it is possible to set the contour interval anywhere from 0.7 mm to infinity.

The object to be studied is placed inside a chamber with a plane-glass window and a double-exposure hologram is taken of the object from outside the chamber. The chamber is partially evacuated between the two exposures so that the refractive index of the air changes. Using the previously given Gladstone–Dale relationship of equation (7.13) and also equation (7.19) above, it is possible to calculate the contour interval. Figure 7.42 shows contours on a compressor blade with a contour interval of 1.5 mm.

7.7.3 Two-wavelength techniques

The final form of holographic contouring considered here is the two-wavelength technique. In this technique, the hologram is formed with two exposures at different wavelengths. It is then reconstructed at just one of the two wavelengths. Once again, the object will appear to be intersected by a family of hyperbolic shells which form the contours. The process is analogous to that described in Sections 5.7.3 and 11.3.4 for obtaining a slow variation of phase in ordinary interferometry.

For two-wavelength contouring, it can be seen from equations (7.2), (7.3) and (7.4) that forming a double-exposure hologram of an object with two different wavelengths for the two exposures will produce fringe orders defined by

$$N = nd \left(\frac{1}{\lambda_1} - \frac{1}{\lambda_2} \right) \approx nd \frac{\Delta \lambda}{\lambda^2} \qquad (7.20)$$

if the wavelength difference is small. It can be seen that, if the refractive index is unchanged between exposures, the fringes will contour d in fixed increments Δh determined by the wavelength change $\Delta \lambda$. Thus

$$\Delta h = \frac{\lambda^2}{n \, \Delta \lambda}. \qquad (7.21)$$

The contour interval achieved by this technique is dependent upon the change in wavelength. The readily available laser sources such as argon have individual laser wavelengths which are spaced by several nanometres, or tens of nanometres. These lead to contour intervals typically of 2–50 μm, which can only be resolved on relatively flat objects. To get larger contour intervals dye lasers can be used (Friesem et al., 1976), but it is notoriously difficult to stabilize their performance to give a known and reproducible contouring effect.

When trying to achieve small contour intervals (<1 mm), a further difficulty arises which makes the optical system more complicated. If the exposures at each wavelength were simply recorded using the same optical system, then when the hologram was reconstructed with a single wavelength, the mean angle of diffraction for each of the holographic images would be different. The two images reconstructed from the hologram would thus be sheared. This would lead to decorrelation, and no contour fringes would be visible. Various devices have been suggested to compensate for this movement; these include tilting mirrors and adjusting the position of the reference source between exposures (Hariharan, 1984) and the introduction of a diffracting element into the reference beam (Varner, 1971). Even with these additions it remains a difficult technique and only works when the object is very close to the hologram plate, or when it is imaged into the plane of the hologram plate. For these reasons the only objects reported to have been contoured with high-definition fringes by this technique are near-planar objects such as coins.

(a) Case Study 12 Large combustion chamber

For large contour intervals (>1 mm) it is necessary to use very closely spaced wavelengths. This is most conveniently achieved by using the longitudinal modes from a laser. The natural laser linewidth from many lasing materials such as argon ion or ruby is very wide. Within this line width a number of longitudinal modes (discrete, narrow laser lines) are supported by the laser cavity. It is usual to place an etalon in the cavity to select just one of these lines, or modes, in order to increase the coherence length of the laser. It is possible to alter the tuning of the etalon so that the laser hops from one mode to another. A hologram taken first at one mode and then at the next will exhibit very wide contours dependent upon the characteristics of the laser, but normally very reproducible.

The object for this study was a large combustion chamber, approximately one metre in diameter. The holograms were recorded using a one-joule pulsed ruby laser. The etalon temperature control was altered so that holograms were recorded with two different longitudinal modes from the laser.

Figure 7.43 shows the combustion chamber contoured by this technique. The object is approximately 1 m in diameter. The contour interval is approximately 14 mm. Using this technique, as can be seen, large objects may be contoured

Fig 7.43 A large combustion liner (1 m diameter) contoured by the two-wavelength method. Contour interval is approximately 14 mm. (Courtesy of Rolls-Royce plc.)

relatively easily. When the contour intervals are large and the wavelengths used are close together, the problems highlighted above disappear. Any shearing of the images is very small and is insufficient to cause decorrelation.

7.7.4 Holographic contouring – new potential

To date there have been no serious engineering applications of holographic contouring. The author surmises that the reasons for this are as follows. The accuracies required in engineering measurement would be typically 5 μm. Fringe-analysis techniques previously available have been accurate to only one-fifth to one-tenth of a fringe. This would imply that contours of 25 μm to 50 μm would be required. For an object with any reasonable depth there would be too many contours to resolve (the hologram could resolve them, but not the imaging system required to transfer them to a computer for analysis). It is for this reason that applications have been restricted to near planar objects like coins.

There is, however, some hope of a revived interest in holographic contouring. This is provided by the development of new phase-stepping approaches (Hariharan, 1984: Breuckmann and Thieme, 1985) to fringe analysis. These provide a resolution of 1/250 of a fringe. It would thus be possible to contour an object with coarse fringes, e.g. 1 mm, and still make measurements to 5 μm. This overcomes the previous resolution problem. As has been demonstrated, producing and imaging coarse fringes on even large objects is a practical proposition.

7.8 CONCLUSION

It is hoped that the foregoing discussion has demonstrated the diversity of holography as a measurement technique. It is also intended to show that, as a result of the development of several vital enabling technologies, holography has made the transition from laboratory curiosity to practical industrial tool. Essential to this process have been reliable pulsed ruby lasers, novel real-time recording materials and cheap, powerful image-processing systems.

The author thanks Rolls-Royce plc for permission to publish this work. Other holographers have gone before him at Rolls-Royce, notably Bernard Hockley, Peter Bryanston-Cross and Phil Storey, and each has contributed significantly to the furtherance of holographic science and to the rich diversity of material available for this book. A new generation of researchers is picking up the mantle, John Brownell and Paddy McKee, and they too have provided material and inspiration for this chapter; they have a bright and productive future ahead of them.

REFERENCES AND BIBLIOGRAPHY

Abramson, N. (1981) *The Making and Evaluation of Holograms*, Academic Press.
Benton, S. A. (1969) Hologram reconstructions with extended incoherent sources. *Journal of the Optical Society of America*, **59**(11) 1545–6.
Benton, S. A. (1980) Holographic displays. *Optical Engineering*, **19**(5), 686–90.

Bragg, W. L. (1912) The diffraction of short electromagnetic waves by a crystal. *Proc. Camb. Phil. Soc.*, **17**, 43–57.
Brandt, G. B. (1969) Image plane holography. *Applied Optics*, **8** (7), 1421–9.
Breuckmann, B. and Thieme, W. (1985) Computer-aided analysis of holographic interferograms using the phase-shift method. *Applied Optics*, **24** (14), 2145–9.
Brown, G. M. (1974) Pneumatic tire inspection, in *Holographic Nondestructive Testing* (ed. R.K. Erf). Academic Press, pp. 355–64.
Brownell, J. B., Millward, J. A. and Parker, R. J. (1989) Nonintrusive investigations into life-size labyrinth seal flow fields. *J. of Engineering for Gas Turbines and Power*, **111**, 335–42.
Bryanston-Cross, P. J. (1986 a) High speed flow visualization. *Progress in Aerospace Science*, **23**, 85–104.
Bryanston-Cross, P. J. (1986 b) The application of holography as a transonic flow diagnostic to rotating components in turbomachinery. *AGARD Conference Proc.*, **399**, 321–2.
Bryanston-Cross, P. J. *et al.* (1981) Interferometric measurements in turbine cascades using image-plane holography. *Journal of Engineering for Power*, **103** (1), 124–30.
BS7192 (1989) British Standard specification for radiation safety of laser products. British Standards Institution.
Caulfield, H. J. (1979) *Handbook of Optical Holography*. Academic Press.
Craig, J. E. (1983) Feasibility study of three-dimensional holographic interferometry for aerodynamics. *NASA-CR-166483*.
Creath, K. (1990) Phase-measurement techniques for non-destructive testing. *Proc. Hologram Interferometry and Speckle Metrology*. Baltimore, Nov., Publ. SEM, pp. 473–9.
Dändliker, R., (1980) Heterodyne holographic interferometry; in *Progress in Optics* (ed. E. Wolf), North Holland, **XVII**, 1–84.
Dändliker, R., and Thalmann, R. (1985) Heterodyne and quasi-heterodyne holographic interferometry. *Optical Engineering*, **24** (5), 824–31.
Dändliker, R., Ineichen, B. and Mottier, F.M. (1973) High resolution holographic interferometry by electronic phase measurement. *Optics Communications*, **9** (4), 413–6.
Decker, A. J. (1981) Holographic flow visualization of time-varying shock waves. *Applied Optics*, **20** (18), 3120–7.
van Deelen, W. and Nisenson, P. (1969) Mirror blank testing by real-time holographic interferometry. *Applied Optics*, **8** (5), 951–5.
Denisyuk, Yu N. (1962) Photographic reconstruction of the optical properties of an object in its own scattered light. *Soviet Physics Doklady*, **7**, 543–5.
Dvorak, V. (1880) Uber eine neue einfache Art der Schlierenbeobachtung. *Ann. Phys. Chem.*, **9**, 502–12.
Ennos, A. E., Robinson, D. W. and Williams D. C. (1985) Automatic fringe analysis in holographic interferometry. *Optica Acta*, **32** (2), 135–45.
Erf, R. (1974) *Holographic Non-destructive Testing*. Academic Press.
Fagan, W. F., Beeck, M. A. and Kreitlow, H. (1981) The holographic vibration analysis of rotating objects using a reflective image derotator. *Optics and Lasers in Engineering*, **2** (1), 21–32.
Friesem, A. A., Levy, U. and Silberberg, Y. (1976) Real-time holographic contour mapping with dye lasers. *The Engineering Uses of Coherent Optics*. Cambridge University Press.
Gabor, D. (1948) A new microscopic principle. *Nature*, **161**, 777–8.
Gabor, D. (1949) Microscopy by reconstructed wavefronts. *Proc. Royal Society*, **197**, 454–87.
Gasvik, K. J. (1987) *Optical Metrology*, Wiley, 67–115.
Gates, J. W. C. (1986) The influence of holography on measurement technology. *J. Phys., E*, **19**, 998–1007.
Green, R. J., Walker, J. G. and Robinson, D. W. (1988) Investigation of the Fourier transform method of fringe pattern analysis. *Optics and Lasers in Engineering*, **8** (1), 29–44.
Haines, K. A. and Hildebrand, B. P. (1967) Multiple-wavelength and multiple-source holography applied to contour generation. *J. Optical Society of America*, **57** (2), 155–62.
Haller, B. R. (1980) The effects of film cooling upon the aerodynamic performance of transonic turbine blades. PhD Thesis. St John's College, Cambridge.

Hancox, J. E. et al. (1989) An evaluation of several techniques for vibration modeshape mapping. *Stress and Vibration: Recent Developments in Industrial Measurement and Analysis. Proc.* SPIE **1084**, 240–52.
Hariharan, P. (1984) *Optical Holography*. Cambridge University Press.
Hariharan, P. (1985) Quasi-heterodyne holographic interferometry. *Optical Engineering*, **24** (4), 632–8.
Hariharan, P. and Oreb, B. F. (1984) Two-index holographic contouring: application of digital techniques. *Optics Communications*, **51** (3), 142–4.
Hariharan, P., Oreb, B. F. and Brown, N. (1982) A digital phase-measurement system for real-time holographic interferometry. *Optics Communications*, **41** (6), 393–6.
Hockley, B. S. (1973) Holographic visualization of large amplitude vibration using reference beam phase modulation. *Journal of Physics E: Scientific Instruments*, **6** (4), 377–80.
Hockley, B. S. and Butters, J. N. (1970) Holography as a routine method of vibration analysis. *Journal of Mechanical Engineering Science*, **12** (1), 37–47.
Hockley, B. S. and Hill, R. J. (1969) Vibration and strain analysis by means of holography. *Aircraft Engineering*, 6–11.
Hockley, B. S., Ford, R. A. J. and Food, C. A. (1978) Measurement of fan vibration using double pulse holography. *Journal of Engineering for Power*, **100** (4) 655–63.
Hung, Y. Y., Hu, C. P., Henley, D. R. and Taylor, C. E. (1973) Two improved methods of surface displacement measurement by holographic interferometry. *Optics Communications*, **8**(1), 48–51.
Jones, D. G. and Parker, R. J. (1987) Optical flow diagnostic measurements in turbomachinery. *Eighth International Symposium on Air Breathing Engines.* Cincinatti, Ohio. AIAA, 607–18.
Jones, R. and Wykes, C. (1983) *Holographic and Speckle Interferometry*. Cambridge University Press.
Keating, P. J. et al. (1984) The holographic storage of study models. *British Journal of Orthodontics*, **11**, 119–25.
Koechner, W. (1976) *Solid State Laser Engineering*. Springer-Verlag.
Leith, E. N. and Upatnieks, J. (1964) Wavefront reconstruction with diffused illumination and three-dimensional objects. *J. Optical Society of America*, **54**, 1295–301.
Longhurst, R. S. (1957) *Geometrical and Physical Optics*. Longman.
McBain, J. C. et al. (1979) Vibration analysis of spinning disk using image derotated holographic interferometry. *Experimental Mechanics*, **19** (1), 17–22.
Mach, L. (1892) Uber einen Interferenz-Refractor. *Z. Instrumentenkunde*, **12**, 89–93.
McKee, V. B. and Parker, R. J. (1989) A numerical simulation of the quasi-heterodyne technique for double-pulsed holographic vibration measurement. *Stress and Vibration - Recent Developments in Industrial Measurement and Analysis, Proc.* SPIE **1084**, 323–34.
Merzkirch, W. (1974) *Flow Visualization*. Academic Press, pp. 62–177.
Moore, C. J. et al. (1981) Optical methods of flow diagnostics in turbomachinery. *Proc International Conference on Instrumentation in Aerospace Simulation Facilities*, 244–55.
Ostrovsky, Yu I., Butusov, M. M. and Ostrovskaya, G. V. (1980) *Interferometry by Holography*. Springer-Verlag.
Parker, R. J. (1978) A new method of frozen-fringe holographic interferometry using thermoplastic recording media. *Optica Acta*, **25** (8), 787–92.
Parker, R. J. (1987) Extraction of 3-D flow data from transonic flow holograms. *Industrial Optoelectronic Measurement Systems using Colernt Light. Proc.* SPIE **863**, 78–85.
Parker, R. J. and Brownell, J. B. (1985) Holographic flow visualization applied to very small flow sections in turbomachinery research. *Optics in Engineering Measurement, Proc.* SPIE **599**, 111–8.
Parker, R. J. (1990) A quarter century of thermoplastic holography. *Proc. Hologram Interferometry and Speckle Metrology.* Baltimore, Nov., Publ. SEM, 217–25.
Parker, R. J. and Gates, J. W. C. (1979) An investigation of the instabilities in photometric standard lamps by holographic interferometry. *J. Phys. E*, **12**(1), 18–20.
Parker, R. J. and Jones, D. G. (1985) Holographic flow visualization in rotating transonic flows. *Institute of Physics Conference Series*, **77**, 141–6.
Parker, R. J. and Jones, D. G. (1987) Industrial holography: the Rolls-Royce experience. *Laser and Optoelectronic Technology in Industry, Proc.* SPIE **699**, 111–26.

Parker, R. J. and Jones, D. G. (1988 a) Holography in an industrial environment. *Optical Engineering*, **27**(1), 55–66.
Parker, R. J. and Jones, D. G. (1988 b) The use of holographic interferometry for turbomachinery fan evaluation during rotating tests. *Journal of Turbomachinery*, **110**, 393–400.
Parker, R. J. and Reeves, M. (1990) Holographic flow visualization in rotating turbomachinery. *Proc. Hologram Interferometry and Speckle Metrology*, Baltimore. Publ. SEM, 500–8.
Powell, R. L. and Stetson, K. A. (1965) Interferometric vibration analysis by wavefront reconstruction. *J. Opt. Soc. Am.*, **55(12)**, 1593–8.
Robillard, J. and Caulfield, H. J. (1990) *Industrial Applications of Holography*. Oxford University Press.
Robinson, D. W. (1983) Automatic fringe analysis with a computer image-processing system. *Applied Optics*, **22** (14), 2169–76.
Robinson, D. W. and Williams, D. C. (1986) Automatic fringe analysis in double exposure and real-time holographic interferometry. *Optics in Engineering Measurement. Proc.* SPIE **599**, 134–40.
Rottenkolber, H. (1985) 25 years of holography - the development of holographic testing. *Industrial Opto-electronic Measurement Systems using Coherent Light. Proc.* SPIE **863**, 134–40.
Saxby, G. (1988) *Practical Holography*. Prentice Hall.
Schorner, J. and Rottenkolber, H. (1983) Industrial application of instant holography. *Industrial Applications of Laser Technology. Proc.* SPIE **398**, 116–22.
Schumann, W. and Dubas, M. (1979) *Holographic Interferometry (from the Scope of Deformation Analysis of Opaque Bodies)*, Optical Science Series **16**, Springer Verlag.
Schumann, W., Zurcher, J. P. and Cuche, D. (1985) *Holography and Deformation Analysis*, Optical Science Series **46**, Springer Verlag.
Shibayama, K. and Uchiyama, H. (1971) Measurement of three-dimensional displacements by holographic interferometry. *Applied Optics*, **10** (9), 2150–4.
de Smet, M. A. (1985) Holographic non-destructive testing for composite materials used in aerospace. *Optics in Engineering Measurement, Proc.* SPIE **599**, 46–52.
Smith, H. M. (1969) *Principles of Holography*. Wiley Interscience.
Smith, H. M. (1977) *Holographic Recording Materials*, Topics in Applied Physics 20, Springer-Verlag.
Stetson, K. A. (1978) The use of an image derotator in holographic interferometry and speckle photography of rotating objects. *Experimental Mechanics*, **18** (2), 67–73.
Storey, P. A. (1983) A study of aero-engine fan flutter at high rotational speeds using holographic interferometry. PhD Thesis. University of Loughborough.
Storey, P. A. (1984) Holographic vibration analysis of a rotating fluttering fan. *AIAA Journal*, **22** (2), 234–41.
Storey, P. A. (1986) Holographic interferometry of rotating components: decorrelation limitations of the double pulsed technique. *Optics in Engineering Measurement, Proc.* SPIE **599**, 66–73.
Suzuki, M. and Yuzo, H. (1983) The present state of holography in Japan. *Industrial and Commercial Applications of Holography, Proc.* SPIE **353**, 74–81.
Syms, R. R. A. (1990) *Practical Volume Holography*, Oxford Engineering Science Series 24.
Takasaki, H. (1970) Moiré topography. *Applied Optics*, **9** (6), 1467–72.
Tanner, L. H. (1966) The application of lasers to time resolved flow visualization. *J. Sci. Instrum.*, **43** (5), 353–8.
Tiziani, H. J. (1983) Holographic interferometry and speckle metrology: A review of the present state. *Proc. Industrial Applications of Laser Technology, Proc.* SPIE **398**, 2–10.
Toepler, A. (1864) *Beobachtungen nach einer neuen optischen Methode*, Max Cohen u. Sohn, Bonn.
Tozer, B. A. and Webster, J. M. (1980) Holography as a measuring tool. *Journal of Photographic Science*, **28**, 93–8.
Tozer, B. A., Glanville, R., Gordon., A. L., Little, M. J., Webster, J. M. and Wright, D. G. (1985) Holography applied to inspection and measurement in an industrial environment. *Optical Engineering*, **24**(5), 746–53.
Trolinger, J. D. (1975) Particle field holography. *Optical Engineering*, **14**, 383–92.
Trolinger, J. D. (1985) Automated data reduction in holographic interferometry. *Optical Engineering*, **24** (5), 840–2.

Trolinger, J. D. *et al.* (1983) Laser diagnostic techniques: a summary. AIAA 16th Fluid and Plasma Dynamics Conference. Danvers, Massachusetts.

Tsurata, T. *et al.* (1967) Holographic generation of contour map of diffusely reflecting surface by using immersion method. *Japanese Journal of Applied Physics*, **7**, 1092–100.

Varner, J. R. (1971) Simplified multiple-frequency holographic contouring. *Applied Optics*, **10**, 212–3.

Varner, J. R. (1974) Holographic and moiré surface contouring, in *Holographic Non-Destructive Testing* (ed. R. K.Erf), Academic Press, pp. 105–47.

Vest, C. M. (1974) Formation of images from projections: Radon and Abel transforms. *J. Optical Society of America*, **64** (9), 1215–18.

Vest, C. M. (1979) *Holographic Interferometry*. Wiley.

Wuerker, R. F. *et al.* (1974) Application of holographic flow visualization within rotating compressor blade row. *NASA-CR-121264*.

Yatagai, T., Nakadate, M. *et al.* (1982) Automatic fringe analysis using digital image processing techniques. *Optical Engineering*, **21** (3), 432–5.

Zehnder, L. (1891) Ein neuer Interferenzrefraktor. *Z. Instrumentenkunde*, **11**, 275–85.

Zelenka, J. S. and Varner, J. R. (1969) Multiple-index holographic contouring. *Applied Optics*, **8** (7), 1431–4.

8
Television holography and its applications

J. C. DAVIES AND C. H. BUCKBERRY

Television (TV) holography is a technique which uses a laser, CCD camera and digital processing to generate holograms at TV frame rate. Several different names have been used to describe it. The earliest is probably 'electronic speckle pattern interferometry' (ESPI), due to Butters and Leendertz (1971). Recently, a proliferation of names has appeared for slight variations on the same theme, such as 'electronic holography' and 'electro-optic holography'. The name TV holography (Ellingsrud and Løkberg, 1989) will be used throughout this chapter. It highlights the real-time operating mode of the technique and also serves as a reminder of its holographic roots.

The authors use TV holography routinely for the analysis of engineering structures. As a result the contents are somewhat biased in this direction, but the scope has been broadened by including examples of work undertaken by other users. The emphasis is on the practical rather than the rigorously theoretical, an aspect that has been treated in considerable depth by several authors (Goodman, 1975; Jones and Wykes, 1983; Kujawinska, 1989). The authors hope that having studied the chapter the reader will understand the operation of TV holography systems, appreciate how they can be used to solve real engineering measurement problems and be in a position to implement them usefully in his or her own particular field.

8.1 LIMITATIONS OF CONVENTIONAL HOLOGRAPHY

Conventional holography is dealt with in detail in the previous chapter, but it is worth restating its underlying principles before considering the electronic version. Holography is an interferometric technique that generally uses a laser as the coherent source of radiation. By means of optical components like mirrors and beamsplitters, laser light reflected from the subject is combined with a second reference beam from the same laser to produce an interference pattern which is recorded by a photographic medium (Fig. 7.1). After photographic processing – developing, fixing and sometimes bleaching – the hologram can be used to reconstruct a three-dimensional image of the original subject, again generally with a laser (Fig. 7.2).

The process can be used to make engineering measurements by exploiting the ability of the hologram to store three-dimensional information. Using a double-

Optical Methods in Engineering Metrology. Edited by D.C. Williams. Published in 1993 by Chapman & Hall, London. ISBN 0 412 39640 8

exposure or a time-averaging technique, it is possible to measure a variety of parameters: displacement, distortion, vibration, shape, density, etc. The information is displayed as fringes striating the holographic reconstruction of the subject. These fringes are contours of equal optical path length and could represent out-of-plane distortion, in-plane displacement, refractive index changes, and so on, depending on the optical arrangement used to form the hologram. Some typical examples are shown in the previous chapter.

8.1.1 The need for holography

The holographic process is based on the phenomenon of interference: phase differences across the two wave fields, one from the subject under study and the other directly from the laser, are converted into density distributions across the emulsion of the photographic recording medium. This phase recording capability of holography enables it to be used to measure changes in the wave field of the light returned from the subject and it is this feature which is of interest to the engineer. Many engineering phenomena generate changes in the shape of a structure which when translated into optical phase are of a suitable magnitude to be recorded on a hologram.

An automobile body shell made from thin sheet steel is subjected in use to excitation from a range of sources: wind pressure, the road, engine self-excitation, etc. Vibrations at the frequencies of interest to passengers are typically in the range 20 to 400 Hz. Outside this range the forces are generally sufficiently reduced to cause no problems.

At 100 Hz, a typical acceleration level would be 0.1 g, approximately 1 m/s^2. Assuming the vibration is simple harmonic, this corresponds to an amplitude of about 2.5 μm or five wavelengths for argon laser light, which is conveniently the degree of distortion that can be recorded holographically. However, many mechanical systems generate optical path length changes outside the range of holographic recording, and it is wise to establish that the measurand is of the correct order of magnitude before attempting any technique as exotic as holography.

Given that the distortion in phase is of suitable size, one has to decide whether holography is the most suitable approach. It may be that there is no other method of making the measurement; examples could be that the structure has to be analysed in a non-contacting manner, or that refractive index variations in a medium cannot be accessed by conventional probes.

Although it may be possible to use conventional means, holographic techniques may be the most cost effective. For measuring the vibration behaviour of a large structure like an automobile body shell, the usual approach would be to use accelerometers, fixed to the body at a range of points and monitored while the structure is subjected to controlled vibrations. At a frequency of 400 Hz, the modes of a body shell will typically give antinodes of 10 cm spatial extent. To monitor this behaviour with adequate spatial sampling requires that the accelerometer measurements are made on a grid of points at intervals of less than 5 cm. A body shell is large and has a very convoluted shape, so that the measurements could require 4000 individual sites. Even using several accelerometers in parallel leaves a task taking several weeks – just to collect the data.

Construction of the mode shapes from the data collected, so that any offending vibration mode can be visualized and tackled, requires further analysis. The phase relationship between the sets of data from each accelerometer has to be established. Generally, the acceleration measurements have to be made at a range of frequencies and the whole set of 4000 records handled at the same time to extract the mode shapes. Proprietary systems exist which are tailored to conduct this analysis on large sets of accelerometer data, but it is an expensive activity and the analysis of a vehicle body shell would take several weeks of skilled effort.

8.1.2 The benefits and limitations of holographic analysis

The attraction of holography is that using the time-average mode of operation described in section 7.5.1, the vibration modes of a body shell can typically be obtained in a couple of days. Furthermore, the whole of the subject can be viewed at once, which can be of great value to the engineer trying to understand the behaviour of a large, complex structure. The potential benefits of holography over conventional techniques can be summarized as:

1. Full field;
2. Non-contacting;
3. Fast;
4. Accurate;
5. Sensitive.

There are, however, difficulties associated with the application of conventional holography in an engineering environment. The most obvious problem is its reliance on a photographic recording medium with a need for wet chemical processing, darkrooms and considerable expertise. An alternative technology, described in section 7.2.3(*b*), uses electrostatic recording on to a 35 mm format slide (reusable for typically 100 holograms). However, it has limited sensitivity, is expensive and constrains the field of view.

The holographic technique requires that during the recording of the hologram the whole interferometric assembly – including the subject under study – is isolated from extraneous vibrations. The stability requirement is made worse by the low sensitivity of holographic recording media. This means that either a high-powered laser has to be used to keep the exposure time short or the measurements have to be carried out under excellent vibration isolation conditions. For a structure the size of an automobile body shell this is not a trivial task; manufacturers have even used a salt mine as a venue for holographic interferometry needing particularly long exposures.

With smaller objects, the technique generally requires the use of a vibration-damped, airmount-isolated optical table on which the holographic interferometer and the subject can be assembled. The whole arrangement frequently is shrouded by a

278 Television holography and its applications

perspex canopy to ensure air currents cause no unwanted optical pathlength changes during the exposure period. The information of interest to an engineer has to be extracted from the hologram by making measurements on the fringes generated at the reconstruction stage, after photographic processing.

The limitations of conventional holography from the engineer's point of view are:

1. Slow;
2. Photography based;
3. High-power laser required;
4. Complex optical arrangement;
5. Complex and expensive facilities;
6. Qualitative analogue output.

Ideally, the behaviour under study should be observed live, as it happens, and the main engineering attraction of TV holography is that it offers this capability.

8.2 PRINCIPLES OF TELEVISION HOLOGRAPHY

8.2.1 Resolution requirements

A photographic recording medium is needed for conventional holography because the data is stored in the form of an interference pattern which is of wavelength scale. A typical 100 mm × 125 mm holographic plate has a resolution capability greater than 3000 line pairs per millimetre and contains some 10^{10} bits of information. A TV camera, nowadays generally having a CCD sensor, typically operates with 600 × 800 pixels and eight-bit intensity resolution. Each image captured by the camera represents perhaps 10^6 bits of information, several orders of magnitude less than that of a conventional hologram. Clearly it is not possible simply to replace the photographic plate with a TV camera. Nevertheless, an optical system using a TV camera can be configured so that it can capture holographic data and enable all the measurements available to conventional holography to be made at TV frame rates. However, the resolution constraints introduced by the camera mean that some information contained in the conventional hologram must be sacrificed.

The reference beam in conventional holography illuminates the photographic plate at an angle of typically 45°. Sufficient space is then available for the subject to be separated from the reference source. As the subject gets larger the reference beam has to be moved further off axis to avoid encroaching on this space. If it is not sufficiently off axis during the formation of the hologram, the image produced will be partially overlapped and swamped by the reconstruction beam. However, if the reference beam is moved further off axis, the structure in the interference pattern generated becomes more finely spaced and the recording medium resolution has to be increased. Figure 8.1 shows schematically the interference fringe spacings produced across the wavefront from a typical object point, and the corresponding spatial Fourier transforms (see section 2.1.1).

Principles of television holography 279

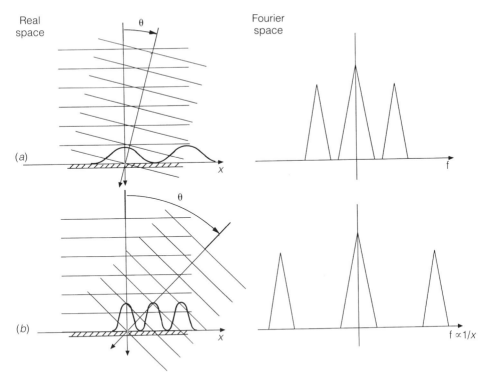

Fig. 8.1 Formation of fringes in a hologram and the corresponding spatial Fourier transforms for (*a*) small and (*b*) large reference beam obliquity angle θ. Number of fringes per millimetre and their spatial frequency *f* are proportional to sin θ.

8.2.2 Forming TV holograms

If the reference beam were to be positioned in the centre of the subject at the reconstruction stage, illuminating along the axis of the holographic system, the resolution requirement for the recording film would be reduced to a minimum. The arrangement would be of little practical use, as the image would be indiscernible in the presence of the high-intensity reconstruction beam. Early Gabor holograms were of this type and it was the development of the off-axis mode of operation by Leith and Upatneiks (1962) that opened up the field of holography. However, the key to recording holographic data with a TV camera sensor is to use the low-resolution mode of operation obtained from an on-axis reference beam geometry in conjunction with an electronic filter or subtraction stage which can eliminate the troublesome reconstruction signal. In ordinary holography, an optical subtraction process is not physically realizable.

The resolution requirements would still be too high for a system based on a TV sensor and the hologram has to have its spatial resolution limited further. This is

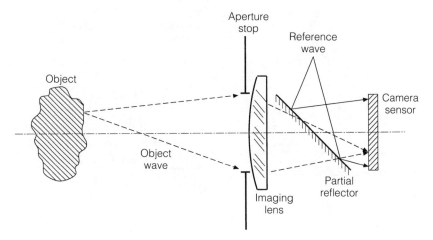

Fig. 8.2 Typical geometry for recording a TV hologram.

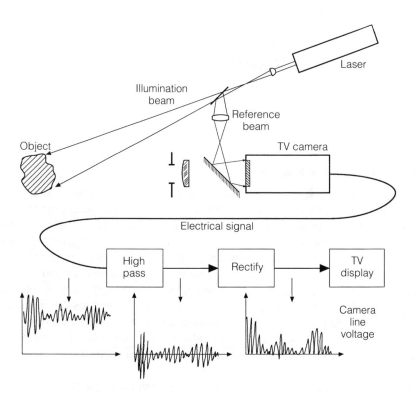

Fig. 8.3 Basic elements of a TV holography system.

achieved by using a lens to form an image of the subject on the sensor. The bandwidth is reduced by having a small aperture on this lens. As the lens is stopped down, the lateral resolution of the system is reduced proportionately. The information requirement can then be degraded from the 10^{10} bits of conventional holography to the 10^6 bits of a TV system. A schematic diagram summarizing the recording geometry is shown in Fig. 8.2.

A complete TV holography system is shown in Fig. 8.3. The reference beam usually has about 5% of the intensity of the subject illumination beam. Having detected the holographic interference pattern, the low-frequency signal due to the on-axis reference beam geometry can be eliminated electronically by high-pass filtering.

The modulation that remains can then be viewed by feeding the filtered signal to a TV monitor. This signal constitutes the reconstructed TV hologram. After high-pass filtering, it is generally squared or rectified to avoid losing negative components. The hologram is generated at TV frame rate, each new image presented to the monitor being the result of filtering a new interferogram collected by the camera. Changing the optical path length between subject and camera will change the interferogram and hence modify the image.

8.2.3 The effect of the aperture stop

A TV holography system can be considered as an assembly of two-beam interferometers, one located at each pixel position. The assembly of pixels, each delivering its own voltage proportional to the incident light intensity, generates a two-dimensional sample of the interference that has taken place between object and reference beams across the image plane of the lens. We shall call this distribution of intensities the **interferogram**. A typical example of the kind of speckle distribution that is sampled is shown in Fig. 8.4.

For each point of the object imaged by the lens, a diffraction pattern with a central spot known as an Airy disc is formed on the sensor (Fig. 8.5). The Airy disc is a consequence of the finite aperture size of the lens used to form the image. The relationship between the lens parameters and the disc diameter σ is

$$\sigma = \frac{1.2 \lambda f}{D}, \tag{8.1}$$

where the symbols are defined in Fig. 8.5. As explained in section 2.1.1, the size of the disc σ is inversely proportional to the diameter of the lens aperture D. A lens of infinite aperture, transmitting all spatial frequencies, would generate a true image point.

A distant point, imaged in argon laser light by a lens of 50 mm focal length operating at $f/32$, gives an Airy disc size of approximately 6 μm. The size of a typical 600 × 800 element CCD array is about 6 × 8 mm, so that the pixel size is approximately 10 μm. Each pixel is then capable of measuring the intensity of one Airy disc, as shown in Fig. 8.5. The diffraction patterns from all the points of the object

Fig. 8.4 Typical appearance of the laser speckle pattern which forms a TV hologram.

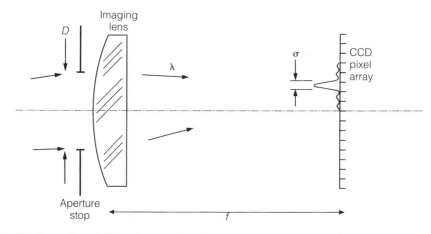

Fig. 8.5 Formation of diffraction spot by an imaging system with a limited aperture.

combine coherently to produce speckles whose size is comparable with that of the individual Airy discs. The result of the coherent addition of the on-axis reference beam to the wave field from the object is still a speckle structure across the CCD sensor, but the speckles have double the size. The intensity generated at each pixel then depends on the relative phase between the image wavelet and the reference beam.

The importance of an on-axis reference beam can now be appreciated. If it illuminates the sensor obliquely, then changes of the image phase relative to it across a pixel will generate a range of fringe phases across the pixel. An overall change of phase between reference and image fields of π radians, which should produce a large

Principles of television holography 283

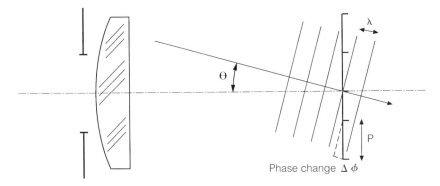

Fig. 8.6 Comparison of reference beam phase variation with pixel size.

change in pixel signal, would in this case generate only a small change. On a sub-pixel scale, one part of the pixel area would become brighter when another part became dimmer. The result is that overall the pixel records little change in total integrated intensity.

If the inclination of the reference beam with respect to the optical axis increases, then the spatial scale of the intensity variations becomes smaller. The modulation measured by each pixel and consequently the signal-to-noise ratio of the system then decrease. Eventually the spatial variations in intensity will be on such a small scale that each sensor will return a signal proportional to the average intensity, which will not modulate sufficiently to be detected when phase variations between reference and image wave fields occur. In the limit of very oblique illumination, we reach the conventional holographic requirement for a spatial resolution of about a half wavelength (0.3 μm). In TV holography, the sensing pixels could in principle be made this small, but a more practical solution is to require the reference beam to be sufficiently on-axis that spatial variations across the image plane occur on a scale greater than one pixel dimension for a standard CCD array.

The reference beam wavefront curvature also needs to be arranged properly. Although the speckle distribution in the image is a random array of amplitudes and phases, any small area comparable with the speckle size contains a little wavefront propagating away from the lens aperture. For the reference wavefront to match these wavelets as well as possible over the whole field, it must apparently diverge from the centre of the aperture.

We now consider the constraint on the reference beam arrangement analytically. If the reference beam is inclined at an angle θ to the optical axis (Fig. 8.6), then across a pixel of width P the phase will vary by $\Delta\phi$ with respect to a wavelet imaged from the object. To keep this phase variation small, we could require that

$$\Delta\phi = \frac{2\pi}{\lambda} P \sin\theta < \frac{2\pi}{10} \qquad (8.2)$$

so that

$$\sin \theta < \frac{\lambda}{10P}. \qquad (8.3)$$

Typically $P = 16\ \mu\mathrm{m}$ so that $\sin \theta < 0.5/160$ radians. Thus the reference beam must be axial to within about 0.2 degree – realizable with modest quality optical assemblies.

8.2.4 Operating modes

The conventional holographic method enables the optical path length changes due to static or dynamic events to be recorded. The photographic emulsion acts as a storage medium, and the frozen-fringe or double-exposure method involves the superposition of two interferograms, separated in time to allow some change to be made to the subject under study.

A CCD sensor can have a similar capability, but only over time intervals shorter than its frame period (1/25 s Euro, 1/30 s USA). At the end of every frame period the pixels are read out and then reset to receive the next frame of data. Two interferograms can be accumulated during a single frame as in the conventional holographic case. For events separated in time by more than a frame period, another means of storing the first TV hologram is needed. Early systems used a video recorder, but modern technology has produced digital frame stores which can collect and store a frame of data from a camera at TV frame rate. Subsequent TV holograms can then be compared with the stored frame.

Conventional holography operates with passive elements, so that signals can only be combined by addition, but with an electronic medium signals can be modified actively. They can be frequency filtered to eliminate unwanted elements and also subtracted, which makes TV holography systems particularly robust when optical noise is present. The frozen fringe technique (section 7.3.1) is accomplished in a TV holography system by subtracting the second hologram, pixel by pixel, from the stored version of the first hologram. The resulting frame of data is high-pass filtered, squared or rectified and displayed. The TV image then contains fringes which delineate equal optical path length differences generated between the two holograms. The electronic processing can all be undertaken by a single image-processing board, installed in the back of a microcomputer and under software control.

8.2.5 Fringe formation

(a) Subtraction fringes

When fringes are formed by subtracting two TV holograms, areas of pixels which have the same voltage in both holograms will subtract to zero and generate dark fringes. Regions with pixels that differ in voltage between holograms will not subtract to zero and will appear bright. Some of the subtractions will generate negative voltages, which cannot be displayed directly on a TV monitor. However, all of the information showing the degree to which the two holograms differ can be displayed by squaring or rectifying the difference signals.

The intensity at any pixel will vary cosinusoidally with phase difference between reference and object wavefronts, just as the intensity at a point in the output of a Michelson interferometer changes if one of its mirrors is moved along the optical axis. Thus regions of pixels where the relative phase has changed not at all or by integer multiples of 360° will always have the same voltage in each of the two recorded holograms and yield dark fringes.

Areas where the phase change is 180° will tend to return large differences, because a pixel which has maximum intensity for the first hologram will have minimum for the second, and vice versa. However, a pixel which is at mid-range for the first hologram will return to mid-range for the second, so it is possible to have dark pixels in a bright fringe region. Furthermore, any pixel at which the subject image happens to contain a dark speckle will have low modulation and always return a small difference value.

The fringes generated by the subtraction of TV holograms are thus not as smooth in appearance as those formed by the conventional holographic process. It is possible to generate pixels of all intensities, including black, in the interior of bright fringes. A typical example is shown in Fig. 8.14(a). The eye can accommodate the dark pixels interspersed within the bright fringes, and has no difficulty in identifying where fringes have been formed, but any automatic analysis algorithm cannot be based on individual pixel values; surrounding pixels must be taken into account. Ways in which this can be achieved are discussed in section 8.5.

(b) Time-average fringes

In this case the light intensity falling on each pixel is time integrated over the framing period of the camera. The signal from a single frame is high-pass filtered, rectified and displayed, so that the display appears brighter in regions where there is a large spread of pixel intensities. This process produces visible fringes in the image of a vibrating subject.

For vibration in a single mode, the phase difference between object and reference fields at each pixel changes sinusoidally at the vibration frequency, possibly with an amplitude spanning several fringe cycles. The effect of this is to reduce the depth of modulation at each pixel if the phase difference is also changed slowly. The process is analogous to that which occurs in time-averaged holographic interferometry (section 7.5.1), and the reduction in modulation as a function of vibration amplitude is again proportional to the square of the J_o Bessel function. Thus the average brightness in regions of the TV display also depends on this function.

The reduction in modulation can be explained as follows. The structure spends most of its time near the two extreme positions. Where the peak-to-peak movement is a multiple of a half wavelength, the pixel intensity distribution will be the same for both positions, so that the integrated effect approximates to the addition of two similar holograms. For other amplitudes, the distributions are different for the two extremes, giving smaller integrated differences between pixels and a lower brightness in the display. As the amplitude of the vibration increases, a progressively smaller fraction of the frame period is made up by contributions from the extreme positions; variations in integrated intensity are hence reduced, so that the fringe contrast tends to become lower.

(c) Addition fringes

A third mode of operation is addition fringe formation. In this case two TV holograms are formed sequentially on the CCD sensor within one frame period – generally by a pulsed laser illuminating the subject with two short pulses of high power. The time interval between the pulses can be so short that random changes in phase due to unwanted influences – vibrations, thermal effects, etc. – can be made insignificant. Time-stationary noise cannot be eliminated in this mode, so the fringes are generally not as well defined.

8.3 SYSTEMS AND THEIR OPERATION

This section deals with the design and operation of TV holography systems, describing the optical arrangements used and how they affect the performance when studying engineering structures.

8.3.1 Optical arrangement

The underlying requirements for the operation of a TV holography system have been described in section 8.2. The reference beam has to be delivered as a smooth wavefront centred on the camera lens aperture, to illuminate uniformly the sensor in combination with the wave field scattered from the object and imaged by the camera lens.

The required operating conditions can be achieved by simple optics, as shown schematically in Fig. 8.3. There have been many variations on this basic theme. For example, one involves introducing the reference beam along the axis of the camera and partially reflecting the image beam rather than the reference beam by a beam-combining cube (Fig. 8.7). Another arrangement uses a small hole drilled in a reference beam combining mirror. The hole is used as the aperture through which the reference wavefront is brought to the camera sensor (Fig. 8.8). In another design a small spherical ball or concave depression is used as the mirror element. The reference beam is then reflected off this and on to the sensor (Fig. 8.9). However, the experience of the authors is that the arrangement of a beam-splitting cube, operating with a reflectance of about 5% and transmittance of 95%, is the easiest to build, align and maintain; they invariably use it.

Systems and their operation 287

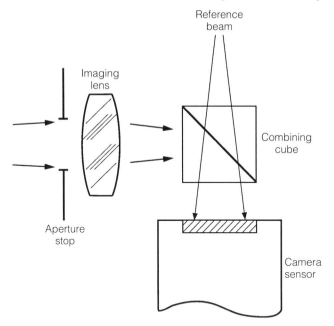

Fig. 8.7 Folded image beam.

Anti-reflection coated optics should be used and, depending on the type of faceplate fitted over the CCD sensor, it may be advisable to have this removed by the camera supplier. If the faceplate is parallel sided or parallel with a face of the beam-splitting cube then multiple reflections can cause unwanted fringes to be generated across the sensor, affecting the modulation of the TV holography signals. Some cameras have an infra-red filter fitted as part of the faceplate structure and it may be advantageous to have this removed – the sensitivity of the camera will generally be increased, typically by a factor of two at a wavelength of 700 nm. The main disadvantage of these precautions is that the sensor loses its protection. It is then extremely delicate and must be treated with great care. It should not be touched and spray cleaning or dipping must be used if it becomes contaminated. The authors have found that it is generally neither advisable nor necessary to remove the faceplate. The images in this chapter have been obtained with a variety of systems, some with faceplates, some without, and there is little to choose between them.

As with any interferometer, it is essential that the optical elements remain fixed relative to each other while data is collected. In the case of a TV holography system this means that the imaging lens, beam-splitting element, reference-beam optics, illumination optics and camera must be fixed securely in space. However, the time period during which this stability must be maintained depends on the mode of operation of the system. In the static subtraction mode, stability is required from the

288 Television holography and its applications

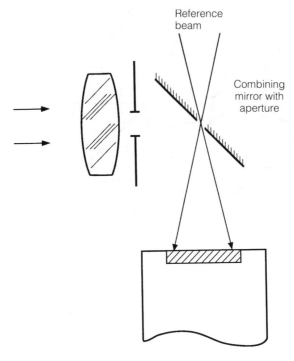

Fig. 8.8 Beam-combining mirror with small aperture.

time the first hologram is collected and stored to the time at which the last hologram is obtained. The system is operating 'live' under these conditions, in a manner analogous to conventional live-fringe holography, and the interferometric stability must be maintained for the duration of the experiment.

In the time-average mode, the situation is different. The sensor of a CCD camera has about 1000 times greater sensitivity than a silver halide emulsion typically used for conventional holography, so that it can record a time-average hologram in a single frame period. The TV holography system is therefore extremely robust; the interferometer need only have dimensional stability for 1/25 s (1/30 s USA). In particular, the subject under study need only remain interferometrically stationary relative to it for this time. Many engineering structures have this degree of stability in a laboratory environment and can thus be examined very easily.

In the addition mode, a ruby laser is typically used to generate two high-power, short (50 ns) pulses. Even if the subject is moving in space by many wavelengths relative to the interferometer, its movement during a pulse is a fraction of a wavelength and a TV hologram will be recorded. A second hologram is formed by the next laser pulse and if the time between the two exposures is less than the frame time the two holograms will be integrated by the sensor. Very unstable objects or fast transient events can thus be studied.

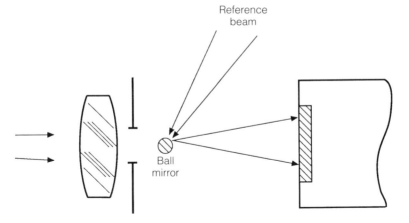

Fig. 8.9 In-line geometry with ball mirror.

This does not mean that the interferometer can itself have relaxed stability constraints. It is easy to generate fringe patterns in such a system, but one must ensure that they relate to the phenomenon of interest and not spurious movements within the interferometer. Pulsed holographic systems are frequently used to study structures under extreme conditions; an operating gas turbine engine is a good example. The measurement environment can be very severe with air and structure-borne excitation present. It is essential under these conditions that the interferometer is extremely stable to ensure that reliable data is collected. It is all too easy to produce fringes but rarely possible to repeat the experiment!

8.3.2 Operating parameters

(a) Lens aperture settings

It would seem ideal to operate with an aperture size that gave a speckle size of pixel dimension. This would ensure that interferometer phase changes would produce maximum intensity changes on a pixel, leading to optimum modulation. For a larger aperture size, several adjacent speckles will overlap on a single pixel, and when the interferometer phase changes due to the subject under study being distorted, the change in intensity at a pixel will depend on how their combination changes. The tendency will be for this to reduce the signal modulation. As the lens aperture is increased and the speckle size decreases, this overlapping condition will become worse and the fringe contrast produced by the system will decrease.

However, another effect counteracts this. As the aperture size increases, the amount of light received from the subject increases. Each lens stop change decreases the point spread function dimension by $\sqrt{2}$ but doubles the intensity. This leads to a better signal-to-noise ratio and improved fringe contrast. The authors have found that for large engineering structures such as vehicle body shells, when gathering light is at a premium, they can operate at $f/1.4$ and still acquire good TV holograms.

(b) Reference beam requirements

As we have seen, the reference beam must appear to emanate from the centre of the imaging lens. An associated concern is that the relative intensities of the reference beam and image at the sensor are arranged to ensure optimum signal modulation. This is to some extent influenced by the axial location of the reference beam. The intensity across its wavefront needs to be uniform and the smaller the exit pupil of the reference beam optics, the further from the sensor it will have to be located in order to ensure that the whole of the sensing area is illuminated evenly. The ratio of reference-to-image intensity for optimum operation depends on the operating mode; static subtraction, time-average, etc.; the topic is dealt with in some depth by Jones and Wykes (1983). The difficulty with modelling the behaviour is compounded by the complex and largely unquantified performance of the cameras and TV systems. A procedure which the authors use to set up a system is outlined in section 8.3.4; it is rather empirical but works satisfactorily.

It is not strictly necessary for the reference beam to have smooth wavefronts, and several workers have used a speckled reference beam formed by optical scattering (Slettemoen, 1980). The results are generally not as good as with a uniform wavefront. This mode of operation occurs in a shearing system (see section 8.3.5).

(c) Polarization effects

For interference to take place, the reference and image beams must have the same polarization. Components which are not of the same polarization will simply raise the overall intensity level at the sensor. As the dynamic range of the sensor is restricted - generally to 256 grey levels - this will reduce the possible modulation depth of the interference signals.

(d) Coherent and time-stationary noise

Coherent noise is a problem faced by every worker in the field of interferometry. It makes coherent optical systems particularly vulnerable to misalignment and damage or dust on the optical components. A scratch or piece of dust present in the system will cause diffraction and the generation of fringes across the whole of the sensor. Care must therefore be taken to design the system to make it resistant to damage and dirt ingress and to make access easy to the parts which cannot be protected completely.

Generally this type of noise is time stationary, its location with respect to the sensor being fixed. In these circumstances static subtraction removes resultant artefacts from the image. The noise remains the same from frame to frame and so is eliminated on a pixel-by-pixel basis by the subtraction process. The effect of the noise is still to reduce the possible modulation, but the subtraction process stops it propagating to the next stages of processing. This can make a very significant difference to the quality of images generated and is one of the reasons why static subtraction images always exhibit better fringe contrast than time-average images.

Because of the noise reduction, the authors operate in a sequential subtraction time-average mode for vibration analysis in order to optimize the fringe visibility on complex engineering structures (Buckberry and Davies, 1990). In this mode, every hologram collected is a time average of the vibration modal behaviour, and is subtracted, pixel-by-pixel, from the previously collected time-average hologram which has been stored temporarily for this purpose. The resulting subtraction time-average fringe map does not suffer from coherent or other time-stationary noise and is easier to deal with subsequently when automatic computer analysis is to be performed (section 8.5).

It is possible to suppress noise artefacts by other means. One way is to filter spatially the image and remove frequencies characteristic of the noise terms. This can be carried out by a procedure of fast Fourier transformation, spatial filtering and inverse transformation, provided there is little signal content at the frequencies being removed (see section 8.5.3(*a*)). However, a subtraction operation will always provide the optimum solution in these circumstances as it is effectively a matched filtering operation.

8.3.3 Fibre optic systems

Fibre optics is a new and valuable development for TV holography systems. Optical fibres offer some important advantages over conventional bulk optics (Davies *et al.*, 1987). They provide a means of steering coherent wavefronts around complex paths, whilst retaining smooth phase and intensity distributions. At the same time a fibre optic arrangement is immune to the dust ingress and mechanical damage which occurs with bulk optical arrangements. Beam-splitting elements can be replaced with their fibre optic equivalent, directional couplers, which are available in a range of split ratios.

As the fibre optic arrangements are simple, they can be made extremely robust and packaged into a small rugged instrument suitable for industrial application. The connection of the interferometer to the laser can be accomplished by using a fibre optic umbilical cable, making it possible to operate remotely from the laser. This is significant if it is necessary to use an argon ion laser which is too bulky to be moved itself. Light can be piped to the interferometer, distributed within it as required and only appear outside the instrument as an expanded beam – an important safety consideration if the instrument is to be used by non-experts. A typical fibre optic arrangement is shown in Fig. 8.10.

The use of fibre optics gives stability and eliminates optical elements like pinhole assemblies which are needed as spatial filters in bulk optical arrangements to deliver smooth wavefronts (see p. 24). The output wavefronts from a single-mode optical

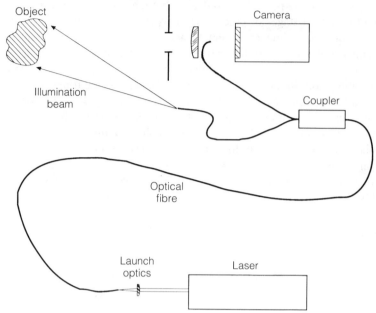

Fig. 8.10 Optical fibre system.

fibre are automatically smooth and require no additional optics, thus being ideal for use in an interferometer.

Fibres offer the opportunity to operate on the intensity and phase of the guided wavefronts by electro-optic means. This feature is discussed in detail in Chapter 11. The techniques for signal modulation can be incorporated into the interferometer with relative ease, thus making available a wide range of operating modes which can expand and enhance the analytical capabilities of the TV holography method. For example, the phase of the reference beam wavefront can be controlled by winding a piece of optical fibre, through which the beam is being guided, around a piezoelectric cylinder. The application of a suitable voltage to the cylinder will cause it to expand and stretch the optical fibre, increasing the path length and introducing a phase delay. The applications for this type of modulation technique are detailed in section 8.4.

8.3.4 Setting up a system

Having built an arrangement to deliver a smooth reference beam along the optical axis of an imaging system as in Fig. 8.2, it is necessary to align the system and set its operating parameters. The initial alignment is best undertaken with the camera output signal passed directly to a high-contrast TV monitor. A suitable object should be positioned in the field of view and brought to a focus on the camera sensor. The features mentioned above need to be considered, namely:

1. Coherence conditions for the laser being used;
2. Aperture size of the lens;
3. Axiality of the reference beam;
4. Reference-to-image intensity ratios;
5. Relative polarizations of reference and image;
6. Stability conditions.

It is best to start with the system under well-controlled and stable conditions. The authors generally set the reference beam geometry by viewing its profile on the TV monitor to ensure it covers the sensor uniformly; the image light should be blocked off for this stage. They, with the image and reference beams both illuminating the sensor, it is possible to view the position of the divergence point of the reference beam wavefront with respect to the image of the object by sighting along the reference beam using the fourth port of the interferometer (i.e. looking into the partial reflector from below in Fig. 8.2).

The aperture can be stopped down in stages and minor adjustments made to the position and axiality of the divergence point to keep it in the centre of the image of the lens iris as seen through the beam-splitting cube. With a little practice it is quite easy to set up the system rapidly and accurately in this way. When the system has been aligned approximately, the interference pattern generated on the TV monitor can be used as a final guide. The phase difference in the interferometer needs to be changed, either by moving the object or probably more conveniently by gently touching one part of the interferometer itself. This will cause the image, which will have a grainy speckled appearance, to modulate in intensity, cycling as the phase difference is changed. It remains to modify the operating parameters detailed above very slightly in an iterative manner until the image on the monitor has highest contrast and cycles in intensity as strongly as possible. Under these conditions the subsequent stages of high-pass filtering and rectification will yield fringes of highest visibility.

If an image-processing board and a desktop computer are used, a word of caution is useful. The computer, image-processing board and CCD camera will all have their own clocks, and only one of these clocks must be used to control the timing of the whole assembly. Generally in the set-up procedures for an image-processing board there is an opportunity for the user to define the master system clock. It is particularly important because when the system is used in a subtraction mode, frames of data must be subtracted from each other with pixel accuracy. That is, the operations must be controlled to a timing accuracy of fractions of a microsecond to ensure that a pixel in one frame is not subtracted from its spatial neighbour in the next frame. If horizontal bright bars are generated under subtraction conditions it is frequently indicative of timing errors in the subtraction process.

8.3.5 Alternative operating geometries

In the study of mechanical assemblies, we are usually interested in measuring how the surface has changed position as a result of some applied force – mechanical,

294 Television holography and its applications

thermal, etc. With the interferometer arranged so that the imaging lens views down the line of the object illumination, or very nearly so, the fringes generated will be contours of equal line-of-sight displacement. The surface may not, of course, move in this direction only. There will be components of movement in other directions also, and these can be observed by viewing the surface along the appropriate direction. As in holographic interferometry, the more general case of different illuminating and viewing directions gives the component of movement along the bisector.

(a) In-plane sensitive systems

It is possible to change the arrangement of the optics and make the interferometer insensitive to line-of-sight displacements of a surface and sensitive instead to some other chosen component (Moore and Tyrer, 1990). If two beams from the same laser are used to illuminate a surface at about 45° to the surface normal, and from opposite sides of that normal, then a system can be arranged to be sensitive only to in-plane surface displacement. The independent reference beam is then not needed. The two beams both generate scattered beams which are collected by the imaging lens (Fig. 8.11), and each can be thought of as acting as the reference for the other. Any line-of-sight movements of the surface change the phases of both beams by the same amount, so that the hologram formed remains unchanged and the system is relatively stable. It is operating in what is termed a local reference beam mode and the fringes generated are contours of equal in-plane displacement.

The application of such a system is restricted, as there are not many engineering problems that permit the optical access required. The measurement of the in-plane

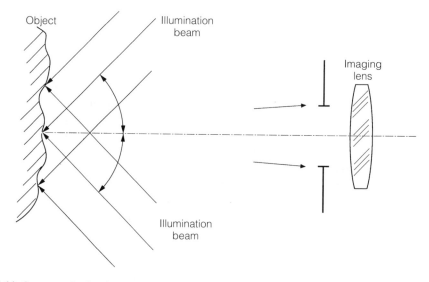

Fig. 8.11 Geometry for in-plane sensitivity.

displacement, and hence the in-plane strain, of flat, notched specimens is one area that is of interest, as these studies can lead to a better formulation of stress intensity factors which are important for characterizing fracture behaviour.

(b) Shearing systems

A closely associated technique is one in which a single beam is used to illuminate the object, but the image is amplitude divided into two equal-intensity images, one being shifted laterally with respect to the other. These two sheared images are brought to a focus on the sensor where each image field again acts as the reference field for the other. A separate reference wavefront is again not needed. The interpretation of the fringes can be understood by reference to Fig. 8.12. The fringe order when two recordings are subtracted depends on change in surface slope rather than surface position. The technique has important applications in the inspection of composite structures, particularly in the aerospace industries. Failure of composite material is generally characterized by a change in local slope. The shearing interferometer generates fringes of equal surface slope change and so is well suited to locating defects.

8.4 MODULATION TECHNIQUES

The operating modes considered so far lead to fringe maps of static distortion or vibration amplitude. The information is qualitative, although it is possible to get a feel for the behaviour by counting the number of fringes between points on the map.

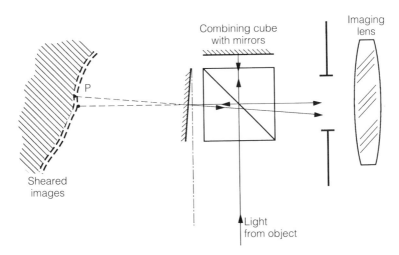

Fig. 8.12 Sheared image geometry. Interference takes place between light effectively scattered from P and P'.

Information at points between fringe peaks is more difficult to establish and at best only an outline picture of the full field behaviour can be gleaned. Furthermore, the fringe map does not contain information about the sign of the deformation or, equivalently, the phase of the vibration. A similar ambiguity exists for all the associated modes of operation: in-plane, shearing, etc.

Further limitations become apparent as soon as a TV holography system is used in engineering situations. However, the limitations have already been studied by the exponents of conventional holography and the task is to transfer their techniques to TV holography, whose electronic nature makes it more amenable to implementation of these techniques.

8.4.1 Limitations of a basic system

The limitations of a basic TV holography system are listed below with a brief explanation of each.

1. *Limited operating stability*. The system needs to be stable interferometrically during its operation, requiring air-mounted anti-vibration facilities.
2. *Qualitative information*. To extract quantitative information the user must estimate the centres of fringe extrema or, with greater difficulty, intermediate fringe values.
3. *Phase ambiguity*. Distortions towards or away from the interferometer yield the same fringe maps; vibration phases are lost.
4. *Visibility loss for high-amplitude vibration*. Time-average Bessel-function fringes have low visibility at high vibration amplitudes.
5. *Vibration travelling waves*. Non-linear vibration behaviour leads to travelling waves which appear as ambiguities in time-average fringe maps. Simple time-average operation leads to adjacent fringe orders overlapping.
6. *Sensitivity limitations*. Phase changes of less than 0.1 wavelengths or greater than 100 wavelengths cannot be seen by eye.

It is possible to avoid these limitations at the price of slightly complicating the system by adding components to modulate the intensity and phase of the optical beams. The following sections detail these modifications and explain how they provide extra information.

8.4.2 Intensity modulation

(a) Chopped illumination

TV holography systems, like all interferometers, require a degree of stability to operate satisfactorily. Conventional holographic systems need highly stable optical tables, because the time to record each hologram may be tens of seconds. TV holography systems have the advantage of a much higher sensitivity detector, making possible frame-rate operation, and a system will operate under time-average conditions in a laboratory environment without an optical table.

In a less well-controlled environment, operation in a time-average mode without an optical table may still be possible by shortening the exposure time. A shutter, generally electro-optic like a Pockels cell, can be used to transmit the laser beam for some fraction of the TV frame period. The system only has to be stable for the duration of the illumination, so the stability constraint is then relaxed. In the limit, with a very small time interval, the system becomes equivalent to a pulsed laser system. However, by shortening the frame period more modestly by means of a Pockels cell, a relatively inexpensive pulsed system can be built which will operate in a workshop rather than in a laboratory environment.

An attractive alternative is to use a CCD camera which has a variable, user-selectable shutter speed and so can operate with a shortened frame period. This is rapidly becoming a standard feature of modern CCD cameras and eliminates the need for a Pockels cell.

(b)   Stroboscopic illumination

We have seen that time-average fringes reduce in visibility at higher vibration amplitudes – they have a Bessel-function variation in visibility as a function of vibration amplitude, as shown in Fig. 7.7. An added complication is that the spacings between fringe peaks are not equal, a problem when the fringes are to be analysed automatically by computer (see section 8.5).

The high-velocity elements of the vibration cycle which cause the reduction in modulation depth can be excluded by allowing illumination at the extremes of vibration only. Unlike the chopped-illumination case described above, there will generally be many separate pulses per frame period, two per vibration cycle. If the duration of each illumination pulse is very short compared to the vibration period, the fringe function can be converted from Bessel to cosinusoidal. The reduction in fringe visibility with increasing vibration amplitude is then eliminated completely, but the price to be paid is a reduction in available light. The authors have found it is possible to use two pulses each occupying 20% of the vibration period and still obtain good cosinusoidal fringes without losing too much light.

A Pockels cell is the most convenient way to control the laser illumination, as in this case the frequency of the switching must be changed to match the vibration. The phase of the signal needs to be variable as well. This is because even if the control signal is taken from the same frequency source as that controlling the vibration excitation equipment, there will inevitably be some phase lag in the system. The phase set at the source will not be the same as the resulting vibration phase of the structure under test. The exact phase lag will also be a function of frequency. To ensure that the illumination is at the vibration extremes it is thus necessary to incorporate a phase-control circuit. This has other important uses, as will be explained in the next section, and so is worthwhile building into the control circuitry for the interferometer.

8.4.3 Phase modulation

(a) Phase stepping

Automatic analysis of interferograms is the way to make the technology more widely usable by engineers rather than optical physicists. The whole of section 8.5 is devoted to the techniques of automatic fringe analysis; this section serves to place it within the wider context of controlling the wavefronts in the interferometer.

The technique of phase stepping as applied to ordinary holography has already been described in section 7.3.4. If a known phase change is introduced between the reference and image wave fields it is possible to eliminate the ambiguity of deformation direction. In its simplest form, the deformation fringes are formed as usual by a static subtraction operation and the interferometer is then moved towards the subject under study by a fraction of a wavelength. Areas of the subject which have been deformed towards the interferometer will effectively move further and their fringe order will increase. Fringes formed by movements away from the interferometer will be identified by decreasing fringe order. It is as if sea level on a contour map were being moved.

It is inconvenient, and rather uncontrolled, to move the interferometer physically. The same effect can be obtained by introducing a phase change electro-optically. One of the most convenient and flexible methods is to use fibre optics. If a fibre guiding the reference beam is stretched, the optical path length increases. The fibre can be stretched accurately by using a piezoelectric cylinder around which the fibre is wound. Application of less than 10V can change the phase by several wavelengths, allowing control by a low-voltage digital-to-analogue converter card driven by a computer.

To undertake automatic fringe analysis, the procedure is taken further. Several versions of the same static subtraction field are collected in a digital framestore, each with a different applied phase change. (Frequently, four frames are used.) From the resulting set of data, the fractional fringe order at each point in the field can be determined; see section 8.5 for a detailed description of this process. The required phase steps can be introduced by fibre optic modulation during the frame flyback period, which enables the frames of data to be collected in rapid succession. Fibre optic modulation gives the same phase change across the whole wavefront, whereas techniques using mirrors can introduce phase variations, leading to errors in subsequent analysis.

(b) Sinusoidal modulation

For sinusoidally vibrating structures being studied by the time-average method, the equivalent of phase stepping can be achieved by phase modulation. The vibration amplitudes at each point in the image can be measured and the vibration phases can be mapped. The principle used is identical to that already described, with one additional complication. As the phase of the wavefront returning to the interferometer from the subject is changing continuously, the deterministic phase introduced into the reference arm must also be changed continuously. In a fibre optic system, the piezoelectric cylinder can be driven by a voltage varying sinusoidally at the frequency of vibration of the structure.

As the modulation amplitude is increased, regions vibrating in phase with the reference-beam modulation will reduce in fringe order, while regions vibrating out of phase will increase in fringe order (Fig. 8.13). Control of the phase of the modulating signal is required, to match the vibration phase of the structure.

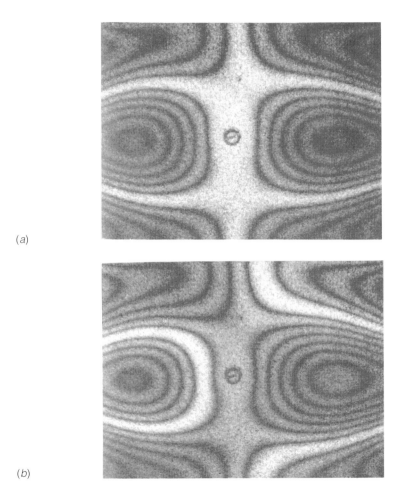

(a)

(b)

Fig. 8.13 (a) Bessel vibration fringes, (b) zero order shifted by sinusoidal phase modulation of reference beam. (Courtesy of Rover Group.)

(c) Heterodyne operation

It is possible to obtain an animated picture of the phase distribution across the structure. If the reference beam phase is again modulated sinusoidally, but at a frequency slightly different from that of the structure, then the time-average fringe map will change at the difference frequency; hence the term **heterodyne operation**. There are now no longer any parts of the image that show stationary phase. The brightest areas of time-average holograms indicate nodal regions of the structure that are not moving at all; under heterodyne operation the brightest regions indicate areas of minimum relative phase change.

The authors use the technique to obtain a rapid understanding of the vibration phase behaviour of complex structures before initiating detailed analysis. Real engineering structures are rarely well behaved; the damping tends to be unevenly distributed, which generates vibration phases other than 0° and 180°, as for a simple linear system. Under these conditions travelling waves are generated, and time-average interferometry will give fringe maps which contain some very confusing features, such as adjacent fringe orders which overlap. Heterodyne operation displays this behaviour immediately.

8.4.4 Combined modulation

(a) Sinusoidal modulation and stroboscopic illumination

For automatic analysis of vibration behaviour, it can be advantageous to convert the Bessel fringes generated by time-average operation into cosine fringes by stroboscopic illumination. Computer analysis is then undertaken in the same way as for the static fringe case. A series of phase-shifted fringe maps is assembled and their relative intensities are analysed for each pixel. For the time-average case, phase stepping has to be replaced by phase modulation, the fringe maps being recorded with different amplitudes of modulation.

(b) Serrisoidal modulation and stroboscopic illumination

Phase modulation undertaken with a serrisoidal instead of a sinusoidal signal generates fringe maps which delineate regions of equal velocity rather than equal position. The ramp rate of the sawtooth is chosen to produce a rate of change of phase equal to that occurring in some region of the image. Stroboscopic illumination restricts the data-collection period to a suitable part of the vibration cycle.

The technique can be used to analyse very large vibration amplitudes and is of interest to noise and vibration engineers as the fringes produced relate directly to the noise-generating mechanism: velocity of the surface. By changing the form of the serrisoidal signal and giving it a frequency which is a multiple of the structural vibration frequency, high-order Bessel-function fringes can be generated. These have the merit of suppressing fringe generation at low vibration amplitudes and enhancing it at high amplitudes (Aleksoff, 1971).

8.5 AUTOMATIC FRINGE ANALYSIS

Whilst TV holography found widespread application to many problems during its first 20 years, very little of the information content of an interferogram was actually used in the analysis, which was largely qualitative and comparative. Even this level of analysis was laborious and required a high degree of user expertise, which was generally unacceptable to industry. Early attempts to analyse fringes by digitizing the centres with graphics tablets had two major problems: first, the accuracy could never be very much better than half a fringe and, second, not enough data points could be obtained. The digitized points were also limited to fringe centres, thereby making it difficult to analyse data on a uniform grid, which would be compatible with other engineering formats such as finite-element models.

8.5.1 Digital processing

Two technological developments occurred in the 1980s that revolutionized the analysis of interferograms. First, with the performance of computers increasing at a seemingly exponential rate, IBM imposed the PC/AT standard on industry. This in turn permitted the development of plug-in boards that performed specific functions. As a result, image-processing boards were available in 1990 for £4000 including a software library of over 200 functions, that then allowed the generation and automatic analysis of TV interferograms in seconds for an image size of 256 000 pixels. The second development, of solid-state charge-coupled-device (CCD) TV cameras, enabled light intensities to be acquired as a two-dimensional array, accurately and consistently.

As a result of this new ability to analyse images, various algorithms for automatic fringe analysis could be studied. This topic is treated more extensively in section 10.1. One of the earliest algorithms was fringe tracking, which mimicked the human approach of following fringe centres, having initially preprocessed the interferograms with some combination of thresholding, histogram equalization and skeletonization. With TV holography interferograms (Button *et al.*, 1985) the speckle noise provides too great an obstacle, but for classical holograms the approach can be successful and is potentially the only way to analyse a fringe field when only one image is available and closed fringes exist (Hunter, 1989).

Phase measurement, as explained in section 10.2, can overcome the problems associated with fringe tracking, and is only limited by the spatial resolution of the camera and power of the computer to handle the volume of data. The earliest reference to the technique is generally accepted to be that of Carré (1966); others followed (Bruning *et al.*, 1974; Sommagren, 1975; Wyant, 1975; Massie, 1980;

302 Television holography and its applications

Takeda *et al.*, 1982; Hariharan *et al.*, 1983; Reid *et al.*, 1984; Dändliker and Thalmann, 1985). Although the theory was established early, practical applications only arose through the development of appropriate technology.

8.5.2 The engineering need

When designing a phase-measuring algorithm for the analysis of any interferometric image field, the first priority must be to create the specification so that solutions to foreseeable problems can be designed in at the start. This ensures computational efficiency and, perhaps more importantly, adaptability to the large variety of images presented in a general engineering holographic laboratory. By considering the images from a variety of objects, it is possible to identify a number of features that need to be accommodated by the algorithm. These are:

1. Fringes of varying spatial width and frequency;
2. Discontinuous or sheared fringes;
3. Fringes of varying contrast;
4. Holes due to either low reflectivity or physical form;
5. Shadows due to illumination angle and surface curvature;
6. Differing fringe orientation;
7. Ambiguous fringe direction;
8. Closed fringes;
9. Faulty camera pixels and digitization noise;
10. Environmental and interferometer stability;
11. Fringe function (Bessel or cosine); and
12. Transient fringes due to thermal or mechanical shock.

These features are general to interference fringes. Interferograms generated by TV holography impose other problems related to the speckle noise, which will be covered later.

8.5.3 Phase-measurement algorithms

(a) Fourier transformation

Any sequence of digitized data can be transformed into the frequency domain using a digital Fourier transform. To construct the Fourier components of an image, the transform must be two-dimensional. The phase values may be obtained without ambiguity from a single image, provided that linear carrier fringes have been introduced in a known direction to remove all closed fringes, as in an ordinary interferometer for visual inspection (see section 2.5.3 and Fig. 10.3(a)). This produces side lobes in the transform, one of which is isolated and inversely transformed.

The Fourier analysis of fringe patterns was originally proposed by Takeda *et al.* (1982), in order to quantify moiré projection fringes representing surface form. The

Automatic fringe analysis 303

theory of the method is given in section 10.2.3. Only one image is required, making it particularly suitable for pulsed images. Given sufficient separation between fringe carrier and background noise, the technique is extremely efficient at noise removal, making the identification of 2π ambiguities very easy. The disadvantage is that if the fringe pattern requires addition of a linear term the object must be sufficiently stable to allow the carrier to be applied.

Kreis (Kreis 1986, 1987; Kreis and Jüptner, 1989) has demonstrated frequently the use of Fourier transformation for the extraction of phase from holograms, showing that it can be done unambiguously given *a priori* knowledge. Preater and Swain (1991) have shown that speckle patterns can be analysed and phase extracted with fringes of very low contrast. Figure 8.14 shows a typical fringe pattern and the resulting phase map, which can subsequently be unwrapped (see section 8.5.5(d)) using a flood procedure. In the Fourier plane (Fig. 8.15), x maps to u, y to v, and the

(a)
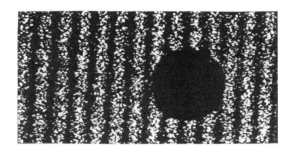

(b)

Fig. 8.14 (*a*) Tilt fringes formed by subtraction process, (*b*) corresponding phase map obtained using Fourier transform method. (Courtesy of R.W.T. Preater, City University.)

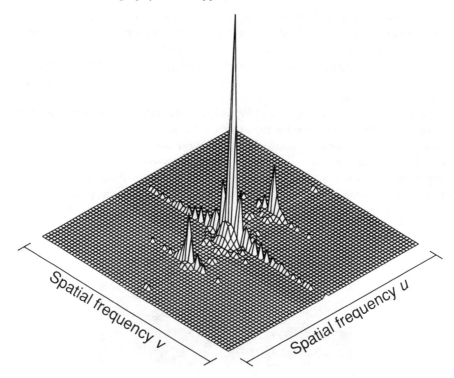

Fig. 8.15 Fourier transform corresponding to Fig. 8.14.(Courtesy of R.W.T.Preater, City University.)

spreading of the lobes is due to the fringe field not filling the complete frame-store data space. The application of this work is described in section 8.6.4.

(b) Phase stepping

The generalized fringe equation is

$$I(x, y) = I_o(x, y) + I_m(x, y) \cos \phi, \tag{8.4}$$

where $I(x, y)$ is the recorded intensity at a pixel (x, y) on the CCD sensor, I_o is the mean d.c. intensity at that pixel, $I_m(x, y)$ is the modulation amplitude and ϕ is the phase difference between object and reference beams. In order to solve for the three unknowns I_o, I_m and ϕ, a minimum of three intensity values is required. Each can correspond to a frame to which has been added, either during the course of exposure or between frames, a known phase shift in addition to the phase ϕ. The frames can be acquired sequentially with the phase shifts applied by translating a mirror by small amounts. Unlike the Fourier-transform approach no carrier fringes are required, provided that a record of the sign is maintained in the calculation of phase ϕ.

If the mirror is stepped between frames, the process is known as **phase stepping**; if the mirror is ramped linearly while successive frames are captured, it is known as **integrated buckets**. Early reports of use in TV holography are given by Creath (1985) for integrated buckets and Nakadate and Saito (1985) for phase stepping. The integrated bucket approach gives somewhat lower modulation, as a result of integrating the sinusoidal intensity change over the frame exposure. Hence, it has not seen much application to TV holography where the modulation depth is already relatively low.

The number of frames N most frequently used by researchers has been 3, 4 or 5. We consider the generalized fringe equation with an additional phase term

$$I_i(x, y) = I_o(x, y) + I_m(x, y) \cos [\phi_d(x, y) + 2i\pi/N] \tag{8.5}$$

where $i = 1$ to N. Here $I_i(x, y)$ is the intensity for frame i, ϕ_d is the phase to be calculated and $2i\pi/N$ is the phase step imposed on frame i. If the number of frames is chosen to be 3 with equispaced steps of 120°, the solution for ϕ_d is (Robinson and Williams, 1986)

$$\phi_d(x, y) = \tan^{-1} [\sqrt{3}(I_3 - I_2)/(2I_1 - I_2 - I_3)]. \tag{8.6}$$

This is the most basic case, considered for holographic interferometry in section 7.3.4. Having calculated the numerator and denominator by simple arithmetic, the inverse tangent of the ratio can be obtained from a look-up table as shown in Fig. 8.16. If a step of 90° is used, an alternative expression (Creath, 1990) is

$$\phi_d(x, y) = \tan^{-1} [(I_3 - I_2)/(I_1 - I_2)]. \tag{8.7}$$

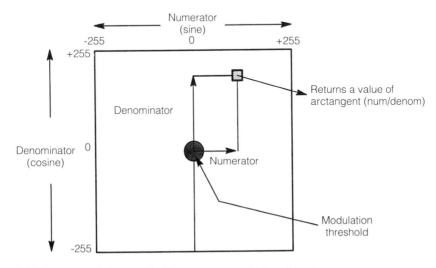

Fig. 8.16 Arctangent look-up table (after Vrooman and Maas (1989))

Use of three frames requires the least computer memory and interferometric stability during the acquisition of data, but because only three samples are taken it is the most error prone. With four frames at an interval of 90°, a solution for $\phi_d(x, y)$ is

$$\phi_d(x, y) = \tan^{-1}[(I_4 - I_2)/(I_3 - I_1)]. \tag{8.8}$$

This uses simple mathematics but more memory. However, as a result of sampling the additional frame, sensitivity to an error in the applied phase step is reduced. With five frames, it is possible to make phase step errors practically negligible. A relationship which achieves this is

$$\phi_d(x, y) = \tan^{-1}[2(I_2 - I_4)/(2I_3 - I_5 - I_1)]. \tag{8.9}$$

As with equations (8.7) and (8.8), the step applied is 90°. This algorithm has been described by Schwider *et al.* (1983) and Hariharan *et al.* (1987). A general formula for any number of steps is given in section 10.2.2.

(c) Max-min scanning

The phase-stepping method solves for the unknowns in equation (8.4) by using a limited number of discrete samples. When interferometers are unstable or are being used in adverse environments with unstable objects, this can lead to significant errors in $\phi_d(x, y)$ which cause propagation of errors when the phase map is unwrapped (see section 8.5.5). These problems are particularly evident with TV holography as it is pushed towards on-line inspection of engineering components on the shop floor.

Vikhagen (1989a) has proposed the max-min scanning technique to overcome this. Rather than deriving $I_o(x, y)$ and $I_m(x, y)$ analytically, they are found experimentally, either by stepping the phase through many small increments or allowing natural temporal drifts to cause at least a 2π change. As the phase varies, the value of $I_i(x, y)$ is read and compared with running records of the maximum and minimum values, which are stored. These can then be equated to $I_o(x, y) + I_m(x, y)$ and $I_o(x, y) - I_m(x, y)$, from which $I_o(x, y)$ and $I_m(x, y)$ can be calculated readily. Then, with the phases steady, the phase at each point is found using

$$\phi_d(x, y) = \cos^{-1}[(I_i - I_o)/I_m]. \tag{8.10}$$

Because the cosine function alone is ambiguous, the phase is found uniquely by recording another image. The principle has also been applied to vibration (Vikhagen, 1989b).

8.5.4 Application of phase stepping to TV holography

TV holograms contain random speckle noise. This can be described analytically by the addition of an extra randomly varying phase term $\phi_d(x, y)$ to equation (8.4), giving

$$I(x, y) = I_o(x, y) + I_m(x, y) \cos[\phi_d(x, y) + \phi_d(x, y)]. \tag{8.11}$$

This means that phase-shifting interferometry as applied to ordinary holographic and moiré systems requires some modification. The following sections will briefly discuss various options.

(a) Subtraction of phase distributions

This approach has two stages. The speckle phases $\phi_d(x, y)$ are calculated first using

$$I_i(x, y) = I_o(x, y) + I_m(x, y) \cos [\phi_d(x, y) + 2i\pi/N]. \tag{8.12}$$

The deformation is then applied to the object and the phase-shift process is repeated to produce a second distribution of phase $\phi_d(x, y) + \phi_d(x, y)$ based upon

$$I_i(x, y) = I_o(x, y) + I_m(x, y) \cos [\phi_d(x, y) + \phi_d(x, y) + 2i\pi/N]. \tag{8.13}$$

The phase $\phi_d(x, y)$ resulting from the deformation is then obtained by subtracting the two distributions (Creath, 1985; Robinson and Williams, 1986; Vrooman and Maas, 1989). This approach requires a significant degree of stability whilst between six and ten frames are acquired, depending on the algorithm used.

(b) Max-min scanning

Max-min scanning was conceived for application to TV holography with its random speckle phases. Having determined the values of $I_o(x, y)$ and $I_m(x, y)$, two single frames can subsequently be acquired, representing the object before and after deformation. This means that the computational requirement is relatively small and the fringes can be non-stationary.

(c) Speckle-correlation fringes

Rather than work with raw speckle patterns as in the above two techniques, an alternative is to phase shift speckle-correlation fringes, formed by subtraction as described in section 8.2.5(*a*). A series of subtraction images is captured, the phase steps being applied to one of the two components. By taking the Fourier transform of each phase-shifted image, it can be band-pass filtered efficiently to leave the fringes free of the speckle noise $\phi_d(x, y)$. This is shown schematically in Figs. 8.14 and 8.15. The phase due to deformation can then be obtained by processing the frames using one of equations (8.6) to (8.9).

The speckle noise is broad band, allowing efficient separation of the fringe frequency component if the fringes have sufficient modulation. The manipulation of the Fourier components is discussed by Kerr *et al.* (1989). The d.c. term I_o can also be removed in the Fourier plane, so that only two correlation fringe patterns are needed for solution of the deformation phase $\phi_d(x, y)$. This approach is also considered by Kerr *et al.* (1989). The disadvantage is the amount of computing required to evaluate the Fourier transforms of several images.

308 *Television holography and its applications*

Rather than transfer the image to the Fourier plane, filtering can be performed by local neighbourhood averaging, otherwise known as convolution filtering (Gonzalez and Wintz, 1988). Many researchers have described this approach (Nakadate and Saito, 1985; Osten *et al.*, 1987; Buckberry and Davies, 1990). Convolving the image can never be as efficient as working in the Fourier plane, due to the finite size of the kernel arrays. If the fringe density approaches the filter array size then, as with the Fourier technique, it becomes difficult to separate the noise. However, by suitable choice of kernels and multiple passes the smoothing achieved can be very significant.

8.5.5 Practical considerations

(a) Fringe density

In phase-measurement techniques, the modulation I_m must be sufficient to be detected by the sensor array and converted to an appropriate voltage. When the fringes are closely spaced, Nyquist sampling theory requires a minimum of two pixels for each fringe. The pixels have a finite size and will therefore integrate the intensity received over a finite range of phase. If this is more than π the intensity will not be strongly modulated, as explained by Creath (1988). An additional effect is seen in the wrapped reconstructed phase; the phase discontinuity is not 2π but approaches π as the Nyquist limit is reached. The unwrapping algorithm must then account for steps of π.

(b) Phase-shifting mechanisms

For phase stepping to be employed, known phase shifts must be introduced by changing the optical path difference between the two beams of the interferometer. Figure 8.17 shows four different methods. Most frequently employed is the pushing of a mirror by a piezoelectric ceramic transducer (Fig. 8.17(*a*)). If it is required that the beam position is not altered, a Porro prism assembly can be used as in Fig. 8.17(*b*). If an interferometer has polarizing optics, a rotating half-wave plate followed by a stationary quarter-wave plate can be employed at the input (Sommagren, 1975), giving orthogonally polarized beams with a variable phase difference (Fig. 8.17(*c*)). When an interferometer is constructed using optical fibres, as described in section 8.3.3, the phase shifting can be achieved intrinsically by wrapping a length of the fibre around a piezoelectric tube and stretching it (Fig. 8.17(*d*)). This technique has the advantage of imparting a uniform phase across the whole wavefront.

If discrete phase steps are applied, any calibrated transducer can be used. When a ramp is applied, nonlinearities must be removed using a waveform generator. Calibrating the phase shifter is an essential requirement and various authors have provided analyses of the errors resulting from incorrect phase steps (Schwider *et al.*, 1983; Chiayu and Wyant, 1987; Creath, 1988). Probably the best approach is to use five frames at 90° intervals and the equation

Automatic fringe analysis 309

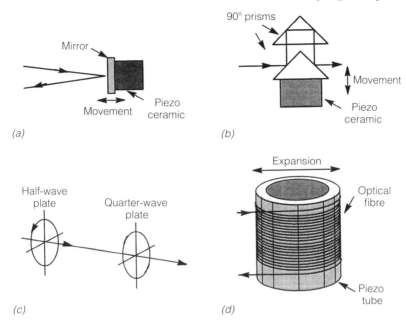

Fig. 8.17 Schemes for phase shifting. (*a*) mirror on piezoelectric translator, (*b*) Porro prism pair and translator, (*c*) rotating half-wave plate, (*d*) optical fibre on piezoelectric cylinder (after Creath (1988)).

$$\phi_N = \cos^{-1}[(I_5 - I_1)/2(I_4 - I_2)] \tag{8.14}$$

to determine the phase step ϕ_N for a known voltage (Cheng and Wyant, 1985). Having applied the phase shifts accurately, it may be necessary to mitigate any errors due to interferometer instabilities by using an algorithm that is insensitive to step size errors, such as the five frame algorithm of equation (8.9).

(c) Low modulation detection

Not all pixels will see sufficient modulation, due to several factors:

1. Dark speckle giving low modulation;
2. Holes on the object;
3. Shadows on the object;
4. Areas off the object;
5. Faulty camera pixels; and
6. Dirt on the camera sensor.

Creath (1985) proposes that the modulation depth be calculated on a pixel-by-pixel basis according to

$$I_m = \sqrt{[(I_4 - I_2)^2 + (I_3 - I_1)^2]/2} \tag{8.15}$$

if four frames are acquired at 90° intervals. If the depth lies below a certain threshold then those pixels can be marked and either averaged with their neighbours or highlighted in some way. Vrooman and Maas (1989) proposed that pixels should be flagged if they lie in the central region of the arctangent look-up table used to calculate the phase, as shown in Fig. 8.16. The radius of the region can be determined experimentally. The authors chose not to modify the flagged pixels by combining with their neighbours, as they generally aided in identifying various features on the object.

(d) Phase unwrapping

As a result of using an arctangent to calculate the phase due to deformation $\phi_d(x, y)$, modulo 2π steps may occur in the phase map. It is necessary to remove these steps in order that a smooth reconstruction of the surface phase may be made. This process has become known as **phase unwrapping**. In TV holography the process is not straightforward, due to noise in the region of the 2π discontinuity from incorrect calculation of speckle phase, speckle decorrelation and low modulation speckles.

The most common approach is based on adding or subtracting an offset of 2π at each pixel if a phase jump greater than π is detected. Starting at the top of any column in the phase map, the offset is set at zero. Scanning downwards, jumps are checked for by examining the difference between adjacent pixels. Whenever one is detected the offset is incremented by $+$ or -2π depending on the sign of the jump. Finally, scanning outwards along each row, offsets are calculated starting with the offset from the scanned column. The resulting map of offsets is sometimes referred to as the **mask**.

This approach works well with the clean fringes of classical holography and moiré. For TV holography the edges can be buried in noise, causing steps to be identified incorrectly. Further, the approach requires that all pixels in the field be valid. If the edges are incorrectly identified, then 'streaks' will propagate in the phase map. More sophisticated algorithms to accommodate this phenomenon have been developed (Nakadate and Saito, 1985; Ghiglia *et al.*, 1987; Huntley, 1989; Vikhagen, 1989a; Vrooman and Maas, 1989; Towers *et al.*, 1990; Osten *et al.*, 1987; Osten and Höfling, 1990; Virdee *et al.*, 1990). Frequently, these algorithms operate in a manner analogous to flood fill algorithms found in computer software graphics packages; from some arbitrary starting point, valid pixels are unwrapped along both rows and columns until bad data is encountered. The algorithm then chooses the next nearest row or column along which it can propagate from a valid data point.

A few points of interest are worth noting from the various routines. Vrooman and Maas (1989) have indicated that they employ a spiral scan in order that more valid pixels are used to determine an edge, and a queue is utilized so that the next optimum direction is taken for propagation. Nakadate and Saito (1985) scan about a pixel to assess the orientation of the gradient. Towers *et al.* (1990) segment the image not only into pixels but also tiles in order that errors are contained. The edges of the tiles are then ranked in statistical order and unwrapped in order of least likelihood of failure. Ghiglia *et al.* (1987) use a cellular automata which involves whole-field iteration over many cycles until a state of 'minimum energy' exists. This approach can also accommodate sheared fringes in the phase map.

8.5.6 Practical examples

(a) Static fringes

The example in Fig. 8.18 shows the analysis of a simple plate in bending. An initial speckle image of the object is recorded in its undeformed state. The deformation is then applied and three further images are acquired as successive TV frames in 3/25 of a second, between each of which the PZT transducer has been stepped during the flyback. By subtracting the first frame from each of the following three and taking the absolute value, three correlation maps are obtained separated by 120° intervals.

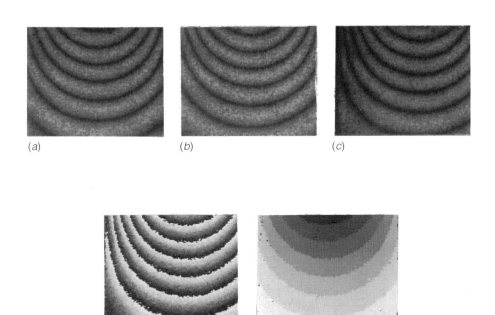

Fig. 8.18 Phase stepping applied to a deformed plate: (*a*) – (*c*) subtraction fringes with 120° relative phase shifts, (*d*) phase map, (*e*) fringe order number mask. (Courtesy of Rover Group.)

These three images are low-pass filtered using a 5 × 5 convolution filter in order to reduce the random fluctuations, yielding the images (a)–(c). Equation (8.6) is then used to obtain sine and cosine maps, which in turn give values for finding the arctangent in a look-up table (Fig. 8.16). This generates a phase map (d), which can then be unwrapped by generating a mask (e) and adding it to the phase map.

(b) Vibration fringes

For fringes that show the time-averaged behaviour of a vibrating object by a Bessel function of some order, the simplest approach to phase measurement is perhaps that chosen by Nakadate et al. (1986). The fringe function is converted from J_o^2 Bessel to $\cos^2$ by using stroboscopic modulation, as described in section 8.4.2(b). A phase-shifter is then made to vibrate at the same frequency as the object with a related amplitude and phase (section 8.4.4(a)). Figures 8.19(a)–(d) show four phase-stepped stroboscopic images of a plate. The procedure is then a repeat of the static case, except that it is possible to use speckle averaging to reduce the degree of low-pass filtering required. Figure 8.19(e) shows the resulting phase map and Fig. 8.20 shows a wireframe of the plate at a vibration extreme.

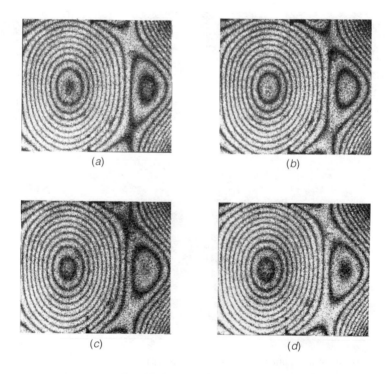

Fig. 8.19 Phase stepping applied to a vibrating plate. (a)–(d) stroboscopic subtraction fringes with 90° relative phase shifts. (Courtesy of Rover Group.)

Automatic fringe analysis 313

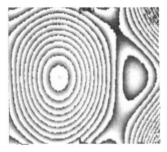

Fig. 8.19 Phase stepping applied to a vibrating plate. (*e*) phase map. (Courtesy of Rover Group.)

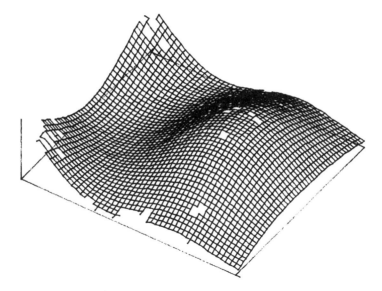

Fig. 8.20 Wireframe plot corresponding to Fig. 8.19 showing amplitude of plate vibration.

An alternative described by Nakadate (1986a, b), Stetson (Stetson, 1990; Stetson and Brohinsky, 1988), Johannson and Predko (1989), Virdee et al. (1990) and Vollesen (1990) is to analyse the Bessel function itself, performing the conversion from Bessel to cosine in the computer rather than optically. This offers greater optical simplicity and light efficiency, but the fringes must be of greater fidelity. The approaches differ, but they all require the addition of a d.c. shift in phase as well as a shift in vibration phase in order to remove the non-linearity between the first and subsequent orders in the Bessel function.

314 Television holography and its applications

Ellingsrud and Løkberg (1989) have developed an approach for the analysis of sub-fringe vibrations, which can be extended to higher amplitudes (see also section 8.6). A set of phase-stepped frames can be acquired from the Bessel function curve, as shown in Fig. 7.22, by working with the linear part of the flank of the zero-order fringe. Many engineering structures (in fact nearly all studied by the authors in the automotive field) exhibit some form of non-linearity in phase between anti-nodes, their phases being other than 0° and 180°. This is a result of either exciting two orthogonal modes simultaneously or there being unevenly distributed damping in the structure. The Ellingsrud approach is currently the only one known to the authors that can cope with this condition.

Figure 8.21 gives an example of the technique applied to a 10 cm mid-range loudspeaker vibrating at 1650 Hz. This is a good example, as it demonstrates the ability of the system to analyse a complex vibration on a surface with very uneven reflectance. The figure shows plots of the vibration phase and amplitude distribution. Loudspeakers have always been basic subjects for holographers; this example would be difficult to analyse by hand.

(c) Transient fringes

Many phenomena worth studying are transient. The Fourier transform approach is applicable if only one frame is available and a linear term can be added to the data. If the event is not occurring too quickly, a modified phase stepping procedure described by Kujawinska (1989) and demonstrated by Buckberry and Davies (1990) can be used. This technique uses speckle correlation as described in section 8.5.4(c), but with the frames acquired in reverse sequence. Three frames are acquired prior to deformation, giving

$$I_i(x, y)) = I_o + I_m \cos [\phi_s(x, y) + 2i\pi/3], \tag{8.16}$$

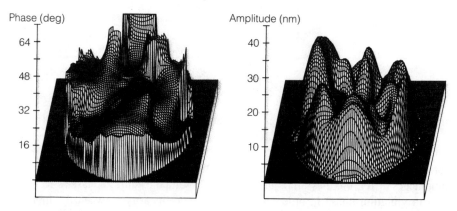

Fig. 8.21 Vibration of a mid-range loudspeaker. (*a*) phase map, (*b*) amplitude map. (Courtesy of S. Ellingsrud, Norwegian Institute of Technology, Trondheim.)

where $i = 1$ to 3, and then only one frame of the event described by

$$I_4(x, y) = I_o + I_m \cos [\phi_s(x, y) + \phi_d]. \tag{8.17}$$

By subtracting the fourth frame from each of the other three and squaring, three correlation fringe maps are generated. Provided correlation is maintained, single frames of the event can continue to be acquired, allowing a time history to be formed. In Fig. 8.22 and 8.23, an example is shown of fringes formed by convection

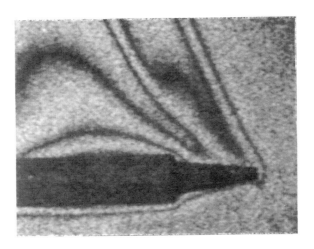

Fig. 8.22 Fringes formed by convection currents in the air around a soldering iron. (Courtesy of Rover Group.)

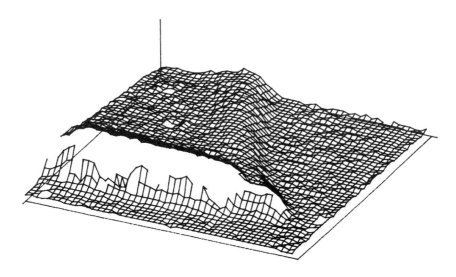

Fig. 8.23 Wireframe plot corresponding to Fig. 8.22.

currents in the air around a soldering iron. Kujawinska (1989) also showed that the random speckle phase term $\phi_s(x, y)$ could be evaluated rather than filtered out, when the frames are acquired in this order. However, this does require the use of a special filter to choose the correct running solution from the two alternatives generated by the equations.

8.5.7 Conclusions

We have given a comprehensive, if not detailed, overview of the fringe-analysis procedures in use today. The approach has been to describe the algorithms in terms of their robustness in solving engineering problems rather than a detailed theoretical analysis. TV holography interferograms represent probably the worst fringe patterns that an algorithm has to analyse, irrespective of problems with the environment and object, but the reader should not be deterred by the task.

Future developments will probably make greater use of the Fourier approach as more powerful processors become available, because of its efficiency in the presence of noise. Unwrapping algorithms will become more sophisticated in order to be faster and more robust, possibly using neural networks. Another development may well be the use of instantaneous phase-measuring interferometers as described by Smythe and Moore (1984), Womack (1984) and Kujawinska and Robinson (1988). These approaches allow fast transient events to be analysed from any fringe field.

Finally, quantified interferograms will form part of some whole system, incorporating perhaps other analytical measurements and computational models. An excellent example, albeit holographic, is the system developed by Toyota Cars (Hyodo and Konomi, 1986; Imai *et al.*, 1986). The authors particularly recommend that these papers should be read to appreciate the usefulness of automatic fringe analysis.

8.6 EXAMPLES OF APPLICATIONS

We shall now see that it is possible to use TV holography in a wide variety of situations and obtain useful engineering data.

8.6.1 Automobile applications

(a) Engines

The authors have frequently been asked to study deformation and vibration in petrol engines, in order to improve their performance and reduce noise and vibration. The loads introduced by the tightening of bolts during assembly have a significant bearing on the fit and wear of components. Figure 8.24 shows how fibre TV holography was used to study the deformation of a cylinder bore. The bore was viewed from the crankshaft end, with the camera looking towards the top of the chamber. Supported through the spark-plug aperture was a 75 mm diameter conical mirror, which was rigidly mounted externally to be independent of the engine

Examples of applications 317

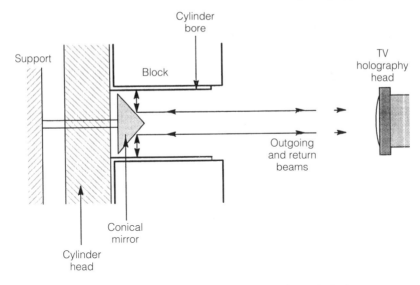

Fig. 8.24 Cylinder-bore deformation measurement by means of a conical mirror.

Fig. 8.25 Phase map resulting from tightening of head bolts, as seen in conical mirror. (Courtesy of Rover Group.)

movement. The camera was thus able to view approximately a 50 mm slice down the bore. In order that known clamping loads could be applied, the bolts were strain-gauged.

Figure 8.25 shows the resulting phase map. In this case the phase map was unwrapped in a circular manner so that the cone could be transformed geometrically as a function of radius to a cylinder. Typical plan and isometric displays of slices

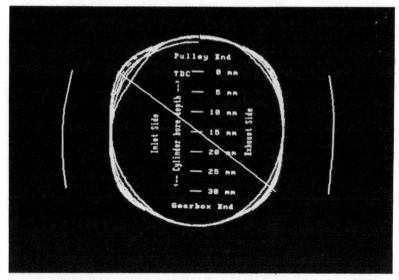

(a)

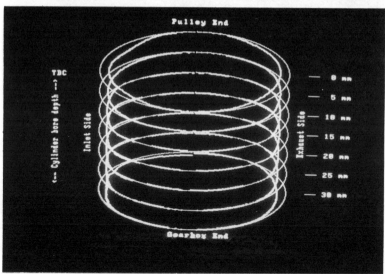

(b)

Fig. 8.26 Distortion of cylinder bore corresponding to Fig. 25. (*a*) Plan and (*b*) isometric views of sections. (Courtesy of Rover Group.)

down the bore are shown in Fig. 8.26. The four-fold deformation can clearly be seen, due to the four bolts surrounding the bore applying a bending moment about the gasket. As a result of the bore becoming square, four gaps are present between the bore and the piston rings, leading to burning of oil and a degradation of the emissions from the engine. A modification to the gasket alleviated the problem.

Examples of applications 319

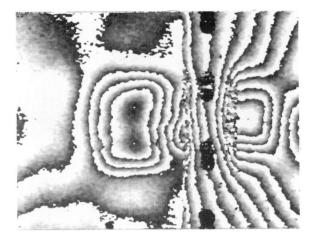

Fig. 8.27 Phase map showing deformation of the exhaust side for a four-cylinder engine when a head bolt is tightened. (Courtesy of Rover Group.)

Fig. 8.28 Phase map showing deformation of a four-valve combustion chamber due to head-bolt tightening. (Courtesy of Rover Group.)

Figure 8.27 shows the deformation of the exterior face of the engine when the head bolts in the above example are tightened. It can be seen that if the bolts are tightened unevenly from one side of the engine to the other, the head tilts about the gasket and local bulging occurs under the head of a bolt. Reaction stresses are created in the block, causing a depression in the water jacket.

A more serious effect of the head bolts being tightened is the deformation of the combustion chamber shown in Figs. 8.28 and 8.29. As a result of the bending

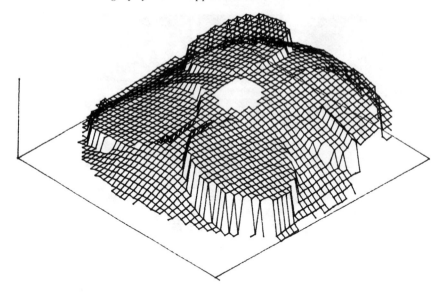

Fig. 8.29 Wireframe plot corresponding to Fig. 8.28.

moment applied about the gasket, the head is bowed towards the crankshaft. The valves, which are held via their stems in a bush, are forced to move differentially to the head casting, giving rise to a gap that may be as large as 125µm at full bolt torque. It can be seen in Fig. 8.29 that the point of contact is nominally the edge of the valve nearest to the spark-plug hole. This lack of contact between the valve and the chamber prevents the valves from losing heat, a vital requirement in lean-burn engines which run comparatively hot in order to burn as much of the nitrous oxide and carbon monoxide as possible. It can also be seen that the left-hand valve in Fig. 8.29 has apparently sunk into its seat. This is physically not possible unless the seat has been cut at an incorrect angle. This was subsequently found to be the case, with a progressively larger error down the length of the engine as the tool drifted off-line.

All the above work was done with the engine bolted to the optical table via its gearbox mountings, the assumption being that the gearbox acted as an infinitely stiff end-constraint about which the engine moved. Engines also act as the primary source of excitation of acoustic noise within the passenger compartment. This may either be transmitted structurally or be airborne. In order to study the natural modes of an engine that may result in this excitation, the authors have constructed a rig as shown in Fig. 8.30. This allows the engine to be held as it would be in the engine compartment of the car, so that it retains the correct degrees of freedom. Single-frequency excitation is provided from an electromechanical shaker attached to the engine via a thin flexible rod known as a **stinger** and a load cell. The frequency input is swept across the range of interest at a constant force and the modes are noted. Figure 8.31, obtained by stroboscopic phase-stepped analysis, shows the first torsional resonance of an engine.

Examples of applications 321

Fig. 8.30 Rig for modal analysis of engine assemblies using TV holography. (Courtesy of Rover Group.)

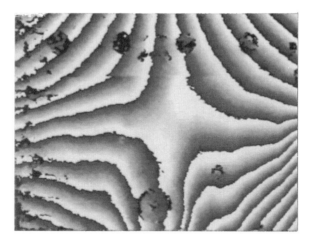

Fig. 8.31 Phase map of the first torsional mode of vibration of a four-cylinder engine at 835 Hz. (Courtesy of Rover Group.)

Not all vibrations are unwanted. Holography has been found to be the ideal tool for deciding where to locate the positions of **knock sensors**. These devices detect the onset of a combustion phenomenon which results from the engine ignition being

322 Television holography and its applications

advanced too far in an attempt to keep exhaust pollutants to a minimum. The signature of knock is a characteristic frequency transmitted to the surface of the engine block, resulting in a vibration which can be detected by a narrow-band accelerometer. This frequency is usually several kilohertz, resulting in antinodes of small spatial size, typically 4 cm across. To find these with an accelerometer is too time-consuming to be worthwhile, whereas holography with its high spatial resolution is the ideal tool. When run in real time using TV holography, the task is trivial. Many sites can be found and ranked in terms of their positional suitability and response to a unit input force across the frequency range of interest. Figure 8.32 shows the modal response of a block at such frequencies.

(b)   Brakes

Brakes can be a serious source of noise, primarily because they usually emit a very clear single frequency which can be distinguished clearly from other noises in the vehicle. Both disc and drum brakes give problems and much work is done to identify the natural resonances of individual components and complete assemblies.

Figure 8.33 shows how the modes of a disc brake can be categorized, plotted as a function of frequency. This plot can be compared with computer predictions for the mode shapes of the disc. The most convenient means of exciting a disc is by using a piezoelectric element bonded to the face. Frequently, because of the symmetry of the disc, orthogonal modes can exist, separated by only a few hertz. If these are excited

Fig. 8.32 Modal response of an engine assembly at 6 kHz. (Courtesy of Rover Group.)

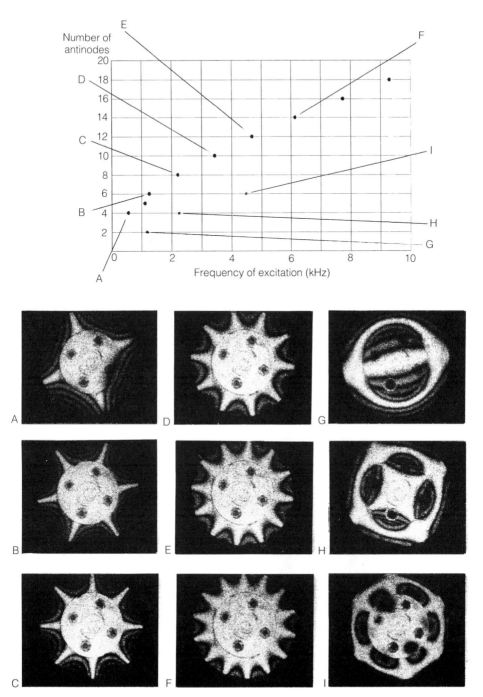

Fig. 8.33 Modal plot for a vented disc brake. (Courtesy of Rover Group.)

324 *Television holography and its applications*

simultaneously because of imperfections in the function generator, combined mode shapes will appear, resulting in travelling waves. Because the structure is highly symmetrical and the damping is comparatively low, modes can easily be disturbed by the addition of only a few grams in mass, making non-contacting techniques desirable. The real operating environment does much to change the structure through both centrifugal stiffening and temperature; Jurid-Werke in Germany have done much work in this area using pulsed TV holography and a derotator (see Section 7.5.3).

Figure 8.34 shows a further example of drum-brake backplate vibration, quantitatvely analysed using stroboscopic phase steps. The brake was studied *in situ* on the vehicle, with single-frequency excitation provided from an electromechanical shaker attached to the rear vehicle subframe. This allowed the assembly to remain undisturbed, as it had been found that the response was sensitive to both the build of the brake and subsequent 'bedding-in'. This again demonstrates the need for a system that is portable and not limited to objects that can be mounted on an optical table. The long arm in Fig. 8.34 is the trailing arm suspension member, by which the wheel is attached to the body.

(c) Body shells

The need for a small, compact head for TV holography can be seen in the following example of a vehicle body shell. It is often necessary to identify regions that are sensitive to excitation from both engine and suspension. It is not practical to mount the vehicle on an optical table or in a vibration-isolated environment. Furthermore,

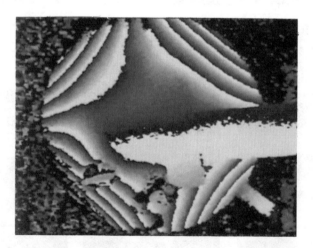

Fig. 8.34 Phase map of the modal response of a drum-brake backplate in situ on a vehicle (Courtesy of Rover Group.)

Fig. 8.35 Vibration amplitude contours overlaid on to the direct image of a vehicle floorpan.(Courtesy of Rover Group.)

the requirement to have line-of-sight access to various parts of the structure precludes the use of any head comparable in size to the commercially available units. A tripod-mounted device can be positioned on the body of a vehicle to observe the interior. Figure 8.35 shows vibration contours overlaid on a floorpan so that the mode can be related to structural features. TV holography allows the rapid assessment of large surface areas whose size would be prohibitive with accelerometers.

8.6.2 *In-situ* measurement of stone degradation

The University of Oldenburg has for several years been using a three-dimensional TV holography system to measure environmental effects on stonework *in situ*. This achievement is remarkable not only because of its technical merit in system design but because of the period of observation, which is several days. The system used for the experiments is based on laser diodes and a CCD camera, with image processing on an IBM/AT machine. Figure 8.36 shows the principal optical setup. One in-plane component (x) and the out-of-plane component (z) of deformation are obtained with this system.

The system was used in a church at Eilsum in the northern part of Germany. The murals in this church are highly deteriorated, showing cracks in the colour layers and the underlying plaster. The observed wall area, 12 cm across, has separated into three regions due to macroscopic cracks. The optical head was suspended on three steel wires from a single peg about 0.5 m above the field of view. Figure 8.37 shows characteristic fringe systems resulting from out-of-plane deformations 6 and 21 hours after the start of the experiment. The three-dimensional plots show the

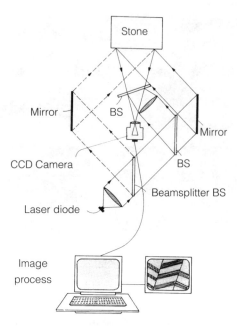

Fig. 8.36 Optical arrangement for combined in-plane and out-of-plane measurement. (Courtesy of K. Hinsch, University of Oldenburg.)

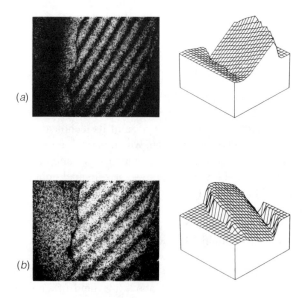

Fig. 8.37 Out-of-plane deformation of church wall section: (*a*) after 6 hours; (*b*) after 21 hours. (Courtesy of K. Hinsch, University of Oldenburg.)

deformation fields evaluated from the corresponding fringe patterns and the comparative tilts of the central region.

It was possible to monitor the wall deformations over a period of 20 days with the same initial reference state. The observed cyclic tilting of the central region over a period of five days is plotted in Fig. 8.38. The periodicity is indicative of a temperature effect. The amount of deformation, about 25 µm per 1 K temperature change, leads to the conclusion that thousands of deformation cycles may drive the deterioration of mural paintings covering such wall sections. In practice, parts of the walls have degraded to such an extent that the only remedy is to replace these regions with a special elastic stone substitute. The requirements for the substitute are that no excessive forces are exerted between the substitute and surrounding stone during hardening, that a good contact prevails between the two media, and that there is only a small difference in the response to environmental stresses.

A further example shows the hardening performance of the substitute over a period of eight days. A hole was milled out of a sandstone sample and filled with the substitute. Figure 8.39 shows the experimental arrangement. Images were captured every 15 minutes showing the z- and x-components of movement. To overcome the problems of decorrelation as the surface hardened, every image acted as a reference state for the following image, so that the fringes represent a 'running' deformation measurement rather than an absolute difference from the initial state.

Several stages were observed over the eight-day period. For the first four hours, the behaviour was plastic and no measurements could be made. During the hardening phase greater stability was observed, and shrinking occurred due to the loss of moisture (Fig. 8.40). Later the amount of shrinkage decreased and the contact

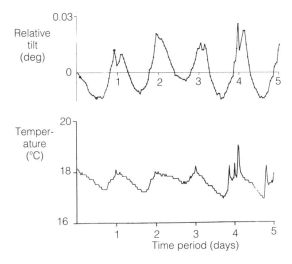

Fig. 8.38 Relative tilt of wall section as depicted in Fig. 8.37 over a period of five days. (Courtesy of K. Hinsch, University of Oldenburg.)

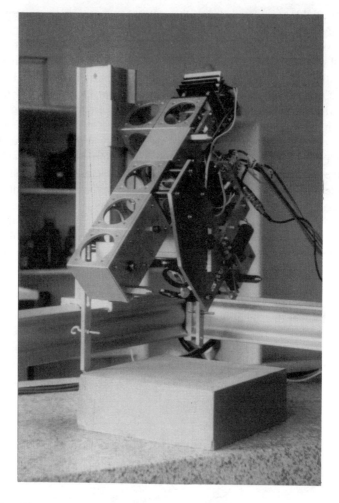

Fig. 8.39 Portable camera used to measure stonework movement. (Courtesy of K. Hinsch, University of Oldenburg.)

between stone and substitute consolidated. After four days the hardening process was complete.

8.6.3 Structural integrity testing

The structural integrity of composites is being assessed by many techniques, including TV holography. To date, no specific method of identifying defects has been found. An example, from Wright Patterson Air Force Base, USA, is a prototype missile fin in which delaminations may occur between the aerodynamic surfaces and

Examples of applications 329

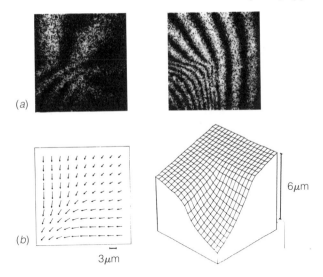

Fig. 8.40 Deformation due to hardening of stone over a period of 10 hours, at 38 hours after start.(*a*) In-plane, (*b*) out-of-plane (Courtesy of K. Hinsch, University of Oldenburg.)

the hat-section core to which they are welded. Holographic recordings were made while the fins were subjected to a random acoustic excitation in a band between 1 and 15 kHz. The lower limit was chosen to exclude the first six resonant frequencies, while the upper limit was found by trial and error.

The principal delaminated regions appear as dark areas. They have an influence on virtually all of the many modes across the entire bandwidth and are clearly visible in real time on the video screen. They have variable amplitudes, giving fringes with orders at least one or two greater than those of the transient mode shapes forming the background. An even clearer picture of the flawed area can be obtained by taking a time-averaged picture of the TV screen over many seconds. It was found that this approach could be applied to many types of structures, as long as a flaw causes a sufficient change of bending stiffness and the modal density in the bandwidth of the random excitation is fairly high.

Some defects are better identified, not by vibration, but by distortion. An example from E. Vikhagen, Norwegian Institute of Technology, Trondheim, shown in Fig. 8.41, is the wing flap of a crashed fighter aircraft stressed by thermal loading. The image has been obtained using the max-min scanning approach to analysis; the resulting phase map is then differentiated in the x- and y-directions to reveal local distortions more clearly. Whilst the honeycomb structure is too fine to be resolved, debonds between the cover ply and the internal structure can be seen at the right-hand edge. A faultless specimen was not available for comparison, so the interpretation of the likely cause (excessive shear stresses at the right-hand edge) could not be verified.

330 *Television holography and its applications*

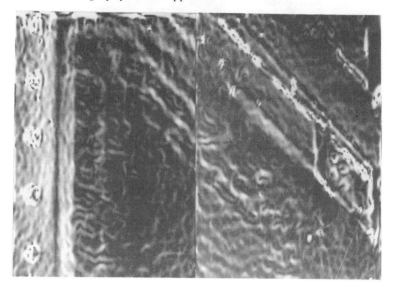

Fig. 8.41 Delamination of a wing from a crashed fighter aircraft, highlighted by displaying the transverse gradient of an out-of-plane phase map. (Courtesy of E. Vikhagen, Norwegian Institute of Technology, Trondheim.)

Rather than derive the gradient of the deformation from an out-of-plane measurement, it can be measured directly using either in-plane TV holography or shearography as described in section 8.3.5. Shearography has gained widespread popularity for this application because of its environmental robustness. A commercial system by Steinbichler Optotechnik is shown in Fig. 8.42, (*a*) being the interferometric head and (*b*) the entire system in which the head is located. The system tests tyres, as in Case Study 4 in section 7.4.1. The tyre is placed under the cover, which is evacuated, causing delaminated regions to swell. The camera is turned to scan the circumference of the inner wall of the tyre. The difficulty with this type of measurement is ensuring that the loading mechanism excites all the defects of interest and then being able to characterize them sufficiently to allow a machine operator to accept or reject the tyres.

8.6.4 Pulsed TV holography

Many structures cannot be analysed using a continuous-wave laser, due to low reflectivity of the surface, or because the disturbance to be measured is transient or the object is not interferometrically stable. These problems have been overcome in normal holography by using pulsed ruby lasers, and more recently Nd YAG lasers. However, their transfer to TV holography has been slow for several reasons. First, the image quality remains poor because of the difficulty in using subtraction routines

Examples of applications 331

(a)

(b)

Fig 8.42 Commercial system for tyre testing: (*a*) shearing camera, (*b*) vacuum chamber within which camera operates. (Courtesy of Steinbichler Optotechnik.)

to remove time-stationary noise. Second, the need for stability and consistency between successive pulses is greater in TV holography in order to obtain good quality. Finally, the pulse-pair repetition rate has until recently been slightly lower than standard video frame rates, preventing realization of the full potential of the system.

Typically, the application of such systems has been to problems that generate fringes in simple ways. Loughborough University has been investigating the use of a Nd YAG laser. Using a diode-pumped system with twin cavities, the stability for successive pulses has been improved considerably, with pulse pairs corresponding to two successive TV frames. The system is used for combined in-plane and out-of-plane measurements. For routine examination of disc-brake vibrations, Jurid-Werke have been developing pulsed TV holography using a highly refined ruby laser. The vibrations are observed via a derotator and the reference beam is introduced by means of a monomode fibre passing through a small aperture in the observation lens.

City University have also been observing rotating discs, but using an in-plane system to record strain distributions around notches. The images are analysed using the single-frame Fourier-transform approach described in section 8.5.3(a). Fringe formation requires two correlated fringe patterns. In this work the laser is fired with extreme accuracy, in order that the rotating component is in exactly the same position every time it is illuminated. Between firings the disc speed is altered, using a high-precision shaft encoder and a sophisticated electronic control system. To obtain complete TV frames the scanning beam of the camera must be turned off when the laser fires and back on again at the start of the next even field. This blanking operation uses the persistence of the camera tube to store the speckle pattern until it can be read into the frame store of the computer.

The optical configuration consists of two cylindrical mirrors which form a radially sensitive in-plane interferometer, shown schematically in Fig. 8.43. The fringes represent in-plane radial expansion with a sensitivity of approximately 0.5 μm. The system operates at any tangential velocity up to about 150 m s^{-1}. The work is aimed at achieving speeds of 300 m s^{-1}, in order to be applicable to aero engines.

Improvement of the poor fringe quality has been the purpose of investigations by the Norwegian Institute of Technology. The aim has been to incorporate the advantages of digital subtraction instead of addition. This has been achieved using a standard interline transfer CCD camera, with the pulses placed on either side of the transfer gate between fields. Each field then contains a single pulsed image, and a pair can be either added or subtracted to form a combined image.

8.6.5 Surface contouring

Recently, two approaches to contouring objects with TV holography have been published. In the first, contour fringes are generated by a change in object illumination angle (Maas, 1990). It is shown that shape is important in determining surface strain on three-dimensional objects, and the measurements of the two can be combined. The example in Fig. 8.44 is a pressurized glass bottle, with strain-field maps corrected for the shape.

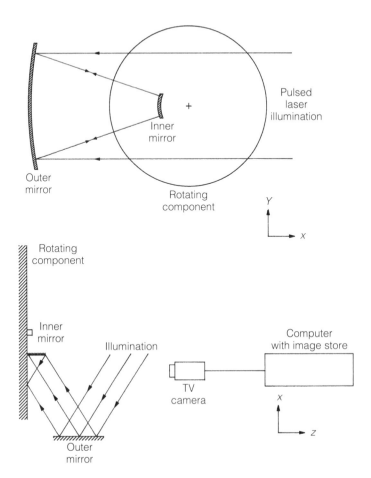

Fig. 8.43 Arrangement for pulsed measurements of radial in-plane movements of rotating objects (after Preater and Swain (1991)).

An alternative approach is to correlate two images resulting from two different wavelengths. This provides a contour interval based upon an effective wavelength $\lambda_1\lambda_2/(\lambda_1-\lambda_2)$, the process being similar to that described for holographic interferometry in section 7.7.3. Usually, the source is a dye laser. Atcha *et al.* (1991) have demonstrated that an alternative is to modulate the wavelength of a diode laser via the injection current. This provides improved flexibility with reduced complexity. A particular advantage of this system is the ability to tune the contour wavelength by real-time observation of the fringes. Figure 8.45(*a*) is the result of operating a laser diode at three wavelengths, making two fringe periodicities visible simultaneously at

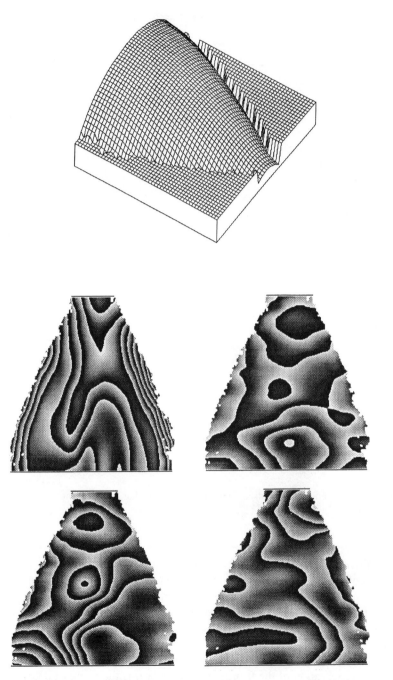

Fig. 8.44 Combination of contouring and in-plane measurement to produce corrected strain field maps. The object was a pressurized glass bottle. (Courtesy of A. Maas, Fokker.)

Examples of applications 335

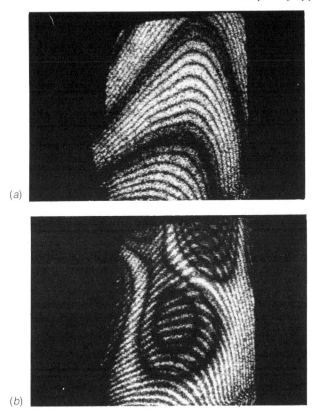

Fig. 8.45 Fringes generated simultaneously on a turbine blade. (*a*) contour fringes at two intervals, (*b*) vibration and contour fringes (Atcha *et al.*, 1991).

TV frame rate. Figure 8.45(*b*) was formed by vibrating the object while driving the laser diode at two wavelengths; surface-form contours and vibration contours are superimposed in real time.

8.6.6 Future developments

There are several areas in which new developments are likely. Solid-state lasers with good beam qualities and long coherence lengths are now available and will be used more in future in TV holography systems. They are rugged, efficient and very small, and thus ideal for incorporation in a portable system. A feature peculiar to laser diodes is the ability to modulate the wavelength, which has enabled intrinsic phase modulation as well as contouring systems to be built, as discussed in section 8.6. Interferometers for differe,nt purposes can be combined in a common optical arrangement, and the authors are developing two such systems. The first is a TV interferometer for combined deformation/vibration and contouring measurement, as described in section 8.6.5. The second is an integrated TV interferometer and scanning laser Doppler system. The two are combined using a fibre optical phase-locked loop; this allows the phase in the TV interferometer to be locked to any point on the surface of the vibrating object, giving a localized reference beam. This not only gives full-field mode shapes but also the modal parameters at points on very unstable objects.

The future of computers and phase measuring is likely to depend on the speeds that can be achieved at acceptable cost. Already, commercial Fizeau interferometers display wrapped-phase maps rather than optical fringes at TV frame rate. TV interferograms are prime candidates for this mode of display, because it removes the speckle structure from the raw fringes. The authors see the break point as being approximately one second. With respect to algorithms, they see spatially synchronous techniques as being of importance, first in order to achieve the ideal of real-time phase maps and second to provide quantitative analysis of transient events.

ACKNOWLEDGMENTS

The authors would like to thank the Rover Group for permission to publish this chapter and the following workers in the field of TV holography and associated areas for contributing material, inspiration and many hours of conversation.

Dennis Ghiglia, Sandia National Laboratories, USA.
Neil Halliwell and John Tyrer, University of Loughborough, UK.
Klaus Hinsch, Claudia Holscher and Gerd Gulker, University of Oldenburg, Germany.
Julian Jones, Jesus Valera and Dave Harvey, Heriot-Watt University, Edinburgh, UK.
Thomas Kreis, Institut für Angewandte Strahltechnik, Bremen, Germany.
Margaret Kujawinska, Technical University, Warsaw, Poland.
Ole Lokberg, Eiolf Vikhagen, S. Ellingsrud and Rudie Spooren, Norwegian Institute of Technology, Trondheim.
Ad Maas, Fokker Ltd, Holland.
Gene Maddux, Wright Patterson Air Force Base, USA.
Richard Preater and Robin Swain, City University, London, UK.
Oliveria Soares, University of Porto, Portugal.
B. Stockley, Test Techniques, Rover Group, UK.
Ralph Tatam and Hashim Atcha, Cranfield Institute of Technology, UK.
Dave Towers and Tom Judge, Warwick University, UK.
Dave Williams, National Physical Laboratory, Teddington, UK.
Ichirou Yamaguchi, Institute of Physical and Chemical Research, Saitama, Japan.

REFERENCES

Aleksoff, C. C. (1971) Temporally modulated holography. *Appl. Opt.*, **10** (6), 1329–41.
Atcha, H., Tatam, R., Buckberry, C. H., Davies, J. C. and Jones, J. D. C. (1991) Surface contouring using TV holography. *Proc. SPIE*, to be published.

References

Bruning, J. H., Herriott, D. R., Gallagher, J. E., Rosenfeld, D. P., White, A. D. and Brangaccio, D. J. (1974) Digital wavefront measuring interferometer for testing optical surfaces and lenses. *Appl. Opt.*,**13** (11), 2693–703.

Buckberry, C. H. and Davies, J. C. (1990) The application of TV holography to engineering problems in the automotive industry. *Proc. SEM Conf. on Hologram Interferometry and Speckle Metrology*, Baltimore, 268–78.

Butters, J. N. and Leendertz, J. A. (1971) Holographic and video techniques applied to engineering measurement. *Meas. and Control*, **4** (12), 349–54.

Button, B. L., Cutts, J., Dobbins, B. N., Moxon, C. J. and Wykes, C. (1985) The identification of fringe positions in speckle patterns. *Opt. and Laser Tech.*, **17** (4), 189–92.

Carré, P. (1966) Installation et utilisation du comparateur photoélectrique et interférentiel du Bureau International des Poids et Mesures. *Metrologia*, **2** (1), 13–23.

Cheng, Y. and Wyant, J. C. (1985) Phase shifter calibration in phase-shifting interferometry. *Appl. Opt.*, **24** (18), 3049–58.

Chiayu, A. and Wyant, J. C. (1987) Effect of piezoelectric transducer nonlinearity on phase shift interferometry. *Appl. Opt.*, **26** (6), 1112–16.

Creath, K. (1985) Phase-shifting speckle interferometry. *Appl. Opt.*, **24** (18), 3053–8.

Creath, K. (1988) Phase-measurement interferometry techniques. *Prog. in Opt.*, **26**, 349–93.

Creath, K. (1990) Phase-measurement techniques for nondestructive testing. *Proc. SEM Conf. on Hologram Interferometry and Speckle Metrology*, 473–9.

Dändliker, R. and Thalmann, R. (1985) Heterodyne and quasi-heterodyne holographic interferometry. *Opt. Eng.*, **24** (5), 824–31.

Davies, J. C., Buckberry, C. H., Jones, J. D. C. and Panell, C. (1987) Development and application of a fibre optic electronic speckle pattern interferometer (ESPI). *Proc. SPIE*, **863**, 194–207.

Ellingsrud, S. and Løkberg, O. (1989) Analysis of high frequency vibrations using TV-holography and digital image processing. *Proc. SPIE*, **1162**, 402–10.

Ghiglia, D. C., Mastin, G. A. and Romero, L. A. (1987) Cellular-automata method for phase unwrapping. *J. Opt. Soc. Am. A*, **4** (1), 267–80.

Gonzalez, R. and Wintz, P. (1988) *Digital Image Processing*, Addison-Wesley, Reading, Mass.

Goodman, J. (1975) *Laser Speckle and Related Phenomena*, Springer-Verlag, Berlin.

Hariharan, P., Oreb, B. F. and Brown, N. (1983) Real-time holographic interferometry: a microcomputer system for the measurement of vector displacements. *Appl. Opt.*, **22** (6), 876–80.

Hariharan, P., Oreb, B. F. and Eiju, T. (1987) Digital phase shifting interferometry: a simple error-compensating phase calculation. *Appl. Opt.*, **26** (13), 2504–6.

Hunter, J. C. (1989) Assessment of some image-enhancement routines for use with an automatic fringe tracking program. *Proc. SPIE*, **1163**, 83–94.

Huntley, J. M. (1989) Noise-immune phase unwrapping algorithm. *Appl. Opt.*, **28** (15), 3268–70.

Hyodo, Y. and Konomi, T. (1986) Mode analyses using laser holography. *Jnl Japan. Soc. Auto. Engnrs*, **7** (3), 44.

Imai, M., Suzuki, S., Sugiura, N. and Saito, H. (1986) Radiation efficiency of engine structures, analysed by holographic interferometry and boundary element calculation. *Soc. Auto. Engnrs*, Paper No. 861411.

Johannson, S. and Predko, K. G. (1989) Performance of a phase-shifting speckle interferometer for measuring deformation and vibration. *J. Phys. E. (Sci. Instrum.)*, **22** (5), 289–92.

Jones, R., and Wykes, C. (1983) *Holographic and Speckle Interferometry*, Cambridge University Press.

Kerr, D., Mendoza Santoyo, F. and Tyrer, J. R. (1989) Manipulation of the Fourier components of speckle fringe patterns as part of an interferometric analysis process. *J. Mod. Opt.*, **36** (2), 195–203.

Kreis, T. M. (1986) Digital holographic interference-phase measurement using the Fourier-transform method. *J. Opt. Soc. Am. A*, **3** (6), 847–55.

Kreis, T. M. (1987) Fourier-transform evaluation of holographic interference patterns. *Proc. SPIE*, **814**, 365.

Kreis, T. M. and Jüptner, W. P. O. (1989) Fourier-transform evaluation of interference patterns: the role of filtering in the spatial frequency domain. *Proc. SPIE*, **1162**, 116–25.

Kujawinska, M. (1989) Analysis of ESPI interferograms by phase stepping techniques. *Proc. FASIG*.

Kujawinska, M. and Robinson, D. W. (1988) Multichannel phase-stepped holographic interferometry. *Appl. Opt.*, **27** (2), 312–20.

Leith, E. N. and Upatnieks, J. (1962) Reconstructed wavefronts and communication theory. *J. Opt. Soc. Am.*, **52** (10), 1123–30.

Maas, A. (1990) Phase Shifting Speckle Interferometry. PhD Thesis, University of Delft.

Massie, N. A. (1980) Real-time digital heterodyne interferometry: a system. *Appl. Opt.*, **19** (1), 154–60.

Moore, A. J. and Tyrer, J. R. (1990) An electronic speckle pattern interferometer for complete in-plane displacement measurement. *Meas. Sci. Tech.*, **1** (10), 1024–30.

Nakadate, S. (1986a) Vibration measurement using phase-shifting time-average holographic interferometry. *Appl. Opt.*, **25** (22), 4155–61.

Nakadate, S. (1986b) Vibration measurement using phase-shifting speckle-pattern interferometry. *Appl. Opt.* **25** (22), 4162–7.

Nakadate, S. and Saito, H. (1985) Fringe scanning speckle-pattern interferometry. *Appl. Opt.*, **24** (14), 2172–80.

Nakadate, S., Saito, H. and Nakajima, T. (1986) Vibration measurement using phase-shifting stroboscopic holographic interferometry. *Opt. Acta*, **33** (10), 1295–309.

Osten, W. and Höfling, R. (1990) The inverse modulo process in automatic fringe analysis-problems and approaches. *Proc. SEM Conf. on Hologram Interferometry and Speckle Metrology*, Baltimore, 301–9.

Osten, W., Höfling, R. and Saedler, J. (1987) Two computer methods for data reduction from interferograms. *Proc. SPIE*, **863**, 105.

Preater, R. W. T. and Swain, R. (1991) Fringe analysis of in-plane displacements on high speed rotating components. *Proc. SPIE*, to be published.

Reid, G. T., Rixon, R. C. and Messer, H. I. (1984) Absolute and comparative measurements of three-dimensional shape by phase measuring moiré topography. *Opt. and Laser Tech.*, **16** (5), 315–19.

Robinson, D. W. and William, D. C. (1986) Digital phase stepping speckle interferometry. *Opt. Comm.* **57** (1), 26–30.

Schwider, J., Burow, R., Elssner, K.-E., Grzanna, J., Spolaczyk, R. and Merkel, K. (1983) Digital wavefront measuring interferometry: some systematic error sources. *Appl. Opt.*, **22** (21), 3421–32.

Slettemoen, G. Å. (1980) Electronic speckle pattern interferometric system based on a speckle reference beam. *Appl. Opt.*, **19** (4), 616–23.

Smythe, R. and Moore, R. (1984) Instantaneous phase measuring interferometry. *Opt. Eng.*, **23** (4), 361–4.

Sommagren, G. E. (1975) Up/down frequency shifter for optical heterodyne interferometry. *J. Opt. Soc. Am.*, **65** (8), 960–1.

Stetson, K. A. (1990) Theory and applications of electronic holography. *Proc. SEM Conf. on Hologram Interferometry and Speckle Metrology*, Baltimore, 294–300.

Stetson, K. A. and Brohinsky, W. R. (1988) Fringe-shifting technique for numerical analysis of time-average holograms of vibrating objects. *J. Opt. Soc. Am. A*, **5** (9), 1472–6.

Takeda, M., Ina, H. and Kobayashi, S. (1982) Fourier-transform method of fringe-pattern analysis for computer-based topography and interferometry. *J. Opt. Soc. Am.*, **72** (1), 156–63.

Towers, D. P., Bryanston-Cross, P. J. and Towers, C. E. (1990) The automatic quantitative analysis of phase stepped interferograms. *Proc. SEM Conf. on Hologram Interferometry and Speckle Metrology*, Baltimore, 480–7.

Vikhagen, E. (1989a) *TV holography in material evaluation*. PhD Thesis, Univesity of Trondheim.

Vikhagen, E. (1989b) Vibration measurement using phase shifting TV-holography and digital image processing. *Opt. Commun.*, **69** (3.4), 214–18.

Virdee, M. S., Williams, D. C., Banyard, J. E. and Nassar, N. S. (1990) A simplified system for digital speckle interferometry. *Opt. and Laser Tech.*, **22** (5), 311–16.

Volleson, J. H. (1990) Quantitative vibration data with time average ESPI and PC-based image processing. *Proc. SEM Conf. on Hologram Interferometry and Speckle Metrology*, Baltimore, 225–31.

Vrooman, H. A. and Maas, A. (1989) New image processing algorithms for the analysis of speckle interference patterns. *Proc. SPIE*, **1163**, 51–61.

Wornack, K. H. (1984) Interferometric phase measurement using spatial synchronous detection. *Opt. Eng.*, **23** (4), 391–5.

Wyant, J. C. (1975) Use of an a.c. heterodyne lateral shear interferometer with real-time wavefront correction systems. *Appl. Opt.*, **14** (1), 2622–6.

9
Moiré methods in strain measurement

C. FORNO

Crown copyright

9.1 CONVENTIONAL MOIRÉ MEASUREMENTS

The evolution of optical techniques since the development of the laser has been particularly significant in the area of holography applied to engineering metrology. For various reasons, such techniques have not been extended to include large-scale applications outside the confines of a well-controlled laboratory environment, where the necessary coherence properties of the laser can be maintained adequately. In the sometimes hostile environmental conditions of an industrial test, techniques based on incoherent or white-light illumination are not so restricted and, unlike holographic interferometry, they can present information concerning in-plane displacements directly.

Such methods are often simple in construction, easy to install and, in comparison with laser-based systems, less likely to interfere with the smooth running of a test programme. In addition, where high-intensity illumination is demanded, for example during the examination of civil engineering structures, white-light sources in the form of floodlights or electronic flash are not generally considered as hazardous – even natural daylight may be an adequate substitute for artificial light.

The main advantage of optical techniques over conventional displacement methods is the potential capability to give a full-field coverage. This feature may be relevant where the anticipated distribution of strain is uncertain and as a preliminary before introducing more sensitive and localized measurement methods, such as strain gauging. Even with the ubiquitous strain gauge, however, there is an alternative optical solution with a comparable sensitivity in the form of moiré interferometry.

9.1.1 The moiré phenomenon

Of all methods using white-light illumination, those based on the moiré fringe phenomenon are most common, although originally the term had no connection with measurement.

Optical Methods in Engineering Metrology. Edited by D.C. Williams. Published in 1993 by Chapman & Hall, London. ISBN 0 412 39640 8

The word moiré has been associated with an exotic silk textile that used to be imported into France from ancient China. A subtle variation in the spacing between threads, introduced during the weaving process, caused an optical interaction to appear in the form of a watered or wavy pattern. Nowadays, the moiré effect is associated with the fringe patterns that are displayed when any two geometric arrays or regularly repeated structures with nearly the same spacing are superimposed. It is no longer specific to moiré silk, owing to the widespread variety of mechanically generated components which exhibit a structure that is repetitive, for example railings, wire fences or the occasional striped shirt on TV.

Rayleigh (1874) gave the first account of the formation of moiré fringes in descriptive form, but Righi (1877) showed in graphic detail how the phenomenon could be made to reveal relative displacements between two superimposed gratings. In the field of experimental mechanics, it was not until 1948 that the potential for detecting and measuring both displacement and strain by the application of moiré was proposed and although the principles involved have changed little, there are several new and interesting developments.

As the majority of engineering measurements relate to displacements in the plane of the surface, it is appropriate to emphasize this aspect after having introduced the principles of some of the simple moiré techniques that are sensitive to out-of-plane movements.

When moiré is applied as a measurement technique, a grating is either attached physically or imaged optically on to the structure to be studied and this is called the **specimen grating**. The second is called the **master** or **reference grating**, because its properties are not modified during the structural testing process. An interaction between the two is produced either by contact, or by imaging the specimen grating on to the reference. There is a wide choice of methods for preparing the specimen grating and its type is dictated by the direction of the measurement vector.

9.1.2 Out-of-plane moiré techniques

In Chapters 7 and 8, holographic techniques for measuring out-of-plane displacements are explained in detail. Moiré provides an alternative, generally of lower sensitivity. In this case, there is no need for a special surface preparation. A grating that is either shadowed or optically projected on to the surface is suitable.

(a) Shadow moiré

Such an intensity-modulated line distribution can be generated in two ways. In the first, shown in Fig. 9.1, a shadow is cast by an obliquely illuminated flat glass reference grating located close to the surface of the specimen. From a viewpoint approximately normal to the surface of the grating, moiré fringes are visible and these are caused by interaction of the reference with its shadow.

At position A, light rays from the source are transmitted through the grating and are cast on to the surface as bright lines. The diffusely reflected line is seen through

Conventional moiré measurements 341

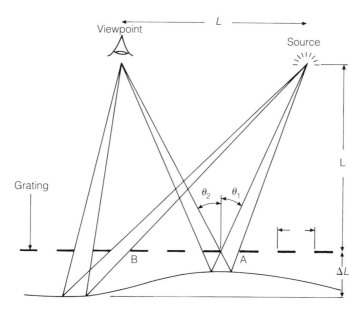

Fig. 9.1 The shadow moiré technique for generating fringes representing height contours.

an adjacent gap in the grating and so local to this region the surface will appear bright. Conversely, the line reflected diffusely at position B is obstructed by an adjacent dark line on the grating and locally a dark region or fringe is observed.

For grating period lenght f, the change in separation between the grating and surface ΔL corresponding to one fringe is related to the source and viewpoint angles θ_1 and θ_2, by

$$\Delta L = p/(\tan \theta_1 + \tan \theta_2). \tag{9.1}$$

For the special configuration shown, where the angle between source and viewpoint is 45° and they are at the same perpendicular distance from the surface, one has

$$\Delta L = p. \tag{9.2}$$

This is a convenient condition, where the moiré fringes represent height contours whose fringe interval is the same as the pitch of the reference grating.

The technique can be employed to determine absolute shape, or to monitor any changes in shape of a diffusely reflecting surface. It is simple to implement; a small filament lamp will provide adequate illumination in most instances. In order to preserve sharp shadows, it is advisable to align the lamp filament with the grating lines.

If the component is curved steeply then the separation between grating and surface will generate a large number of contour fringes. In addition, the shadows will

342 *Moiré methods in strain measurement*

become blurred if the gap is excessive. For such cases, it is practical to employ a curved grating whose shape approaches that of the component (Marasco, 1975).

(b) Projection moiré

Where the implementation of a glass grating in shadow moiré is impractical, for example with large structures or restricted access, the comparable technique of projection moiré is appropriate. Here, the optically generated surface pattern is derived from an obliquely projected image of a fine grating. The surface pattern is then imaged by a camera at a position corresponding to the shadow moiré viewpoint (see Fig. 10.8). Fringes are produced by comparison of the live image with a pre-recorded master grating placed in the focal plane of the camera.

Much of the published literature on applications of shadow and projection moiré deals with biostereometrics, especially in the area of detecting and correcting scoliosis (Pekelsky, 1986). This approach has proved so successful in Canada that nationwide screening for spinal deformities in five-year-old children has been adopted. Some novel early work (Miles and Speight, 1975) describes an application to measure the meat content of cattle.

(c) Reflection moiré

If the surface is specularly reflecting then the technique of reflection moiré becomes suitable. In this, the interaction between the reference grating and its distorted appearance when reflected by the surface generates fringes which are contours of partial slope. Details of this somewhat specialized process and its applications in experimental mechanics can be found in Ritter (1982).

9.1.3 Imaged moiré

In exceptional circumstances, the structure may be so large that even projection techniques are not feasible, because of problems with illumination. One such example, reported by Kearney and Forno (1989), and probably the largest moiré experiment ever conducted, was the study of buckling behaviour inside a 30 m diameter liquid methane storage tank belonging to British Gas. Here the measurement area, consisting of one-third of the circumference of one of the ring bays, exceeded 30 m in length and 3.5 m in height (Fig. 9.2).

Since the anticipated mechanism of buckling involved little in the way of in-plane displacements, the usual projected pattern was replaced with a pattern of coarse stripes printed on paper. This was pasted with PVA adhesive to the inside surface of the tank. On scaffolding positioned close to the wall of the tank, three camera systems were mounted 10 m apart so as to view the whole of the papered section. Each consisted of a 35 mm single-lens reflex camera mounted in front of and in line with a field lens and video camera (Fig. 9.3). An initial or pre-test condition of the patterned surface was recorded on high-contrast film. A sharply resolved image is achieved

Conventional moiré measurements 343

Fig. 9.2 5000 tonne liquid methane tank (British Gas).

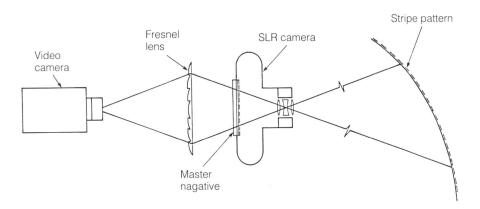

Fig. 9.3 Camera system for viewing internal deformation of tank.

over the full width of the focal plane if it is aligned according to the Scheimpflug condition, which requires that the object and focal planes should intersect in the plane which passes through the centre of the lens and is normal to its axis. The negatives were processed and mounted, emulsion outward, on to glass slides.

Each negative was replaced in its corresponding camera focal plane so that, when viewed through the camera back, moiré fringe patterns were seen from the superimposition of the negatives and real images of the pattern. The video cameras, behind the still cameras, allowed remote inspection and recording of the fringe patterns from the safety of a cabin outside.

344 Moiré methods in strain measurement

As the internal pressure of the vessel was reduced, spherical bulges began to develop at intervals along the inside wall. Figure 9.4 shows an example of a fringe pattern recorded close to the termination of the test, at the lowest pressure state. The wall is almost completely filled with large, spherical bulges whose radial deformations, in some places, exceed 40 mm. Any further reduction in pressure would have resulted in the joining of adjacent bulges and probably a dramatic collapse of the tank.

In this experiment, the displacements Δr were calculated from measurements on the fringe pattern using a relationship similar to equation (9.1), namely

$$\Delta r = p/\tan(\cos^{-1} Hf/2Rh), \qquad (9.3)$$

for one fringe, where p is the pattern pitch and f the camera lens focal length. R is the radius of the cylinder, H the actual height of the wall and h the local image height.

9.1.4 In-plane measurement analysis

A comprehensive treatment of the formation of moiré fringe patterns can be complex, but in comparison with other optical techniques the actual derivation of the appropriate engineering parameters is exceptionally simple.

When using moiré for the determination of in-plane displacements and strains, movements of the surface are correlated with movements in a pattern that is physically applied to the specimen, for example by printing on to its surface. Fringes are produced by superposition of a reference grating.

Figure 9.5 demonstrates the formation of a moiré pattern of this type. If we refer to the line separation or pitch of the specimen grating as p_1 and the reference grating pitch as p_0, then it can be shown by geometry that the perpendicular spacing of the fringes resulting from the superimposition of the two is

Fig. 9.4 Example of a moiré pattern showing internal spherical dilation.

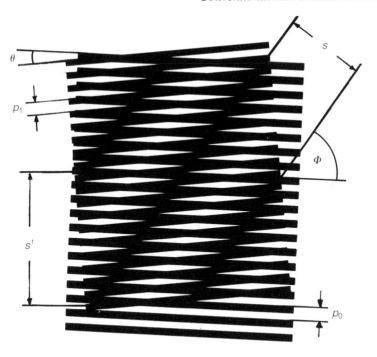

Fig. 9.5 The formation of moiré fringes.

$$s = p_0 p_1 / [(p_0 \sin \theta)^2 + (p_0 \cos \theta - p_1)^2]^{1/2}, \tag{9.4}$$

where θ is the angle between the gratings (see, for example, Chiang (1978) or Durelli and Parks (1970)).

The orientation of the fringes with respect to the reference grating is given by

$$\sin \phi = p_0 \sin \theta / [(p_0 \sin \theta)^2 + (p_0 \cos \theta - p_1)^2]^{1/2}. \tag{9.5}$$

In most practical situations, the gratings are aligned with θ close to zero, when equation (9.4) reduces to the simple expression

$$s = p_0 p_1 / (p_0 - p_1). \tag{9.6}$$

An alternative, directly related expression uses the terminology of **spatial frequency**, which introduces the parameter of lines per unit length on the gratings u_0, u_1 and the number of moiré fringes u_m per unit length, where $u_0 = 1 / p_0$, etc. Equation (9.6) then becomes

$$u_m = u_0 - u_1, \tag{9.7}$$

which simply states that the moiré fringe frequency is the difference between the spatial frequencies of the gratings.

(a) Linear strain

For the determination of strain, the pitches of both reference and specimen gratings are generally assumed to be identical with a value of p_0. Then, under the application of some deformation process, the specimen grating changes to p_1. The linear strain measured along a direction normal to the grating lines, the **principal direction**, is then given by

$$\epsilon^L = (p_1 - p_0) / p_0. \tag{9.8}$$

Here ϵ^L is the Lagrangian description of strain, relating the extension to the initial length. However, measurements are carried out on the moiré pattern as it develops and the Eulerian description of strain referred to the final length is therefore applicable, namely

$$\epsilon^E = (p_1 - p_0) / p_1. \tag{9.9}$$

Combining equation (9.9) with equation (9.6), we have

$$\epsilon^E = p_0 / s. \tag{9.10}$$

The distinction between ϵ^E and ϵ^L is significant only where large strains are involved in which case, as a preferred engineering measurement, it may be necessary to convert strain values using the relationship

$$\epsilon^L = \epsilon^E / (1 - \epsilon^E). \tag{9.11}$$

For the linear strain derivation from crossed gratings, it is necessary to measure the spacing of the fringes in directions corresponding to the principal directions. Consequently, equation (9.10) is more commonly expressed as

$$\epsilon_{xy} = p_0 / s_{xy}. \tag{9.12}$$

(b) Shear strain

Apart from linear strain, values for shear strain can be obtained directly by measurement of the angle of rotation θ, which is derived from the fringe spacing s' in a direction parallel to the grating lines. For small angles and when p_0 is close to p_1, provided that there has been no rigid-body rotation, equation (9.4) reduces to

$$s' = p_0 / \theta, \tag{9.13}$$

and hence the shear strain is given by

$$\gamma = p_0 / s'. \tag{9.14}$$

In practice, the orientation of the reference grating cannot be guaranteed and for a unidirectional grating, any rigid-body rotation may be misinterpreted as a shear effect.

The use of crossed orthogonal gratings for both object and reference will resolve the ambiguity. Any rigid-body rotation will generate two similar, superimposed crossed fringe patterns and from the fringe spacings s_x, s_y measured in orthogonal directions, the absolute shear strain is determined as

$$\gamma = p_0 \, (s'_x + s'_y) / s'_x s'_y. \qquad (9.15)$$

It is important to note that there is a sign change for either s_x or s_y so that the rigid-body motion component is eliminated. In addition, the directions of measurement of s'_x and s'_y are normal to the corresponding principal directions of the gratings, i.e. s'_x is the spacing of fringes measured along the x-direction from a grating with lines parallel to the x-direction.

9.1.5 In-plane measurement techniques

The following section gives some examples of metrological applications of moiré involving in-plane deformation. Again, there are alternative methods for applying different surface patterns and ways of interrogating these by comparison with a master pattern.

(a) Contacting grating

The most direct method uses a reference in contact with the object grating (Boone *et al.*, 1982). This can take the form of a stencilled pattern (see section 9.4.2), or other light-weight surface layer which does not influence the material properties of the component. Sometimes referred to as **transmission moiré**, this method has advantages, including the possibility of real-time measurement, but also some limitations, for example a relatively small inspection area defined by the grating dimensions. In addition, errors can be introduced by imperfect contact between the gratings. Nevertheless, the method is adequate for quantifying strain beyond the elastic region and the formation of the moiré pattern is both simple and direct. If the three-dimensional shape of the component is modified during testing then the reference grating can be supported on a flexible film so as to accommodate the new form (section 9.4.2(*c*)).

(b) Self-imaged grating

If the object is flat, then the master need not be in contact in order to generate good contrast fringes. At discrete planes, separated from the surface by a distance d, the reference grating is self-imaged according to the relationship

$$d = np^2/\lambda, \qquad (9.16)$$

where n is an integer and λ is the mean wavelength of the light. A specimen grating in coincidence with one of these planes will interact with this self-image to produce a moiré fringe pattern. Although the plane closest to the grating, corresponding to $n = 1$, offers the best fringe visibility, discernible patterns can be resolved at more distant planes.

(c) Photographic moiré

In difficult conditions, for instance where the object is heated, the interaction between gratings can be achieved by relaying an image of the specimen grating so that it is in superposition with the master. Theocaris (1964) described an imaging technique for studying elasto-plastic strains in steels. He used a relatively coarse grating (8 lines/mm), photo-etched on to the surface. A similar process has been described in section 9.1.3 for the measurement of out-of-plane displacements, but the approach is more relevant to in-plane displacements in conjunction with the technique of moiré photography.

If a photographic record is made of the specimen grating in its initial state, then this can be used as a reference grating with which subsequent records of deformed states are compared.

9.2 HIGH-RESOLUTION MOIRÉ PHOTOGRAPHY

An apparent omission in the library of techniques for deformation measurement was revealed following the disastrous collapse of the London high-rise building, Ronan Point, and the failure during construction of a box girder bridge at Milford Haven. At the time, classical methods such as surveying or levelling were providing point-by-point information about displacements, but no method existed for monitoring small displacements over the complete surface of large objects.

This omission did much to provide the impetus for developing a low-cost high-resolution moiré photography (HRMP) technique especially for use in situations involving the testing of large civil engineering structures.

Since the sensitivity of the conventional photographic moiré method is proportional to the pitch of the surface pattern, there should be an advantage in using a camera that exhibits a high resolution over the complete image plane, even if the object is deep or curved. Unfortunately, for a conventional camera this presents a serious difficulty.

A camera lens fitted with a circular iris has a depth of field that varies inversely as the square of the numerical aperture, but because of diffraction, the aperture cannot be reduced without degrading the resolution. A compromise therefore has to be made between resolution and field depth. Consequently, the conventional form of moiré photography is usually associated with a low sensitivity to displacement. The problem can be solved, however, by using a simple mask installed inside the camera lens.

Burch and Forno (1982) deal comprehensively with the theoretical performance of masked apertures; only the basic properties will be described here. Perhaps the best-known example of aperture masking is Michelson's stellar interferometer, in which the front of a large telescope is covered by two widely separated small apertures. The

High-resolution moiré photography 349

main advantage of this instrument is the excellent resolution which it provides in one direction over a narrow range of spatial frequencies; its second property, namely its increased depth of focus, is hardly of interest to the astronomer. When a masked aperture is used in conjunction with a camera lens, however, both these properties offer distinct advantages over the unobstructed circular aperture.

Most cameras need to achieve good image contrast over a wide range of spatial frequencies, but a much more selective response may prove quite satisfactory for moiré grid photography, because all that is necessary is to record faithfully those frequencies that are contained in the periodic grid pattern. The 'tuning' of the response must not be made too sharp, however, because the pattern is not perfectly regular – indeed, it is the departures from regularity which generate most of the desired information displayed as moiré fringes.

9.2.1 Tuning the lens response

(a) Slotted apertures

The method of modifying the camera lens response, so that only a selected bandwidth of spatial frequencies is resolved, involves replacing the conventional circular aperture with an opaque mask bearing two orthogonal pairs of slots (Fig. 9.6). Apart from reducing the effective aperture, the presence of the slots alters the modulation transfer function of the lens. The standard triangular form for a circular aperture,

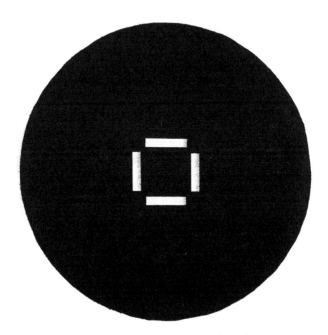

Fig. 9.6 Slotted masked aperture.

350 *Moiré methods in strain measurement*

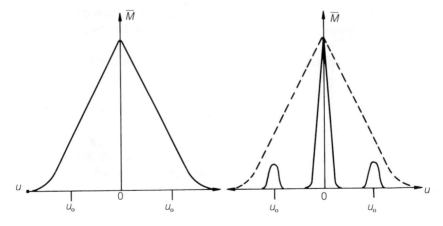

Fig. 9.7 Optical transfer function $\bar{M}$ for circular and slotted apertures.

covering a range of spatial frequencies, is converted to a narrow triangle centred at zero spatial frequency, together with two smaller side bands. These side bands are centred on the chosen tuned frequency with no response to intermediate frequencies. For polychromatic light the response is somewhat reduced in amplitude, but remains selective (Fig. 9.7).

With a slot separation d, the tuned frequency of u_0 lines per millimetre resolved by the lens-mask combination is simply

$$u_0 = d/R\lambda, \tag{9.17}$$

where λ is the effective wavelength set by the spectral response of the film and the predominant wavelength of the illumination – for electronic flash illumination and the recommended film, Agfa Gevaert 10E56, λ is centred on 500 nm. Here R is the distance from the exit pupil to the image plane which, for the majority of situations other than close-range photography, is the same as the focal length.

In the object field, there is a corresponding tuned distance D at which a specimen grating of pitch p is resolved, given by

$$D = pd/\lambda. \tag{9.18}$$

So that the tuned frequency is not critically selective, the width of the slots is made approximately one-seventh of their separation. This also allows for a tolerance of almost 10% on other parameters, including grating pitch, distance from the camera and an angular misalignment of the grating by approximately ±7°.

(b) Depth of field

In the object space, the corresponding depth of field depends on the nominal target distance and the focal length of the modified camera lens. Calculations have been

High-resolution moiré photography 351

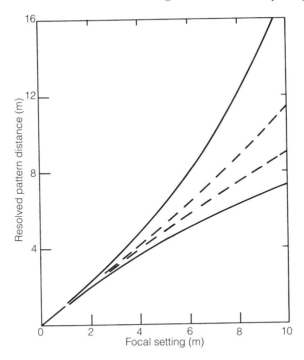

Fig. 9.8 Depth of field of masked lens compared with the same diameter conventional aperture (dashed).

made for the recommended Asahi Pentax $f/2$ Takumar lens for which the measured focal length is about 57 mm. The results, plotted in Fig. 9.8, confirm that the double-slot system provides the advantage of a depth of field of the order of three times that from a conventional aperture.

(c) Camera lens modification

Through extensive testing of different modern camera lenses, the most satisfactory objective, in terms of total image format coverage at 300 lines per millimetre, is the standard Pentax 55 mm $f/2$ Super Takumar. Unfortunately, this lens is no longer produced, although it is not difficult to find in second-hand camera shops. If it cannot be obtained, the masking of an alternative lens should not be carried out until a check has been made for astigmatism (Burch *et al.*, 1974).

In calculating the dimensions for an aperture to be embedded inside the lens, from equation (9.18), a scale factor of 0.695 is applied to both slot width and separation to compensate for magnification effects produced by the first three elements of the Takumar. The final four-slotted aperture, shown to a scale of 2.7:1 in Fig. 9.6, has a slot separation of 5.9 mm and is designed for use with object distances of more than

352 *Moiré methods in strain measurement*

Fig. 9.9 Pentax camera fitted with modified Takumar lens.

1 m. It can be made from thin paper sprayed matt black. Slots can be cut simultaneously in pairs by means of two pairs of bonded knife blades glued together and separated by a spacer.

With the Takumar lens it is not difficult to obtain access to the iris diaphragm. By unscrewing the slotted ring at the rear of the lens, the back elements can be lifted clear as a single assembly, exposing the iris. The diameter of the mask is 29.5 mm so that it rests on the shoulder of the iris mount clear of the iris leaf sprockets. After orientation, the mask can be located with adhesive at a few points around the periphery. Figure 9.9 shows the lens with its mask mounted on the Pentax camera.

In this mask design, the tuned response is centred around 300 lines per millimetre, with a bandwidth of 30 lines per millimetre, and this offers an approximate order of magnitude higher sensitivity to displacement compared with conventional moiré photography. For certain standard camera lenses exhibiting good freedom from astigmatism, the tuned response can be made as high as 600 lines per millimetre.

9.2.2 Photographic recording

(a) Recording material

When the technique was originally developed, only a few 35 mm films were available with sufficiently high resolving power. Currently, this is still the case and Agfa Gevaert 10E56 remains the most suitable photographic material for recording the 300 lines per millimetre detail resolved by the modified Pentax camera. The emulsion is sensitive in the blue region of the spectrum (hence the effective wavelength of 500 nm stated in the previous section) and exhibits a speed equivalent to approximately 2 or 3 ASA when processed in high-contrast, speed-enhancing developer. Unlike other conventional 35 mm films, the 10E56 emulsion is coated on to a polyester base and this provides the best dimensional stability, other than glass.

There are, however, one or two features that affect the accuracy obtained with high-resolution moiré photography. They are related to characteristics of the recording film. First, there is a time-dependent displacement of the film within the camera and second, an effect of humidity on the substrate material. A third effect, which is apparent only over long-term investigations, is latent image regression, whereby the recorded image fades progressively while the film remains unprocessed (Forno, 1980).

A method for reducing the first two effects and improving the mechanical stability of the film uses a transparent registration glass 1 mm thick mounted in the image frame, on to which the film is clamped. The pressure plate inside the rear cover of the camera is fitted with a threaded rod which passes through the back of the camera. A nut on the outside provides a means of alternately clamping and releasing the springs on the pressure plate while exposing and then winding on the film. By covering the pressure plate with thin black cloth, a uniform pressure over the film is achieved.

Humidity and thermal effects on the film can produce an overall contraction or expansion of the recorded image which could be misinterpreted as change in the overall strain level. As it is not always practical to control humidity during an experiment, a method which compensates for it has been adopted. A reference pattern is positioned alongside the structure that is being examined. Any humidity changes are then revealed as expansion fringes on the reference surface, and a compensation can then be applied to the measured deformation of the structure.

(b) Exposing and processing the film

Illumination requirements when studying structures up to a maximum size of 5 m can be met with the typical 1.5 kJ output of conventional studio flashlamps. Two are usually required, positioned on either side of the camera to avoid specular reflections from the surface of the structure. Alternatively, continuously operating floodlights or, for studies conducted in the open, natural ambient daylight can be adequate. It is important, however, to confirm that no displacement or vibration occurs during the exposure time of about a second required for daylight.

Recommended development of 10E56 is in a high-contrast developer such as Kodak D19, after which the film is rinsed, fixed and washed. The black and white amplitude image is then converted to a phase image by bleaching, using Kodak R10 (stock solutions A: ammonium dichromate 20 g, sulphuric acid 14 ml, water to 1 litre: B: sodium chloride 45 g, water to 1 litre. Use: 1 part A + 1 part B + 5 parts water). This treatment enhances the diffraction efficiency of the recorded grating structure and ameliorates the effects of density variations caused by non-uniformity of the object pattern or the incident or reflected illumination. As a precaution against a darkening of the bleached image which occurs with prolonged exposure to bright light, the negative needs to be stabilized in a dilute solution of potassium iodide. Processed films should be cut into manageable lengths about seven frames long and kept flat, wrapped in paper.

354 *Moiré methods in strain measurement*

9.3 APPLICATION OF MODIFIED CAMERA

There is a variety of surface patterns which are tailored for different types and sizes of engineering structure. Figure 9.10 indicates the optimum tuned distance for the camera together with the corresponding field coverage for different specimen gratings.

The pattern usually consists of a regular array of dots running horizontally and vertically across the surface, and at the tuned distance it will be demagnified to appear as 300 lines per millimetre in the image plane. Photographs are then made of the structure whilst it is being deformed. Each negative record can take the form of a double exposure of two different states, or individual states can be recorded on separate film frames.

After the film has been processed, the recorded pattern detail, in the form of two orthogonal phase gratings, is examined in a spatial filtering system. This enables the moiré fringe information to be isolated as the two first-order diffraction patterns. Each negative generates an x- and a y-displacement map, dark fringes appearing where movements of odd multiples of one-half the pattern pitch have occurred, so as to cancel the resolved image of the grating in these regions.

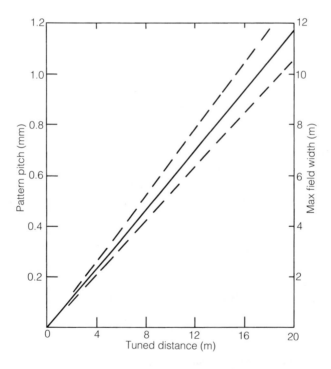

Fig. 9.10 Tuned distance and field coverage for object patterns of different pitch.

9.3.1 Spatial filtering of moiré fringe patterns

(a) Double exposures

Spatial filtering systems allow a separation of the pattern on the surface of the object from surrounding unwanted detail. For the double-exposure records, the direction of illumination required for the separation process is not critical. Monochromatic light or white light that has been filtered spectrally can be used. A schematic representation of a system is shown in Fig. 9.11. In this, the image of the surface pattern is viewed in the light diffracted by either the x-oriented array in the specimen grating or the y-array. One or other can be selected by rotating the prism or negative through $90°$. The prism also serves the purpose of redirecting the diffracted light along the axis of the system, so that the optical components can be mounted on a straight bench.

(b) Single-exposure pairs

With the comparison of two single-exposure records in superposition there may be some phase errors introduced into the moiré pattern by parallax unless a collimated illumination system is used. Such an arrangement, shown in Fig. 9.12, is also relatively simple in construction but demands critical alignment of the negatives in a jig having x- and y-translation, together with rotation between the negatives about the optical axis.

Although the use of separate exposures may entail some loss of accuracy, because of dimensional changes in the negatives, it does offer important advantages. Not only does the technique generate deformation data from any two recorded states, but the direction of movement of the fringes during alignment defines unambiguously the

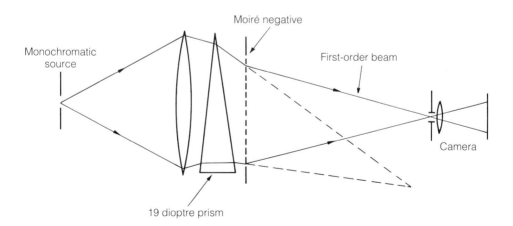

Fig. 9.11 System for reconstructing moiré patterns from a double-exposure negative.

356 Moiré methods in strain measurement

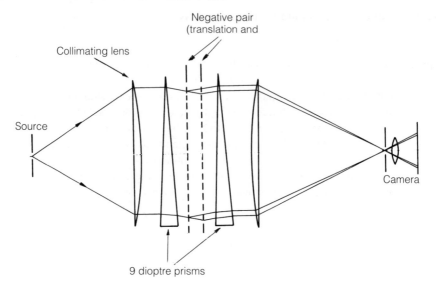

Fig. 9.12 Reconstruction system for a pair of single-exposure negatives.

direction of the deformation. For example, in examining the x-displacement map, a small clockwise rotation is given to the second state negative with respect to the reference state. If the moiré fringes appear to move towards the negative y-direction when the second state is displaced in the positive x-direction, then tension is indicated. In addition, this procedure can be used to record a deformation programme of an indefinite length, which might include the removal of the camera and its replacement at a later stage. Errors in the repositioning of the camera usually introduce straight fringes, due to changes in the demagnification factor of the lens and for these a simple compensation can be applied at the analysis stage.

Another virtue of single-exposure comparison is that it provides a very sensitive means of detecting the presence of cracks when the fringes are manoeuvred slowly across the field of view. The cracks are conspicuous as small phase discontinuities on either side of a fringe minimum.

9.3.2 Fringe-pattern analysis

(a) Automatic analysis

In automatic analysis of the moiré fringe information by the method of phase-stepping, described in the previous two chapters and section 9.6.3, phase shifts can be introduced by in-plane translation of one of the negatives. This approach may suffer from frictional effects between the negatives, but a more precise control of the phase steps is obtained by tilting the complete negative sandwich through small calibrated angles. A slight foreshortening of the aspect ratio of the image produced by the tilt does not appear to introduce significant errors.

(b) Manual analysis

A simple device for analysing moiré fringe patterns can be made from a piece of thin, transparent plastic sheet. This is a template marked with a series of cursor lines on either side of a small central hole which is used with a hard copy of a fringe pattern (Fig. 9.13). The separation between lines is preset to a specified fringe spacing corresponding to a given strain level, derived from equation (9.6), and the hole provides a way for marking on the print the mean position for a point on the strain contour. By manoeuvring the template across the print whilst attempting to fit the marked lines to match adjacent fringes, contours of different levels of strain can be defined over the complete image.

(c) Range of strain measurement

At 300 lines per millimetre, there are approximately 10^4 lines resolved and recorded along the x-direction. If the complete width is utilized as the equivalent of a single strain gauge, the fundamental sensitivity, corresponding to the presence of two moiré fringes on either side of the field, is $1 : 10^4$ or 100 microstrain. Clearly, using the whole field as a single gauge does not provide information about local deformations. As a rule of thumb, the field is divided into 10 units and so the corresponding minimum sensitivity is $1 : 10^3$ or 1 millistrain.

Manual interpretation, either by digitization or the template method, allows an estimation to 0.1 of a fringe. This results in a minimum strain level accuracy of 100 microstrain over the subdivided field. With automatic analysis, the phase determination is in the region of 0.02 of a fringe, which implies a regional strain resolution of 20 microstrain. However, instability of the recording film may produce errors between frames and this figure cannot be justified over distances greater than 4 mm on the image. The maximum level of strain that can be accommodated by HRMP is limited to the appearance of approximately 100 fringes across each subdivided field

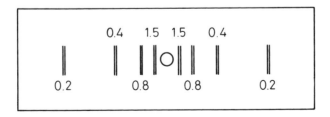

Fig. 9.13 Representation of a template used for manual analysis of strain.

358 *Moiré methods in strain measurement*

area, which corresponds to a level of 10% strain. A higher density of fringes will inevitably lead to some confusion.

9.3.3 Examples of engineering structures

Many different structures have been studied during the course of the development of HRMP. They fall into three categories:

1. Small structures, 20–150 mm wide, which include nuclear reactor components and welded joints, tested at high temperatures;
2. Small-to-medium structures up to 1.5 m wide at ambient temperatures, including plastics and composite materials and small masonry units;
3. Large-scale objects, 1.5–30 m wide, including civil engineering structures such as buildings, bridges and pressure vessels.

Each type has required a modified approach, largely as a result of the differing material properties. For Type 3, for example, because the influence of a surface coating is negligible it may be quite thick, such as a printed pattern on paper; whereas for a soft plastic material, a tenuous pattern is more appropriate. It is also important to use as much of the film format as possible so as to gain the highest displacement sensitivity in proportion to the size of the structure.

Before detailing all the current surface-preparation techniques, an example of an application to study the deformation of a wooden structure will be described briefly.

9.3.4 Deformation measurement in a timber structure

Work in the Institute for Stress Analysis at EPFL, Lausanne, has shown that the method is ideally suited to the study of structures constructed from wood. This results from two properties of timber. It can be subjected to a relatively high level of strain without suffering permanent deformation and, because of its anisotropic fibrous nature, a non-uniform behaviour will often occur, sometimes resulting in particularly interesting fringe patterns.

One of the structures examined is shown in Fig 9.14. It consisted of two short horizontal beams or joists of solid pine supported by a central column, approximately 250 mm wide. The column was of multi-layered pine construction, with the wood grain running perpendicular to the front surface. Resting on cutaway shoulders on the column, the beams were linked across the top by a steel strap attached with nails. On either side, two hydraulic rams fitted with load cells provided a clockwise and anticlockwise couple to the beams about the centre up to a load of 20 kN.

For this study, the required surface pattern took the form of contact copies of a 20 lines per millimetre crossed grating, produced photographically by the method described in section 9.4.1(*c*). A standard PVA wood glue was used to attach the contact copies over the central T region of the structure. Illumination was provided by a 1 kW quartz-iodine spotlight and, with the modified camera at approximately

Fig. 9.14 Test applied to a wooden 'T' structure – prior to the application of the surface pattern.

1 m from the surface, records were taken over an exposure time of 4 s (2 s for each state of a double exposure).

For loads up to a maximum load of 20 kN, a sequence of records was made which included the final unloaded condition. Negatives were processed and examined in the appropriate spatial filtering system from which photographs were taken of the x- and y-displacement fringe maps. An example of the reconstructed pair derived from the comparison between the unloaded condition and the state at approximately 10 kN is shown in Fig. 9.15.

There is a vast amount of information which can be interpreted visually from such patterns concerning the structural and mechanical properties of the wood. Some discrimination is therefore required in providing only relevant information to the engineer. Values for the x and y strain, as well as shear strain, were determined locally by the template method across the central column only, there being little evidence of deformation in the beams. Compressional x-strains below the regions of contact of the beams with the column, displayed in Fig. 9.16, rise steeply and are responsible for the small Poisson's ratio y-extension appearing in the top half of the column.

As the load was increased, the displacement patterns became complex, showing considerable rotation of the beams without significant strain (Fig. 9.17). In the column, the measured strain exceeds 2% and the fringes display an interesting feature

360 *Moiré methods in strain measurement*

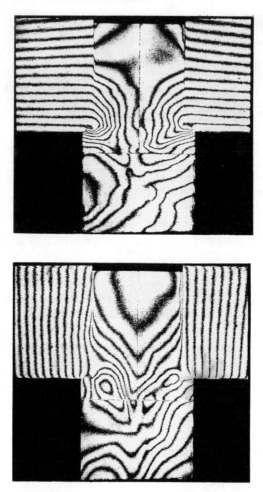

Fig. 9.15 *x*- and *y*-displacement maps of the 'T' structure under load (one fringe represents 50μm displacement).

of slope discontinuities running through the central sections. Not only do these follow the joints between the individual sections of wood blocks, but they also run along lines emanating from the corners of the shoulders. They could be considered as slippage planes parallel to the layered fibrous structure, resembling lines of dislocations in crystals.

Such features are not unusual throughout the many studies carried out using the technique. Rather than presenting the details of different studies, emphasis will be given to the numerous approaches adopted for particular structures.

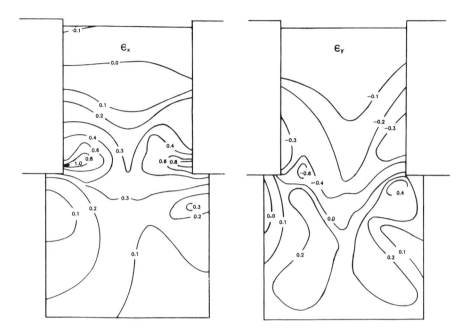

Fig. 9.16 *x*- and *y*- strain distributions over the central column.

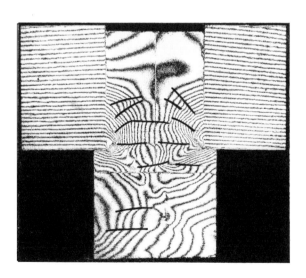

Fig. 9.17 *x*-displacement pattern at a load of 17 kN.

9.4 TYPES OF SURFACE PATTERN

The type of pattern used on timber is clearly unsuitable for use on structures such as large buildings or metal components at high temperatures. However, it is its versatility that makes the technique attractive in all fields of engineering. Using the identical recording system, only the surface pattern is changed for different structures. The various materials and their compatible patterns are given in Table 9.1, while Table 9.2 summarizes the pattern types and their resistance to unusual environmental conditions. By defining a particular material and test condition, the appropriate pattern may be identified.

9.4.1 Printed patterns

(a) Paper

The most effective method for treating large areas of a structure surface is by wall-papering with paper printed with an array of dots. The pattern is derived from a negative, commonly referred to as a **film tint**, obtained from a printing supplier (Haynes Graphic Arts, Station Road, New Malden, Surrey KT3 6JF, England).

It is economic to produce ink-printed litho copies of the tints onto sheets of paper (A3 size or larger and of weight 120 gsm). A range of rulings is available from 2.5 to 6 dots per millimetre, but for a coarser pattern intended for structures 10 m wide a dot pattern negative has been made from a sheet of wire mesh.

Table 9.1 Patterns associated with different materials

Metals	Masonry	Plastics and composites	Others
Non-ferrous 1–10	Concrete 1–4	PVC, etc. 5, 10	Textiles 11
Steel 1–10, 12	Brick 1, 2	Carbon fiibre 6	Card 4
	Timber 1–3	Rubber 5, 9	Graphite 5
	Clay 5		Skin 4

Key:
1. Printed paper patterns;
2. Printed paper patterns subjected to adverse weather conditions;
3. Photographic prints;
4. Direct printing;
5. Stencilled patterns;
6. Protected stencilled patterns;
7. Explosive-resistant patterns;
8. Random patterns – scratched;
9. Random patterns – retroreflective paint;
10. Stripping film;
11. Untreated materials.

Table 9.2 Environmental resistance of patterns listed by test condition

Internal	External	High temp or ambient	Abrasive	Explosive.
1	2	–	–	–
3	3	–	–	–
4	4	4	–	–
–	6, 7	5	6	7
–	8	8	8	–
9	9	–	9	–
10	10	–	10	–
11	–	–	–	–

Key as in Table 9.1.

Standard paperhanging paste is a satisfactory adhesive for most surfaces. Adhesive is applied to the structure first, followed by the sheets of dampened paper. This procedure allows for a certain amount of manoeuvrability in positioning and produces an intimate contact with the surface. As an aside, the author has found that this approach works well in the domestic situation for hanging all forms of wall and ceiling papers.

Figure 9.18 shows the Victorian canal bridge of section 4.6.2 prepared in this way, together with an example of a y-displacement map derived from a load test of the structure. The fringes represent displacement contours of 0.83 mm with small movements conspicuous in the main piers, as well as large vertical displacements in the span. From this study, the complete sequence of load states up to 2300 kN was examined. One record at a load of 200 kN revealed the initiation of cracking which eventually led to failure of the bridge.

Water-based adhesives are not recommended for use on steel unless the surface has been painted first with a primer, otherwise rust stains will mark the surface of the paper.

(b) Protection from adverse weather conditions

Paper coatings, prepared as described in the above section, may be afforded some protection against the elements with a proprietary transparent damp-proofing or water-proofing liquid such as Polycell Dampcure. Figure 9.19 shows a 15 m wide facade of a pair of houses that were wallpapered and then protected, for an investigation into subsidence of the building over a three-month period.

Many other large structures have been prepared with printed paper, from civil-engineering structures, including components for the Hong Kong and Shanghai Bank in Hong Kong, to small components such as conventional pavement slabs.

364 *Moiré methods in strain measurement*

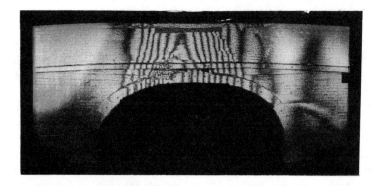

Fig. 9.18 Unprepared brick-built Victorian bridge and a vertical displacement map recorded at a heavy load (one fringe represents 0.83 mm displacement).

(c) Photographic prints

Where the required dot density is between 6 and 40 per millimetre, printing does not produce consistently good reproduction. Contact copies of crossed gratings on resin-coated photographic paper will transfer the finer detail faithfully if the emulsion is separated from the paper base. As the plastic-coated emulsion layer is thin, shearing effects through its thickness do not occur and the true deformations of the underlying structure are reproduced in the surface pattern.

The thickness of the emulsion layer can be controlled as it is peeled away by altering the angle between it and the supporting base. If the paper is held vertically and the emulsion layer is separated whilst maintaining an angle greater than 90°, the

Fig. 9.19 Facade of a house treated with patterned paper and tested for subsidence.

emulsion can be removed with a minimum amount of attached paper. This layer can be bonded directly on to the surface of the structure, or the remaining paper fibres can be removed completely by washing (Sevenhaijsen, 1981).

Photographic copies of a 20 lines per millimetre pattern were employed in a study of yielding in the cylindrical walls of a mild-steel chemical containment vessel which was subjected to a drop test. A range of strain values from 0.01% to 10% was revealed, which demonstrates the effective adhesion of epoxy resin with this type of pattern.

(d) Photographic stripping film

A finely detailed pattern can be made by contact printing a grating on to stripping film and bonding the emulsion layer to the component. It has the disadvantage of a carrier layer and some additional preparation is necessary to enhance the pattern contrast, but the coverage can be considerably greater than that produced by stencilling.

Huntley and Field (1990) have reported the use of HRMP in high-speed deformation studies. Patterns at 150 lines per millimetre, prepared on stripping film, were attached to polymethylmethacrylate specimens with cyanoacrylate adhesive. The modified camera lens used was an Olympus 80 mm $f/4$ macro objective.

During the photography, steel ball projectiles were fired at the edges of the specimens and high-speed electronic flash was triggered at approximately 1 μs intervals following impact. As the recorded fringe patterns represented frozen states, without the possibility of applying fringe displacement analysis methods, the authors carried out automatic fringe analysis by the Fourier transform method (see section 10.2.3).

(e) Direct printing

Some materials, such as thin card, will not accept a surface layer without a significant modification of their mechanical properties. Others may require intermittent inspection of the surface during testing, for example in order to monitor the progress of cracks in concrete. In these cases a direct printing method is appropriate. Card can be litho printed with an ink pattern, or a rubber stamp may be used on concrete. Although the individual printed areas may be small when produced by this method and appear somewhat piecemeal, the quality is improved by the recording process (see section 9.5.4).

9.4.2 Stencilled patterns

(a) Production

Where a structure is small and composed of a soft or elastic material, a fine pattern which is free from any supportive layer will allow the uninhibited transfer of the material strain through to the surface. Such a pattern can be produced by means of stencilling through one of a range of fine metal meshes available from Buckbee Mears Co., St Paul, MN 55101. The mesh, which in its least expensive form is made of nickel, is fixed temporarily to the object with a non-drying adhesive such as diluted detergent solution. After it is registered, the excess fluid is removed with a tissue and once the surface appears to be dry it is ready to be sprayed with a white pigment.

For both ambient and high-temperature applications, a suitable stable pigment is titanium dioxide (BTP Tioxide Ltd, Cleveland, England). A suspension of an 'organically' treated version is made up in ethyl alcohol in an approximate weight ratio of 1:10. Spraying is carried out with an airbrush so as to produce a fine controllable mist and until the surface appears completely white. When the mesh is gently lifted away, the pattern is revealed as white blocks of pigment in relief against the surface (Fig 9.20). To improve the pattern contrast and the perfection of coverage, it is important to make sure that the initial surface finish of the specimen is smooth and polished and the illumination used during recording is directed obliquely at the surface.

At ambient temperature, the stencilled pattern method has been employed in the characterization of low-modulus plastic materials (Dean *et al.*, 1984). In this way, the influence on mechanical behaviour through local heating effects and the added strength of the bonding adhesive when using a conventional resistance strain gauge are completely avoided, providing for a high level of confidence in the calculated levels of strain. Moreover, the local strain involved in testing plastics is generally in excess of 0.1% and so, as with timber, the resulting moiré patterns are often dramatic and can be interpreted with relative ease.

One high-temperature application of this pattern was in the measurement of total mechanical and thermal strain in welded nuclear reactor components. Here, a welded

Types of surface pattern 367

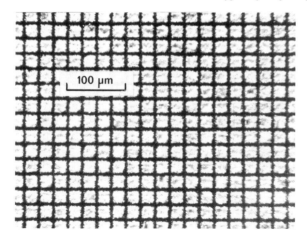

Fig. 9.20 Appearance of stencilled pattern with 40 lines per millimetre.

plate was taken through its operating temperature range of 20 °C – 550 °C. Analysis of the pair of reconstructed fringe patterns indicated that severe shearing stresses were introduced at the weld interface by the temperature change (Webster *et al.*, 1981). Trials above 800 °C have been successful and there is unpublished information indicating that stencilled patterns have withstood 1500 °C.

(b) Protected stencilled patterns

The following protective treatment is intended for use where only a mild abrasion resistance is required. During the testing of carbon fibre composite materials, for example, they are often examined by contacting acoustic scanning methods concurrently with moiré photography.

The surface is first prepared with a stencilled pattern, as in the above section, and it is then sprayed completely with a 10% solution of shellac lacquer in ethanol. Other diluted varnishes may be suitable, but the solvents present in some may affect the behaviour and integrity of the resin in composites and should be avoided. Figure 9.21 shows an example of a composite material that has been loaded cyclically to failure, together with a fringe map showing the indications of fatigue during the test.

(c) Explosive-resistant patterns

The moiré method is attractive for measuring explosively induced deformation, because high levels of strain are often encountered in this type of test. A specially robust pattern has been developed for such circumstances. The pattern is again stencilled using the nickel mesh, but with the white oxide pigment suspended in a

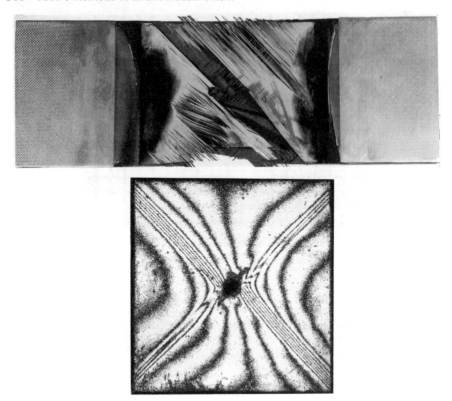

Fig. 9.21 Failed composite test specimen and a fringe pattern recorded during a cyclic loading test.

dilute epoxy resin medium. One resin that is readily available is the twin-tubed Araldite (Ciba-Geigy) and a solution can be made up from an initial mixture of approximately 1:1 resin to acetone, which is further diluted 1:5 with ethyl alcohol. The mixture is first lightly sprayed on to one side of the mesh which is then gently attached to the polished structure. With the remaining adhesive solution, a paint is made up by adding 10% of the oxide and this is sprayed over the stencil.

Stretching the mesh is usually unavoidable during its removal and so it cannot be reused. The surface should be warmed, or left until the resin is completely cured, before testing the structure. Figure 9.22 shows the method applied to a 10 mm thick steel plate that has suffered an explosive deformation exceeding a strain level of 20%.

In view of the large plastic strains and gross out-of-plane displacements, HRMP is generally unsuitable for explosive studies. Instead, transmission moiré, in conjunction with a reference grating on film which is pressed onto the specimen, has proved effective.

Types of surface pattern 369

Fig. 9.22 A steel plate treated with an explosive-resistant pattern.

9.4.3 Random patterns

(a) White-light speckle

The property of the lens–mask combination in generating a small interference pattern surrounding the image of each bright point offers an alternative approach to regular patterning. With random patterns, consisting of fine, bright detail against a dark background, a quasi-periodic pattern is recorded at the tuned frequency, for any camera–specimen distance. Using the same double-exposure reconstructing system, moiré fringe patterns can be obtained which are very similar to the full-field patterns of laser speckle photography as discussed in section 10.1.1 (Archbold *et al.*, 1970). Although the contrast and clarity of the fringes are inferior to those of the regular pattern process, this **white-light speckle** method does permit the examination of re-entrant or steeply curved structures where it would not be possible to treat the surface with one of the standard arrays (Forno 1975). The main disadvantage of this pattern type is, however, the same as that of laser speckles; if the deformation is sufficient to produce many fringes in the reconstruction, the fringe contrast will be degraded by speckle decorrelation. There are also problems with superimposing single exposures; a loss of correlation occurs across the base layer of the separated images and the fringe contrast is considerably degraded.

370 *Moiré methods in strain measurement*

(b) Scratched surfaces

For metal objects, the surface can be first polished and then scratched in two orthogonal directions by means of a coarse abrasive cloth. If the illumination is orientated at approximately 45° to the pattern and at a high incident angle, the scratches behave as fine linear sources against a dark ground. The method works well on near flat surfaces and has been found suitable for some high-temperature applications where the stencilled pattern method (section 9.4.2) is unsuitable. Added protection from the effects of high-temperature oxidation is obtained if a layer of nichrome is vacuum-evaporated over the scratched surface.

(c) Retroreflective Paint

The individual glass beads in retroreflective paint (Codit reflective liquid, 3M Company) produce very bright points of light when viewed from a position close to the illuminating direction. Considered as a whole, they represent a random distribution of points containing all spatial frequencies up to the reciprocal separation of the closest beads. The paint may be applied by brushing or spraying and offers a more uniform pattern coverage than the scratching process. Figure 9.23 demonstrates an application involving the inside panel of a car door subjected to heavy vibration at 165 Hz. Here the fringes are produced by recording from an

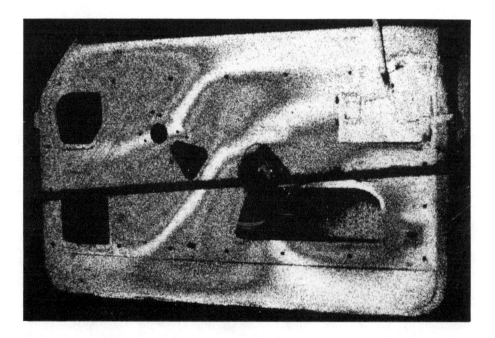

Fig. 9.23 Vibration mode in a car door displayed by white-light speckle photography.

Types of surface pattern 371

oblique viewpoint, so that any out-of-plane displacement is converted to a transverse displacement in the image and the bright nodes are clearly visible where there has been no motion.

(d) Untreated materials

Occasionally, a material may exhibit a built-in pattern; for example, some textiles are woven so that the warp and weft threads are equispaced. Such materials can generate visually interesting patterns from relatively low loads because of the lack of rigidity and the way in which orthogonal movements are independent. By reducing the size of the spatial filter in the reconstructing system, the bandwidth of transmitted spatial frequencies can be restricted, a process that often reveals information about weaving methods and flaws in the material.

Figure 9.24 shows an example pair of reconstructed fringe patterns and strain distributions from a silk-screen textile possessing 4.8 threads per millimetre.

Fig. 9.24(*a*) Fringers and strain contours from a silk-screen material (vertical direction).

372 *Moiré methods in strain measurement*

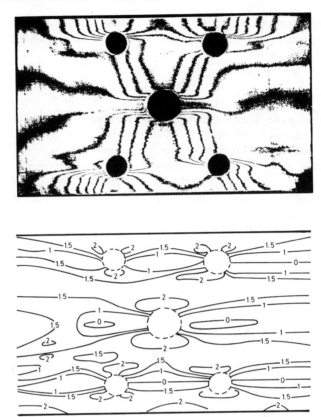

(b)

Fig. 9.24(b) Displacement fringe patterns from a silk-screen material and the associated strain distributions (horizontal direction).

9.5 ADDITIONAL FEATURES OF THE TECHNIQUE

There are a number of extensions of moiré photography, such as pattern frequency multiplication and changing the displacement sensitivity direction. Probably the two most useful features for engineering applications have been optical differentiation to generate contours of strain directly and the analysis of three-dimensional deformations.

9.5.1 Optical differentiation

The process of local differencing or differentiation of fringe values in order to determine strain can be carried out optically (Durelli and Parks, 1970). Two records of the deformed pattern are made with a small translation introduced between them. It can be either predetermined and fixed in a double-exposure record or adjustable, in

Additional features of the technique 373

which case it is applied at the reconstruction stage with two identical single exposures. The resulting moiré fringes represent contours of constant partial derivative of displacement, or strain, in the direction of the translation. A strain contour ϵ_x, defined by the order N_p of the fringe, is related to the translation Δ_x in the x-displacement map by

$$\epsilon_x = N_p/\Delta_x. \tag{9.19}$$

Although a double exposure will display a unidirectional strain field, a pair of single exposures can be displaced and interrogated in both directions with complete freedom in the amount of applied translation.

There are two limitations to this approach for exhibiting strain fields. It is useful only where the anticipated strain levels are high (in excess of 1 millistrain) and the fringes define an average strain over a gauge length equivalent to the translation distance. It should also be noted that the initial undeformed pattern must be uniform in spacing and orientation. Unlike other conventional moiré methods, however, the strain contour map is displayed free from the original grating detail. Figure 9.25 displays the strain distribution around a fatigue-induced crack in an aluminium alloy plate. The contours immediately outside the crack tip represent strain contours at approximately 0.05% intervals.

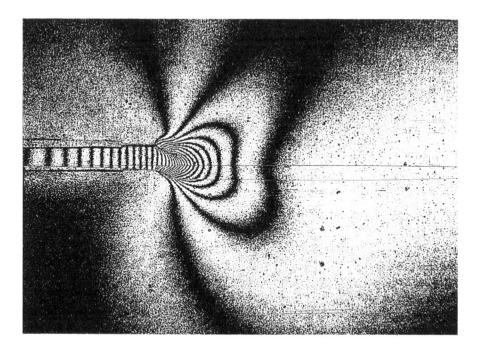

Fig. 9.25 Direct display of strain contours by optical shearing.

9.5.2 Three-dimensional movement

The moiré fringe information obtained with the high-resolution camera relates solely to lateral shifts in the recorded image. These shifts correspond to object movements at right angles to the line of sight of the camera, so that in the centre of the field there is no sensitivity to displacement in the z-direction. Away from the centre, there is a sensitivity to z-movement which increases linearly with the tangent of the field angle.

Information regarding only z-displacement can be extracted by differencing the records obtained with two separate cameras arranged symmetrically on either side of the z-axis. As with photogrammetry (Chapter 4), the sensitivity of this technique depends on the convergence between the cameras and, because of the restricted depth of focus, the maximum sensitivity obtainable over the whole field is approximately one third of that for x and y.

Alternatively, by adding fringe values from a pair of photographs, any z-movement that has occurred can be eliminated, leaving twice the values for the x- and y-displacements. For small structures, the effects of z-movement can also be removed optically during recording by using a telecentric field lens, but aberrations may become troublesome (Dean *et al.*, 1984).

9.5.3 Pattern frequency multiplication

Some regular specimen gratings can be resolved by the camera at distances other than that set by the fundamental tuned response of the lens. If the bright elements in the pattern are narrow relative to the width of the dark elements, then at one-half the optimized object to front focus distance, the lens will generate a modulated image corresponding to twice the pattern line frequency. Thus a 40 lines per millimetre pattern can be recorded as though it displayed 80 lines per millimetre, resulting in double the sensitivity to displacement.

The method can be extended to the third and fourth harmonics of the specimen grating, although the effects of lens aberrations when working at shorter distances become pronounced and the regions of the field resolved adequately are smaller.

An alternative method of multiplication, suggested by Cloud (1976), employs a mask for the camera lens which has been redesigned to resolve actual harmonics present in the specimen grating. By this means a sensitivity multiplication of four was achieved, although at 79 lines per millimetre the selected tuned frequency of his modified Goerz lens was considerably less than is possible with the Takumar lens fitted to a Pentax camera.

9.5.4 Optical regeneration of defective gratings

The property of the slotted mask in generating a modulated image offers a somewhat unexpected advantage over conventional photographic moiré. Where a specimen grating is defective, through damage or poor fidelity introduced during preparation, the defect may be filled-in by the modulation present in adjacent grating structures.

There is always an associated loss of overall image resolution because of this effect, but as the process does not extend further than three or four resolved line widths, the actual degradation in visual fidelity of the fringe pattern is not noticeable.

9.5.5 Displacement sensitivity direction

Without altering the surface pattern, it is possible to change the orientation of the displacement sensitivity. The regular grid array possesses additional periodicities at different orientations to the primary array direction. For instance, at 45° to a 40 lines per millimetre pattern an array of 57 lines per millimetre is displayed. This can be recorded by the modified camera, if the slots are aligned in this direction and the camera-object distance is reduced by a factor of $1/\sqrt{2}$.

High-resolution moiré photography still finds application in all fields of strain measurement. The equipment is unusually inexpensive for such a versatile method, the major disadvantage being that information is recorded on film – a material that may not readily be available in the future. One can foresee the development of CCD cameras containing arrays that are compatible with the 10 000 by 7000 resolved points available with the photographic technique and this will offer the possibility of real-time investigations at high sensitivity.

9.6 MOIRÉ INTERFEROMETRY

In all moiré processes, the sensitivity to displacement (and therefore strain) is directly related to the frequency of the reference grating. With very fine gratings, however, the phenomenon of optical diffraction becomes a major problem. Unless the reference grating is in intimate contact with the object, or very close to one of the self-imaged planes (section 9.1.5(*b*)), different orders of diffraction will overlap in the gap separating the two and the effect will be to degrade the contrast of the moiré fringe pattern. With grating frequencies higher than 40 lines per millimetre the effect becomes so serious as to prevent practical measurements.

9.6.1 Two-beam interference

If instead of the physical reference grating being of solid construction, as is the case with transmission moiré or shadow moiré, an optically generated grating is used, then much higher displacement sensitivities are possible by the technique of fringe multiplication (Post, 1968). This method utilizes the harmonics contained in a square-wave specimen grating with sharp edges. One particular harmonic is selected and interrogated with an optically generated reference grating whose spatial frequency corresponds to that of the chosen harmonic. The reference, usually referred to as the **virtual grating**, is generally produced by interference between two overlapping beams from a collimated laser.

376 Moiré methods in strain measurement

With high-quality specimen gratings, interference between the 50th diffracted order or 50th harmonic has been achieved, giving a sensitivity multiplication of ×50. In general, gains between 8 and 10 are more commonly realized due to a combination of reduced efficiency and unreasonable background noise present in the highest orders of diffraction. Although the application of this method can extend to quite large areas of specimen (100 mm × 100 mm), it is generally suitable for transparent gratings, either mounted on transparent objects or replicated in a transparent material such as silicone rubber.

Moiré interferometry elegantly combines conventional moiré with interferometry whilst dramatically improving the visual quality of the generated fringe pattern. This technique originated a long time ago, two of the earliest references being Boone (1970) and Cook (1971). The test component is coated with either photographic film or photoresist and exposed to a laser-generated virtual grating. Some deformation is introduced in the component and a second exposure is made. The resist is processed and, under diffuse illumination, dark fringes appear over the surface where there have been displacements between the exposures equivalent to odd multiples of one half of the pitch of the grating.

(a) Fresnel's bi-prism arrangement

A simple 45° Porro roof-prism can be used to generate the two-beam pattern by bisecting and recombining a collimated laser beam (Fig 9.26). In the space behind the prism, where the divided beams overlap, a virtual grating with pitch f is formed, where

$$f = \lambda/2 \sin \theta. \tag{9.20}$$

With a crown-glass prism and using the blue wavelength from an argon laser, the

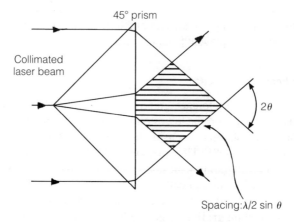

Fig. 9.26 Prism arrangement for generating a two-beam interference pattern.

fringe spacing is equivalent to a displacement sensitivity of 0.52 μm per fringe.

Provided that the photoresist coating receives both components of the illumination, it is possible to examine the behaviour of a specimen whose shape departs considerably from a flat surface. However, the moiré fringe pattern represents a permanent record of a single deformation event.

Wadsworth *et al* (1973) suggested exposing the resist once, processing the surface grating, and replacing it in the virtual grating. In this way, the deformation behaviour could be monitored continuously. This procedure forms the basis of the technique which is now called moiré interferometry. Even in this initial research, experiments were conducted on carbon fibre composites, a material that currently forms a major part of experimental studies using the process.

Preparation of specimens with photoresist, involving either spraying, spinning or dipping is not easy to control and probably prevented the early adoption and development of the technique. It was not until the 1980s, with the introduction of a convenient method for applying a grating in the form of a replica from a master grating together with the use of a Lloyd's mirror system for producing a two-beam interference pattern, that moiré interferometry was converted to an acceptable metrological tool.

(b) Lloyd's mirror arrangement

Figure 9.27 illustrates the principles of this two-beam interferometer. The specimen, bearing a replicated grating, usually with a spatial frequency of 1200 lines per millimetre, is held close to the plane mirror and normal to its surface. Light from a He-Ne laser is expanded and collimated by a lens or parabolic mirror and directed

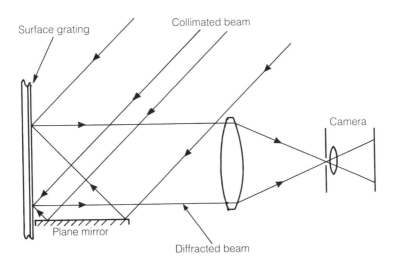

Fig. 9.27 Moiré interferometry principle using a Lloyd's mirror system.

378 Moiré methods in strain measurement

onto the specimen in such a way that half directly illuminates the grating and half is reflected on to it. Again, the two overlapping beams generate an interference pattern of light and dark lines with pitch defined by equation (9.20). If the pitch of the specimen grating is twice that of the virtual grating and the alignment of the two is close to being parallel, interference occurs between the gratings to produce a moiré fringe pattern in light which is propagated normal to the surface. Employing a specimen grating of 1200 lines per millimetre is convenient in that the orientation of the incident beams is approximately 45°.

There is a distinction which should be recognized between an interaction between two amplitude gratings, as in the case of shadow, photographic and other forms of classical moiré, and moiré interferometry. In the former, the formation of fringes is explained as resulting from the principles of geometric obstruction of light by the reference grating. Moiré interferometry can also be understood as a true interference phenomenon; the two wavefronts returned along the same direction by first-order diffraction from the specimen are distorted because of the deformations present in the surface grating. These wavefronts combine and interfere to form a fringe pattern.

(c) Three-mirror interferometer

For the determination of both x- and y-strain components, a similar three-or four-mirror arrangement is required to generate two sets of virtual gratings orthogonal to each other. There are alternative methods, using semitransparent wedges, transmission gratings and incoherent or even white-light illumination, but a Lloyd's mirror arrangement and He-Ne laser illumination offers the least complicated combination. A ray diagram for the three-mirror system is shown in Fig. 9.28. Mirrors A and B reflect two beams that generate a set of fringes aligned with the x-direction. The direct beam and that from mirror C give the y-aligned set.

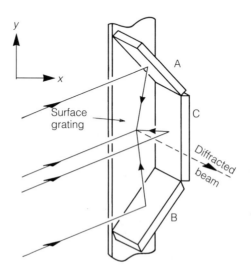

(a)

Fig.9.28(a) Propagation of rays in a three-mirror system.

Moiré interferometry 379

(b)

Fig. 9.28 (*b*) a compact three-mirror interferometer.

Accompanying the ray diagram is a photograph of a compact interferometer developed at Warsaw Technical University. As the instrument is attached to the test specimen, small movements normal-to-surface and rotations have little effect, although the asymmetry of the single central mirror cannot accommodate axial rotations without introducing an apparent strain. These can be avoided, not only by restricting the movement of the specimen, but also by detaching the interferometer from the surface or employing a fourth mirror and face-on illumination. Alternatively, an area of reference grating can be attached locally to the specimen within the field of view of the interferometer. Fringes that appear on the reference can be interpreted as the rotational component of the specimen, enabling the absolute deformation to be calculated.

9.6.2 Surface Grating

There are more than twenty articles which are recommended reading in this subject. Post (1987) references most of them and gives a comprehensive treatment of all the major aspects of moiré interferometry.

In the replication of a grating on to the specimen, epoxy resin is the most commonly used material. Silicone rubber is also suitable and will withstand some

380 *Moiré methods in strain measurement*

degree of heating (Kang *et al.*, 1990). There is adequate energy reflected from silicone rubber to produce good contrast fringes without the need to coat the grating with an evaporated layer of aluminium. Figure 9.29 shows an example of an *x*-displacement map produced by a five-hole alloy specimen, prepared with a rubber grating and subjected to load.

Applications of moiré interferometry have been as diverse as those of high-resolution moiré photography, especially in the area of material behaviour. A few examples include adhesive joints (Davidson, 1987), carbon fibre composites (Post, 1991), dovetail joints in Nimonic materials (Ruiz *et al.*, 1985) and aluminium alloys (Kang *et al.*, 1987). It is possible to reveal strain information directly from the

Fig. 9.29 Example of an *x*-displacement map generated by the three-mirror interferometer (one fringe represents 0.42μm).

displacement-map interferograms by a process analogous to that described in section 9.5.1 (Weissman, and Post 1982), and the methods involving optical shearing of the moiré pattern (Patorski *et al.*, 1987) offer strain contouring in real time.

The examination of the behaviour of materials at high temperature is a recent development. For this, it is necessary to protect or replace the combustible resin grating with a heat-resistant layer. Alternatively, by ion etching through the photo-resist layer the substrate can be marked lightly with the imprint of the grating. A protective layer of precious metal can then be evaporated over the surface. Such **zero-thickness gratings** are attached directly to the substrate and are therefore immune to shearing effects through their thickness (Ifju and Post, 1991).

A simpler approach that will withstand temperatures of 600 °C for a few hours involves chemical deposition of nickel from solution directly on to the resist grating. If the specimen is heated slowly through the decomposition temperature of the resist then the grating structure is preserved and the metal film will adhere strongly at high temperatures. It should also be possible to prepare such coatings on small areas of large structures by applying the solutions with a brush. Preliminary trials with different vacuum-evaporated metals for small components operating above 800 °C show that platinum and nichrome are both effective.

The extension of moiré interferometry to the microscopic examination of materials (Han and Post, 1991) has the potential for revealing local displacements on an extremely small scale. For example, observation of the movements between fibres and matrix in fibre composite materials has been demonstrated. By using new methods involving fringe shifting and immersion fluids on the grating a sensitivity to displacement of a few tens of nanometres is claimed.

The problems associated with vibration have until recently remained a difficult limitation to moiré interferometry. Guo (1991) suggests attaching a transparent phase grating of slightly lower frequency to the specimen grating. Then the optimum two-beam pattern is generated in the gap between the two from the first orders of diffraction produced by the phase grating. Two beams of collimated light at a small angle to each other provide the illumination for the system. As the virtual pattern moves with the specimen and the external beams occupy almost the same path, movements or convection effects on the specimen or surrounding air do not disturb the fringes significantly.

9.6.3 Moiré fringe analysis

The phase-stepping method mentioned in section 9.3.2(a) is suitable for the interpretation of most interference patterns, including those from moiré interferometry. It involves capturing the whole field intensity distribution of the interferogram in at least three conditions, with the phase of the pattern shifted between each state. The choice of the number of steps depends on the type of error present in the measuring instrument. In the case of HRMP, changing the phase of the fringe map is carried out by tilting the pair of superimposed negatives in the spatial filtering system (section

9.3.1(*b*)). In moiré interferometry, a controlled tilt to an optically flat glass plate, inserted in one of the interferometer beams, provides accurate phase steps. Salbut and Patorski (1990) describe a polarization-stepping technique that involves a rotation of a polarizing filter directly in front of the recording camera.

A variation in the displacement sensitivity of the instrument, for example due to non-uniform thickness of the film base in comparing negatives from HRMP, can be accommodated in a four-step method. A five-step routine allows for errors in asymmetry of the fringe waveform, such as might be introduced by a non-linear detector. The images are captured in a frame store and the intensity values are used to calculate the phase independently at each picture element. The phase information, representing displacements over the complete field, can be presented as an intensity-coded distribution with 'jumps' where the original pattern changes by a whole fringe value of 2π. It is possible to derive strain information at this stage, but a process of unwrapping the phase map produces a continuous grey scale without discontinuities.

Smoothing the data before and after unwrapping eliminates some of the artefacts which may degrade the next process, the determination of strain. For this, there are at least two methods. A calculation of slope using several adjacent pixels on the phase map yields local values for strain. Alternatively, a calculation of the difference in intensity between two pixels at a chosen separation is the equivalent of determining the first difference of displacement, which is strain. A scaling factor defined by the distance between the pixels, referred to the specimen, is required to determine the absolute strain value.

As well as phase-stepping methods, Fourier-transform techniques and single-phase approaches are now common. These aspects are covered in section 10.2.3. Procedures for analysis are under development at many institutes and the prospects for generating the type of data an engineer will readily accept are both promising and exciting.

Although the original impetus for its development was for large structures, high-resolution moiré photography has lent itself to a variety of measurement problems at realistic levels of deformation and sometimes in hostile environments. By virtue of whole-field examination, critical regions can be located – a necessary preliminary if the advantages of more sensitive displacement techniques such as moiré interferometry are to be exploited.

ACKNOWLEDGEMENT

As said in the Preface, Professor Jim Burch has played a pioneering role in many aspects of optics, especially in the field of deformation measurement. The two techniques described here are no exceptions. Some of the early experiments in double-exposure moiré interferometry were carried out under his supervision and the author was fortunate in being able to work with him during the development of moiré photography.

References

Archbold, E., Burch, J.M. and Ennos, A.E. (1970) Recording of in-plane surface displacement by double-exposure speckle photography. *Opt. Acta* **17**(12), 883–98.
Boone, P.M. (1970) *Nouveaux Revue d'Optique Appliqueé*, **1**(2), 10.
Boone, P.M., Vinckier, A.G., Denys, R.M. and Sys, W.M. (1982) Application of specimen-grid moiré techniques in large scale steel testing. *Opt. Eng.* **21** (4), 615–25.
Burch, J.M. and Forno, C. (1982) High resolution moiré photography. *Opt. Eng.*, **21**(4), 602–14.
Burch, J.M., Forno, C. and Tanner, L.H. (1974) A lens-hologram test for the symmetric aberrations of a camera system. *Opt. and Laser Tech.*, **6**(3), 109–13.
Chiang, F.P. (1978) Moiré methods of strain analysis, in *Manual on Experimental Stress Analysis* (ed. A.S. Kobayashi), SEM, Prentice Hall.
Cloud, G. (1976) Slotted apertures for multiplying grating frequencies. *Opt. Eng.*, **15**(6), 578–82.
Cook, R.W.E. (1971) Recording distortion in irregularly shaped objects. *Opt. and Laser Tech.*, **3**(2), 71–73.
Davidson, R. (1987) Laser moiré interferometry as applied to adhesive joints. *Proc. SPIE* **814**, 479–98.
Dean, G.D., Forno, C., Johnson, A.F. and Dawson, D. (1984) Determination of the strain distribution in a circular plate under uniform pressure for biaxial stress tests. *Polymer Testing* **4**, 3–29.
Durelli, A.J. and Parks, V.J. (1970) *Moiré Analysis of Strain*, Prentice Hall, Englewood Cliffs, N.J.
Forno, C. (1980) Film characteristics in high resolution moiré photography, *Opt. Eng.*, **19**(6) 908–10.
Forno, C. (1975) White-light speckle photography for measuring deformation, strain and shape. *Opt. and Laser Tech.*, **7**(5), 217–21.
Huntley, J.M. and Field, J.E. (1990) High resolution moiré photography: application to dynamic stress analysis, *Opt. Eng.*, **28**(8), 926–33.
Kang, B.S., Kobayashi, A.S. and Post, D. (1987) Stable crack growth in aluminium tensile specimens, *Exp. Mech.* **27**(3), 234–45.
Kearney, A.M. and Forno, C. (1989) A large scale deformation study using moiré photography. *Photogrammetric Record*, **13** (74), 217–24.
Marasco, J. (1975) Use of a curved grating in shadow moiré. *Exp. Mech.*, **15**, 464–70.
Miles, C.A. and Speight, B.S. (1975) Recording the shape of animals by a moiré method, *J. Physics E*, 773–7.
Pekelsky, J.R. (1986) Automated contour ordering in moiré topograms for biosterometrics in *Biosterometrics 85* (ed. A.M. Coblentz and R.E. Herron), *Proc. SPIE*, **602**, 6–17.
Post, D. (1968) New optical methods of moiré fringe multiplication, *Exp. Mech.* **8**(2), 63–8.
Post, D. (1991) Moiré interferometry: advances and applications, *Exp. Mech.* **31**(3), 276–80.
Lord Rayleigh. (1874) On the manufacture and theory of diffraction gratings. *Phil. Mag.*, **47**, 193–205.
Righi, A. (1877) Sui fenomeni che si producono colla sovrapposizione die due reticoli e spora alcune lora applicazioni: *I. Nuovo Cimento*, **21**, 203–7.
Ritter, R. (1982) Reflection moiré for plate bending studies. *Opt. Eng.*, **21**(4), 663–71.
Salbut, L. and Patorski, K. (1990) Polarisation phase shifting method for moiré interferometry and flatness testing. *App. Opt.* **29**(10), 1471–3.
Ruiz, C., Post, D. and Czarnek, R. (1985) Moiré interferometric study of dovetail joints. *J. Appl. Mech.*, **52**(1), 109–14.
Sevenhaijsen, P. J. (1981) Two simple methods for deformation demonstration and measurement. *Strain*, **17**(1), 20–4.
Theocaris, P.S. (1964) The moiré method in thermal fields. *Exp. Mech.*, **4**, 223 – 31.
Wadsworth, N., Marchant, M. and Billing, B. (1973) Real-time observation of in-plane displacements of opaque surfaces. *Opt. and Laser Tech.*, **5** (3), 119 – 23.
Webster, J. M., Hepworth, J. K. and Forno, C. (1981) Strain determination in a transition joint at high temperature. *Exp. Mech.*, **21**(5), 195 – 200.
Weissmann, E.M. and Post, D. (1982) Full-field displacement and strain rosettes by moiré interferometry. *Exp. Mech.*, **22**(9), 324–8.

Selected papers on shadow moiré

Creath, K. (1988) Phase-measurement interferometry techniques. *Progress in Optics* **26**, 349-93.
Kujawinska, M. and Wojciak, J. (1991) High accuracy Fourier transform fringe pattern analysis. *Opt. and Lasers in Eng.*, **14**(4), 325-39.
Meadows, D. M. *et al.* (1970) Generation of surface contours by moiré patterns. *Appl. Opt.*, **9**, 942-7.
Pirodda, L. (1982) Shadow and projection moiré techniques for absolute and relative mapping of surface shapes. *Opt. Eng.*, **21**(4), 640-9.

Selected papers on phase-stepping method for fringe analysis

Reid, G. T., Rixon, R. C. and Messer, H. I. (1984) Absolute and comparative shape measurement of three-dimensional shape by phase measuring moiré topography. *Opt. and Laser Tech.*, **16**(6), 315-19.
Takasaki, H. (1970) Moiré topography, *Appl Opt.*, **9**, 1467-72.

Selected papers on moiré interferometry.

Bowels, D. F. *et al* (1981) Moiré interferometry for thermal expansion of composites. *Exp. Mech.*, **21**(12), 441-7.
Czarnek, R. (1991) Three-mirror four-beam moiré interferometer and its capabilities. *Opt. and Lasers in Eng.*, **15**(2), 93-101.
Guo, Y. (1991) Vibration insensitive moiré interferometry system for off-table applications. *Proc. SPIE conf. Photomechanics and Speckle Metrology.*
Han, B. and Post, D. (1991) Moiré interferometry with increased sensitivity. *Proc. SPIE conf. Photomechanics and speckle metrology.* Submitted to *Opt. Eng. and Exp. Mech.*
Ifju, P. and Post, D. (1991) Zero-thickness gratings for moiré interferometry. *Exp. Tech.*, **15**, 2.
Kang, B. S., Wang, F.X. and Liu, Q. K. (1990) High temperature moiré interferometry for use to 550 °C. *Proc. SEM Conf. Hologram Interferometry and Speckle Metrology*, Baltimore.
McDonach, A. *et al.* (1983) Improved moiré interferometry and applications in fracture mechanics, residual stress and damaged composites. *Exp. Tech.*, **23**(2), 20-4.
Patorski, K., Post, D., Czarnek, R. and Guo, Y. (1987) Real-time optical differentiation for moiré interferometry, *Appl. Opt.*, **26** (10), 1977-82.
Pirodda, L. (1988) Improvements to the technique of moiré interferometry. *Opt. Eng.*, **27**(3), 219-24.
Post, D. (1987) Moiré Interferometry, in *Manual on Experimental Mechanics* (ed A. S. Kobayashi), SEM, Prentice-Hall.
Post, D. and Baracat, W. A. (1981) High-sensitivity moiré interferometry - a simplified approach. *Exp. Mech.* **21**(3), 100-4.
Post, D., Czarnek, R. and Smith, C.W. (1984) Patterns of U and V displacement fields around cracks in aluminium by moiré interferometry, in *Applications of Fracture Mechanics to Materials and Structures* (ed. G. C. Sih *et al.*), Martinus Nijhoff, Hingham, Mass., pp. 699-708.
Weissman, E. M. and Post, D. (1982) Moiré interferometry near the theoretical limit. *Appl. Opt.*, **21**(9), 1621-3.

10
Automatic analysis of interference fringes

G. T. REID

Crown copyright

It can be seen from the preceding chapters that interferometry is one of the successes of applied optics. As well as being a fruitful topic of academic research, interferometric methods have made important contributions to optical testing (Malacara, 1978), the medical sciences (van Bally, 1979) and several branches of engineering (Erf, 1974; Theocaris, 1969). In many spheres of application, interferometric results must be expressed in such a way that non-specialists can comprehend them or in such a way that they may be combined with other, non-interferometric, measurements. In such cases, interferograms must be analysed so that the results can be presented in the required numerical or graphical form.

In the early days of interferometry, fringe-pattern analysis was carried out manually. Thomas Young was presumably one of the first practitioners of fringe analysis when, in the early 1800s, he measured the spacing of interference fringes in order to calculate the wavelength of light (Peacock, 1855). During the 1960s a number of electronic aids to fringe analysis were developed (Dew, 1964; Dyson, 1963). These devices allowed a substantial improvement in the accuracy with which fringes could be located within an interferogram but the overall analysis remained essentially manual.

During the last few years, however, the wide availability of digital image processing equipment has prompted a number of research groups to investigate the possibility of automatically analysing interferograms, thereby removing much of the tedium from the process. Techniques for interferogram analysis have reached the point where they can provide useful results in an ever-increasing range of applications, thereby extending the effectiveness of interferometry as a practical measuring tool.

At a time when an increasing number of people are making use of automatic fringe-analysis equipment, it is appropriate to summarize in this book the developments which have taken place during recent years. In supplementing the coverage in the three preceding chapters, this chapter aims to provide an introduction to the literature on the subject and to present a cross-section of the capabilities of various analysis techniques. For convenience, the terms **interferogram** and interference **fringe pattern** will be used to describe the fringe-like patterns which are generated by a variety of optical systems, whether or not such patterns are actually generated

Optical Methods in Engineering Metrology. Edited by D.C. Williams. Published in 1993 by Chapman & Hall, London. ISBN 0 412 39640 8

by interferometric means. Broadening the definition of interference is justified in the context of this chapter since the image-analysis procedures are essentially unaffected by the mechanism of fringe generation.

10.1 INTENSITY-BASED METHODS

10.1.1 Dedicated systems

When interferometric techniques are used for repetitive calibration, non-destructive testing or inspection, it is sometimes possible to design a comparatively simple fringe analysis system. After identifying the characteristics of the fringe pattern which are peculiar to the application, the capabilities of the analysis system can be confined to those required for the measurement in question (Ennos *et al.*, 1985; Seemuller, 1982; Snyder, 1980).

Der Hovanesian and Hung (1982), for example, dramatically simplified the fringe analysis task in a moiré fringe-shape comparator (a device which compares the three-dimensional shape of manufactured components with that of a master component) by imposing constraints upon the optical and mechanical design of their apparatus. Using a projection moiré method, they ensured that the image of a manufactured component was uniformly dark if that component was identical to the master. If the component differed from the master, however, bright fringes appeared in the image and, consequently, the overall intensity of the image increased. Photoelectric detection of the image then allowed the presence or absence of fringes to be deduced from the size of the photocurrent, a manufactured component being rejected if the photocurrent exceeded some threshold value.

Despite the variety and complexity of the interferograms which are produced during holographic non-destructive testing, a number of effective means of simplifying the fringe analysis have been proposed. When using holographic interferometry to study the plastic deformation of pressure vessels, Tichenor and Madsen (1978) found that they could identify defective vessels on the basis of fringe density. After digitizing an interferogram, the image was segmented into adjacent blocks of 64×64 pixels. The acceptable number of fringes appearing in each block was determined from a knowledge of the behaviour of good quality pressure vessels. Analysing the interferogram involved counting the number of fringes in each block of pixels to compile a fringe-density distribution over the image. If the fringe-density distribution departed significantly from that obtained from a good-quality vessel then the test vessel was classified as defective.

Robinson and his co-workers have described a number of specialized fringe analysis procedures (1983a, b) for use in engineering applications of interferometry. Two of these are described in section 7.3.4, and we now describe two more. In holographic interferograms produced when testing honeycomb panels, debrazing of the honeycomb produced groups of nearly circular fringes of a particular size and fringe density, as shown in Fig. 10.1. Robinson developed a method of searching

Fig. 10.1 Holographic interferogram showing faults in a brazed honeycomb panel. (Courtesy of National Physical Laboratory.)

interferograms automatically for the presence of these pre-specified faults. The method involves counting the number of fringes appearing on several horizontal lines through the image. When a comparatively large number of fringes appear on a given line, a flaw is presumed to exist on that line. Short vertical scans about the line are then used to search for the location of the flaw by looking for a comparatively large number of fringes in the vertical direction. Having identified the probable existence and location of a flaw, the system carries out a further check by counting the number of fringes along each of four short vectors, angularly spaced by 45° centred on the probable flaw site. If the same number of fringes appear on each of the four vectors then the existence of the flaw is confirmed and the flaw site is marked with a small cross.

As well as the techniques of speckle interferometry described in Chapter 8, speckle patterns can be used to make measurements of surface movement without

388 *Automatic analysis of interference fringes*

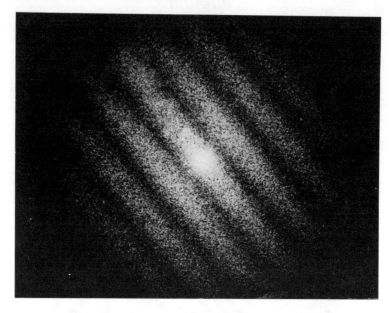

Fig. 10.2 Fringes formed by double-exposure laser speckle photography. (Courtesy of National Physical Laboratory)

invoking optical interference. If a laser-illuminated surface is photographed before and after a lateral deformation, each area of the image contains two identical but slightly displaced random speckle patterns. To analyse the recording, a laser beam is projected through small areas of the photographic film. The wavefronts scattered from corresponding pairs of points then combine to form straight equispaced fringes in the far field, as shown in Fig. 10.2, whose spacing and orientation describe the displacement vector at the place on the deformed object. Determining the spacing and orientation of the fringes would be straightforward were it not for the high level of noise which is inherent in these interferograms. Some degree of filtering or averaging is therefore required before an accurate analysis of the pattern can be performed. Although it may be possible to apply curve-fitting or low-pass filtering techniques, the predictable nature of the fringe pattern allows a less complicated approach (Kreitlow and Kreis, 1979; Robinson, 1983b).

The speckle photograph is digitized and intensity values are summed along each of a series of radial vectors which pass through the centre of the pattern. Assuming that a bright fringe always passes through the centre, the vector giving the largest intensity summation will be most nearly parallel to the fringes. Iteratively reducing the angular separation of the scan vectors allows the fringe direction to be determined. Intensity values are then averaged along the direction of the fringes, thereby reducing the two-dimensional fringe pattern to a one-dimensional function which, in consequence of the averaging, has a comparatively low level of noise. The period of this function is, of course, equal to the spacing of the fringes. It is also possible to compress the fringe pattern from a plane to a line optically by means of a cylindrical lens (compare the application in section 2.4.6), but this requires the lens to be set at the correct orientation (Kaufmann *et al.*, 1980).

When analysing the comparatively noise-free fringe patterns produced during the interferometric measurement of gauge blocks, as described in section 5.6.2, Pugh and Jackson (1982) required to determine the phase displacement, shown in Fig. 10.3(*a*), between the fringes on the gauge and those on the reference flat. Since the fringes were known to be parallel and equispaced, the phase displacement could be calculated from a knowledge of the positions of the fringe minima alone. These were determined from the two-step process illustrated in Fig. 10.3(*b*). In the first step, a first estimate of the location of a fringe minimum is obtained by finding the point where the gradient of the fringe intensity is zero. In the second step, a quadratic function is fitted to intensity data in the region of the first estimate. The turning point of the quadratic function is the final estimate of the fringe minimum. The commercial version of the instrument can locate minima to an accuracy of 0.01 of the fringe spacing by this method.

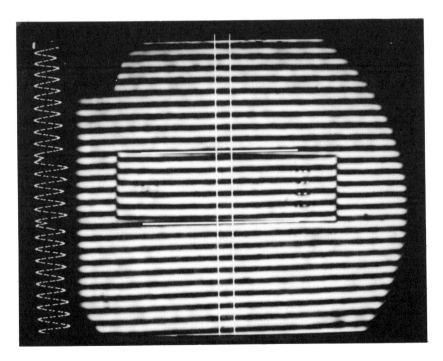

Fig. 10.3 (*a*) Interference pattern generated during interferometric measurement of gauge blocks. (Courtesy of National Physics Laboratory.)

390 *Automatic analysis of interference fringes*

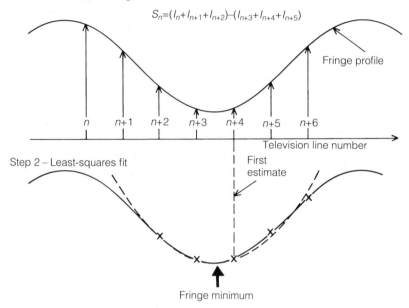

Fig. 10.3 (*b*) Process for locating fringe minima.

10.1.2 Fringe tracking

Perhaps the most obvious approach to fringe pattern analysis involves identifying and subsequently tracking fringes in an interferogram. This approach can be quite general purpose, requiring little or no prior knowledge of the fringe pattern characteristics, and has the important feature of allowing simple verification of the fringe-tracking algorithm by comparing a graphical representation of the tracked fringes with the original interferogram.

A number of fringe-tracking methods have been proposed (Augustyn, 1979; Bergquist, 1982; Button *et al.*, 1985; Chambless and Broadway 1979; Choudry, 1981; Cline *et al.*, 1982; Funnell, 1981; Nakadate *et al.*, 1983; Trolinger, 1985; Yatagai and Idesawa, 1982; Yatagai, 1991). Although these methods differ in detail, they usually rely on the following procedure.

1. Filter the image (Kreis and Kreitlow, 1979).
2. Either (i) fit curves to the intensity data with a view to interpolating between the fringe centres (Schemm and Vest, 1983), or (ii) identify and track the intensity maxima and/or minima with a view to skeletonizing the pattern and thereby minimizing the amount of data which must subsequently be processed (Yatagai, 1991).

3. Number the fringes interactively (Yatagai, 1991) or automatically (Cline *et al.*, 1984).
4. Calculate the measurement parameter from the fringe pattern data.

When analysing the high-contrast, noise-free interferograms which are produced in many applications of classical interferometry, it is often possible to proceed directly to the fringe-tracking or curve-fitting stage without filtering the image. Many other types of interferogram, however, contain comparatively high levels of noise and/or exhibit quite severe variations in fringe contrast and therefore require pre-processing. When the spatial frequency of the noise is significantly higher than that of the fringes (e.g. when speckle noise is present) then low-pass filtering of the interferogram can be performed (Pratt, 1978, pp. 319–21). Low-pass filtering is carried out by replacing the intensity value at each pixel by the average, weighted average or median of the intensities measured at the pixel and its neighbours. When the noise has a much lower spatial frequency than the fringes (e.g. when the workpiece has been unevenly illuminated) then high-pass filtering can be used. In this case, the image can be segmented into blocks containing comparatively large numbers of pixels and the average intensity of each block calculated (Becker *et al.* (1982), for example, used blocks of 32×32 pixels). A smooth curve is then fitted to the values of average intensity and this curve is subtracted from the interferogram to generate a more uniform background level.

When the noise has approximately the same spatial frequency as the fringes then a different approach must be adopted. If the noise is known to be **stationary** (i.e. movements of the fringes across the field of view are not accompanied by movements of the noise) then a high level of noise suppression can be obtained by combining two interferograms of opposite phase (Kreis and Kreitlow, 1979). When an interferogram is shifted in phase by π radians, the resulting dark fringes occupy the previous locations of bright fringes, and vice versa.

Stationary noise is unaffected. If one interferogram is subtracted from the other, therefore, the noise content will subtract to give zero while the fringes will combine to give a higher contrast than in the original. Alternatively, if the phase-shifted pattern is added to the original, the fringes add to give a uniform grey level and the noise is reinforced. Stationary noise can then be suppressed by dividing the original interferogram by the sum of the original and phase-shifted patterns.

If it is planned to extract data from points on the interferogram which do not coincide with the maxima or minima of fringe intensity then a continuous curve can be fitted through the intensity data (Hot and Durou, 1979). In the interest of speed, however, most authors have rejected this computationally intensive approach in favour of fringe tracking and subsequent fringe skeletonizing so as to minimize the amount of information which must be processed during the evaluation of the measurement parameter. Yatagai (1991) for example uses a two-dimensional maxima detection method which is performed within a 5×5 matrix of pixels. The central pixel is defined as lying on a skeletal line if a series of mathematical conditions are satisfied simultaneously along several directions. In principle, a point lies on the skeleton when it and its neighbours satisfy these conditions for any one direction. In practice, a more reliable result is obtained by ensuring that the conditions are satisfied simultaneously along two or more directions and by thinning the skeletal lines after the entire image has been processed and all likely skeletal points have been identified.

An alternative technique (Becker and Yu, 1985; Marshall *et al.*, 1987) combines thresholding, thinning and tracking operations to identify fringe centre lines. First, local threshold values are calculated for blocks of pixels over the fringe pattern. An array of threshold levels is formed either by finding average intensity values within

Fig. 10.4 Successive thinning of a thresholded fringe pattern to identify the fringe centre lines (courtesy of S. J. Marshall *et al.*).

Intensity-based methods 393

blocks of selectable size or by calculating a running average intensity over the field. A **shrink-expand operator** is then used to remove noise which appears as isolated bright or dark pixels. The fringe pattern is then subjected to repeated passes of a thinning operator which erodes the fringes from all sides, while maintaining continuity, until only the fringe centre lines remain. At this stage, the presence of noise on the original fringe pattern may result in gaps in the fringes, spurious branches emerging from fringes or spurious interconnections between fringes. A tracing routine is then used to refine the fringe pattern. Each fringe is followed separately and when a gap or branch is encountered, a set of rules is invoked to determine the most probable path of the fringe. Figure 10.4 shows this technique in operation.

Unambiguous numbers must now be assigned to the fringes in order to complete the interferogram interpretation. If the operator has sufficient prior knowledge of the interferogram then the fringes can be numbered interactively. This procedure might consist of the operator identifying a fringe with a light pen or cursor and then typing the fringe number into the computer. Hunter and Collins (1990) have carried out semi-automatic numbering of fringe patterns showing the density variations in the flow past a turbine blade in a transonic cascade. This application of holographic interferometry is described in section 7.6. Figure 10.5(*a*) shows an original interference pattern. In Fig. 10.5(*b*) the blade area has been masked with the aid of a cursor and the fringes have been converted to binary black or white, with due

(*a*)

Fig. 10.5 (*a*) Holographic interference fringes formed in the flow past a blade in a transonic cascade (Courtesy of Nuclear Electric.)

394 *Automatic analysis of interference fringes*

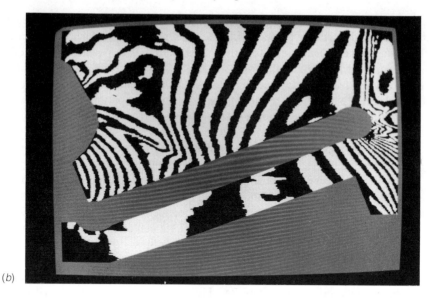

(b)

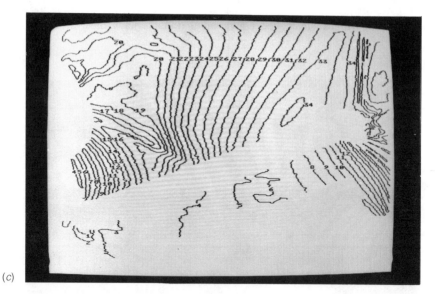

(c)

Fig. 10.5 (b) Binarized fringe field, (c) tracked and numbered fringe field (courtesy of Nuclear Electric).

allowance for local mean grey levels. For Fig. 5.10(c), the boundaries have been tracked and the resulting lines ordered by counting automatically between operator-defined points. This procedure required between 10 and 20 minutes, compared with one to two hours for totally manual analysis.

Interactive fringe numbering can be quite hazardous when complicated interferograms are under analysis. An erroneously numbered fringe can lead to substantial errors in the final calculation of the measurement parameter. In many circumstances, it is possible to overcome the fringe-numbering problem by introducing a substantial degree of tilt to the interferogram (Varman, 1984) so that, as shown in Fig. 10.18(a), the fringes become essentially parallel with the measurement parameter being encoded as a deviation from straightness of the fringes. In this case, the fringe number increases by unity as we move from one fringe to the next across the picture and fringe numbering can be carried out automatically.

When the above technique is impractical, perhaps because sufficient tilt cannot be introduced, then a more sophisticated approach (Jaerisch and Makosch, 1973; Livnat et al., 1980) is required. The three moiré fringe contour patterns shown in Fig. 10.9(a) have been shifted in phase by successive amounts of one-third of a cycle. In consequence of the phase shift, the fringes have moved with respect to the highest point on the surface. If the fringes can be identified in two successive interferograms then the fringe numbering can be performed automatically after noting the direction of the displacement of each fringe (Idesawa et al., 1977). Having numbered the fringes, the interpretation of the interferogram is complete and the measurement parameter may be calculated.

10.2 PHASE-BASED METHODS

10.2.1 Electronic heterodyning

We have seen that, by shifting the phase of an interferogram through appropriate amounts, we can achieve either a high degree of noise rejection or the ability automatically to assign order numbers to the fringes. In either case, we have been forced to conclude that a single interferogram does not always contain sufficient information to allow automatic evaluation of the measurement parameter. With this conclusion in mind, a number of research groups have constructed modified interferometers which allow heterodyne detection of the interferogram. These heterodyne interferometers, although more complicated than their conventional counterparts, are constructed around the requirements of automatic fringe analysis rather than around the requirements of visual interferogram interpretation. In consequence, heterodyne interferometers are free from many of the restrictions of conventional systems.

Heterodyning has been employed in a wide variety of interferometric techniques (Dändliker et al., 1976; Dändliker, 1980; Dändliker and Thalmann, 1985; Indebetouw, 1979; Lavan et al., 1975; McKelvie et al., 1978; Nakadate, 1985; Perrin and Thomas, 1979; Reid, 1983; Stricker, 1985), but the important features of the method can be described with reference to the heterodyne interferometer shown schematically in Fig. 10.6 (Massie, et al., 1979; Massie, 1980). Bragg cells (see sections 6.1.2 and 11.4.3) are placed in both beams of the interferometer to impose unequal frequency shifts $\Delta\omega_1$ and $\Delta\omega_2$ on the interfering laser beams. Denoting ω as the optical

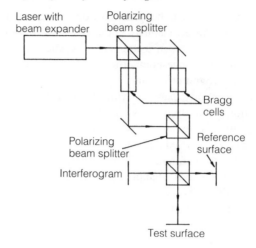

Fig. 10.6 Schematic diagram of a heterodyne Twyman – Green interferometer.

frequency of the light leaving the laser, E as the optical field amplitude and ϕ as the optical phase, we can represent the two interfering laser beams, at time t, by

$$E = E_{01} \cos[(\omega + \Delta\omega_1)t + \phi_1] \tag{10.1}$$
$$E = E_{02} \cos[(\omega + \Delta\omega_2)t + \phi_2]. \tag{10.2}$$

A photoelectric detector lying within the interference field will generate a photocurrent i of the form

$$i = <E_1 + E_2>^2, \tag{10.3}$$

where $<>$ represents time averaging due to the finite bandwidth of the detector. Expanding equation (10.3) we get

$$\begin{aligned} i = &E_{01}^2 < \cos^2[(\omega + \Delta\omega_1)t + \phi_1]> + E_{02}^2 < \cos^2[(\omega + \Delta\omega_2)t + \phi_2]> \\ &+ E_{01}E_{02} < \cos[(2\omega + \Delta\omega_1 + \Delta\omega_2)t + \phi_1 + \phi_2]> \\ &+ E_{01}E_{02} < \cos[(\Delta\omega_1 - \Delta\omega_2)t + \phi_1 - \phi_2]> \end{aligned} \tag{10.4}$$

Assuming that the bandwidth of the detector exceeds $(\Delta\omega_1 - \Delta\omega_2)$ but is much less than φ, then only the last term of equation (10.4) will contribute to the a.c. photocurrent. Under this condition, therefore, the phase $(\phi_1 - \phi_2)$ of the a.c. photocurrent which is generated at a given pixel is equal to the phase of the interferogram at that pixel. Massie et al. (1979) used an image dissector camera and an electronic phase meter to acquire values of phase at several thousand pixels within the interferogram, phase values being obtainable at any point within the interferogram and not only at fringe centres. A mechanically scanned photoelectric detector can provide essentially the same capability, the reduction in scanning speed being offset by potentially higher scanning accuracy.

Heterodyne interferometry allows considerable simplification of the fringe analysis software. Since the values of phase are largely unaffected by stationary noise and variations in fringe contrast, and since the phase values increase or decrease in sympathy with an increasing or decreasing fringe order number, many elements of a conventional fringe analysis routine are unnecessary. Furthermore, since phase values can often be obtained with an accuracy approaching $2\pi/1000$, heterodyne fringe analysis can provide an improvement in accuracy of between two and three orders of magnitude over conventional systems.

10.2.2 Phase stepping

Since heterodyne fringe analysis relies on the generation of an a.c. photocurrent, a non-integrating photoelectric detector (i.e. a detector with a bandwidth which exceeds the frequency $(\Delta\omega_1 - \Delta\omega_2)$ with which the fringes sweep over the interferogram) must be used. In consequence of the available frequency-shifting techniques, most authors have reported values of $(\Delta\omega_1 - \Delta\omega_2)$ in the range 5–100 kHz (Drain, 1980, pp.164–81). The majority of digital image-processing equipment, however, has been designed to take data from cameras which acquire only 25 images per second. Only by substantially reducing the value of $(\Delta\omega_1 - \Delta\omega_2)$ can the benefits of the heterodyne method be combined with the convenience of using widely available video image processing equipment.

It can be seen from sections 7.3.4, 8.5 and 9.6.3 that one of the most important developments in automatic fringe analysis was the advent of the phase stepping or quasi-heterodyne method (Bruning *et al.*, 1974; Bruning, 1978; Cheng and Wyant, 1985a; Halioua *et al.*, 1985; Prettyjohns, 1984; Prettyjohns *et al.*, 1985; Srinivasan *et al.*, 1984; Wyant *et al.*, 1983; Yatagai, 1984; Yatagai and Kanou, 1984). The method has been reviewed by Creath (1988). By introducing discrete shifts in the position of

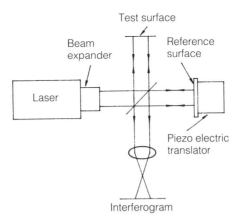

Fig. 10.7 Schematic diagram of a phase-stepping Twyman – Green interferometer.

the fringes, this technique allows the a.c. signal required for phase measurement to be acquired from a number of discrete measurements, the bandwidth requirement for each measurement being low enough to allow the data to be acquired by a video camera. The Twyman–Green interferometer of Fig. 10.7 represents one of the simplest implementations of the technique. This interferometer has been adapted for phase stepping by mounting the reference mirror on a computer-controlled piezoelectric translator. In operation, an interferogram is detected by the camera, digitized and read into computer memory. The phase of the reference beam is then changed by $2\pi/k$ by moving the reference mirror through a distance $\lambda/2k$, where λ represents the wavelength of the laser light.

A second interferogram is then read into computer memory. This process is repeated until, in the general case, the reference mirror has moved through $(k - 1)$ equal steps and a total of k images have been acquired so that k intensity levels have been stored at each pixel location. To calculate the phase $\phi(x, y)$ at pixel location (x, y) the Fourier series coefficients $\alpha_1(x, y)$ and $\beta_1(x, y)$ are evaluated using

$$\alpha_1(x, y) = \sum_{N=1}^{k} I_N(x, y) \cos(2\pi N/k) \qquad (10.5)$$

$$\beta_1(x, y) = \sum_{N=1}^{k} I_N(x, y) \sin(2\pi N/k) \qquad (10.6)$$

where $I_N(x, y)$ is the intensity level at pixel (x, y) for image number N. The phase is then given by

$$\phi(x, y) = \arctan[\beta_1(x, y)/\alpha_1(x, y)]. \qquad (10.7)$$

As in the heterodyne method, phase values are available at every pixel location, concave and convex features are unambiguously resolved, static noise is suppressed and the accuracy of measurement is significantly better than that obtained from intensity-based techniques (Schwider *et al.*, 1983; Cheng and Wyant, 1985b). Several potentially useful variations to the above technique have been proposed. Morgan (1982) has described a phase-calculation method which can tolerate time-dependent perturbations of the fringe pattern during the period of data acquisition, while Greivenkamp (1984) has described a phase-calculation method which allows unequal phase steps to be used. Other variations to the technique allow rapid data acquisition (Smythe and Moore, 1984) and continuous, rather than discrete, shifting of the reference beam phase (Wyant, 1982).

The phase-stepping technique has been incorporated in a wide range of interferometric methods. In the projection moiré contouring system described in section 9.1.2 and shown in Fig. 10.8 (Reid *et al.*, 1984, 1986), the phase-shifting process is carried out by translating the projection grating in its own plane. If, for example, the fringe phase $\phi(x, y)$ is calculated from three interferograms, so that $k = 3$ in equations (10.5) and (10.6), then projection moiré contour patterns such as those shown on a golf ball in

Phase-based methods 399

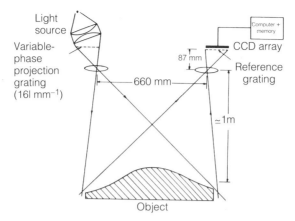

Fig. 10.8 Schematic diagram of a phase-stepping projection moiré system.

Fig. 10.9(*a*) are digitized for three equally spaced positions of the projection grating. In this case the projection system used a laser beam, the ball being placed at one of the self-imaged planes explained in section 9.1.5 (Rodriguez-Vera *et al.*, 1991).

The phase $\phi(x, y)$ is calculated, as before, from equation (10.7). Figure 10.9(*a*) also shows the resulting phase map, the effectiveness of the process in reducing intensity noise being apparent. Three-dimensional surface co-ordinates can then be

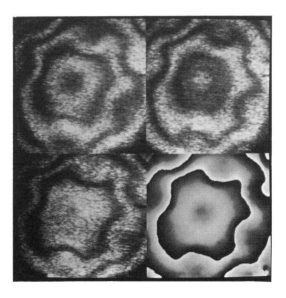

Fig. 10.9 (*a*) Phase-stepped projection moire contour fringes on a golf ball and resultant phase map (courtesy of Loughborough University).

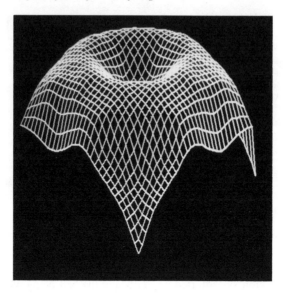

Fig. 10.9 (*b*) Wire-mesh profile of ball surface (courtesy of Loughborough University).

calculated from the phase values through a knowledge of the contour interval and the image magnification. Figure 10.9(*b*) shows a wire-mesh diagram of the golf ball of Fig. 10.9(*a*); the diagram was generated from the co-ordinate data obtained from the phase map. A similar phase-measuring approach can be taken in the shadow moiré contouring method, also described in section 9.1.2 (Reid *et al.*, 1985; Reid, 1986).

Dändliker and Thalmann (Dändliker *et al.*, 1982; Dändliker and Thalmann, 1983, 1986) and Hariharan *et al.* (Hariharan *et al.*, 1983; Hariharan, 1985) have employed phase-stepping methods for the automatic analysis of holographic interferograms. For surface contouring by dual-reference-beam holographic interferometry as introduced in section 7.3.4 (Dändliker *et al.*, 1976) the object is recorded twice on a holographic plate H with different object and reference beams for each recording. The hologram is then reconstructed simultaneously, as shown in Fig. 10.10, with reference beams R_1 and R_2 to generate a holographic contour map of the object. Phase stepping is then carried out by using a piezoelectric translator to change the path length of one of the reference beams.

Thalmann and Dändliker (1985a, b) propose an interesting method of controlling the piezoelectric translator. Since the two reference sources are placed close together the holographic plate is illuminated by a pattern of quite widely spaced interference fringes. A twin photodiode, lying beside the hologram plate as shown in Fig. 10.10, detects the centre of one of the fringes and an integrating feedback loop stabilizes the fringe pattern by driving the piezoelectric translator from the photodiode signal to compensate for path length changes induced by, say, vibration. Phase stepping is

Phase-based methods 401

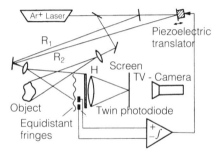

Fig. 10.10 Schematic diagram of a phase-stepping holographic interferometer with fringe stabilization (courtesy of R. Dandliker and R. Thalmann).

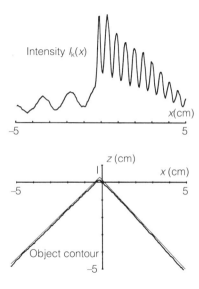

Fig. 10.11 Results from phase-stepped holographic contouring: (*a*) intensity profile of contour fringes on two sides of a cube, (*b*) evaluated object shape (courtesy of R. Dändliker and R. Thalmann).

then carried out by translating the photodiode pair through some fraction of the fringe spacing so that the piezoelectric translator moves through a corresponding fraction of the wavelength of illumination to return a broad fringe to the centre of the photodiode pair. Stabilization of the fringe pattern and well-controlled phase stepping are thereby achieved. Figure 10.11 shows the result of measuring a horizontal line across the edge of a cube using the above method. We see that, even when the fringe contrast undergoes a significant change, the surface profile of the object is reproduced correctly.

402 *Automatic analysis of interference fringes*

Hariharan *et al.*, (Hariharan *et al.*, 1983; Hariharan 1985) have extended the application of phase-stepping techniques to include some of the variations on holographic interferometry. After including a piezoelectric translator in the reference (i.e. reconstruction) beam of a two-index holographic contouring system (Erf, 1974, pp.139–44), phase stepping was used to retrieve phase values from an interferogram which, with a contour interval of approximately 200 μm, depicted the shape of a wear mark on a metal surface (Hariharan and Oreb, 1984). Also in the field of shape measurement, a phase-stepping version of two-wavelength holography was used for the quantitative measurement of the shapes of aspheric surfaces. Wyant and his co-workers (Wyant *et al.*, 1984) found that the two-wavelength method provided phase values which were repeatable to at least 1/100th of the equivalent wavelength–which, in this case, was 9.47 μm.

Combining phase stepping with stroboscopic holography allowed Hariharan and Oreb (1986) to measure the instantaneous displacement of a metal plate which was vibrating at over 200 Hz. The apparatus is shown schematically in Fig. 10.12. In preparation for the measurement, the resonant frequencies of the object were determined using a hologram of the stationary object and visual observations of the real-time, vibration-induced fringes with continuous illumination. After selecting the mode to be studied, the electro-optic modulator (EOM) and its associated polarizer and analyser were used to produce periodic pulses of illumination which could be synchronized to the vibration cycle of the object. A new hologram of the vibrating object was then recorded using stroboscopic illumination. This hologram was reconstructed using continuous illumination and phase stepping of the reference beam by means of the piezoelectric translator; this allowed analysis of the real-time fringes obtained from the stationary object. Figure 10.13(*a*) shows phase data obtained from

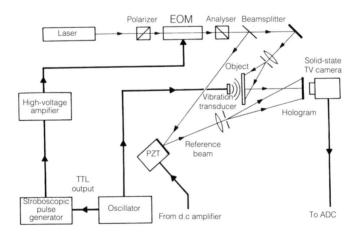

Fig. 10.12 Schematic diagram of an experimental arrangement for phase-stepping stroboscopic holography (courtesy of P. Hariharan and B.F. Oreb).

such a procedure, while Fig. 10.13(b) shows the result of fitting a polynomial surface to the phase data.

Television holography, as described in the previous chapter, relies on both electronic and optical phenomena to generate a fringe pattern (Jones and Wykes, 1988). The fringe-generation mechanism is therefore quite different from that in other interferometers which generate fringes by purely optical means. In the basic case, the speckle field from an illuminated object interferes with a reference beam which is derived from the same laser light source. The resulting speckle interference pattern is detected by a TV camera, digitized and stored. After deformation of the object a second speckle interference pattern is generated and detected by the camera. One pattern is then digitally subtracted from the other. Since speckle fields are used as information carriers, a high level of noise is inherent to the technique and intensity-based fringe analysis is difficult.

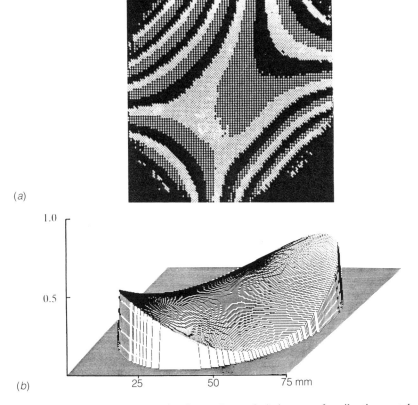

Fig. 10.13 Results from analysis of a stroboscopic hologram of a vibrating metal plate. (a) phase distribution in the fringe pattern, (b) surface representing vibration amplitude distribution (courtesy of P. Hariharan and B.F. Oreb).

404 *Automatic analysis of interference fringes*

As seen in section 8.5, phase-stepping techniques have been applied successfully to the analysis of TV holography patterns (Creath, 1985; Nakadate and Saito, 1985; Robinson and Williams, 1986), but the method of phase evaluation described by these authors is slightly different from that of the previous examples. The phases of points in the first speckle field are calculated from equation (10.7) after a number of stepwise changes have been made to the path length of the reference beam, just as in the previous examples. Phase values from the second speckle field are calculated in the same way. Subtracting one set of phase values from the other gives a measure of surface deformation in terms of the phase differences. Apart from the discontinuities which occur when the phase difference is a multiple of 2π, the phase difference is related linearly to deformation so that a series of **phase fringes**, such as those shown in Fig. 10.9, can be generated. The intensity of these phase fringes is proportional to the phase difference, so the fringes have a saw-tooth intensity profile from which the direction of the surface deformation can be inferred. Figure 10.14 shows the deformation of a diaphragm due to excess air pressure on the rear side. It was obtained using a fast microcomputer fitted with an inexpensive six-bit video digitizer board (Virdee *et al.*, 1990).

The effect of vibration is to reduce the amplitude of modulation at each speckle, in proportion to the zero-order Bessel function. Vibration fringes can therefore be generated by calculating the amplitude as $\sqrt{(\alpha_1^2 + \beta_1^2)}$ instead of the change of the phase arctan β_1/α_1; Fig. 10.15 shows a vibration mode of a diaphragm displayed in

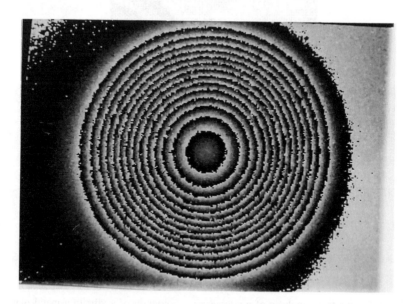

Fig. 10.14 Phase-stepped fringes from a laser speckle interferometer showing bulging of a diaphragm (courtesy of National Physical Laboratory).

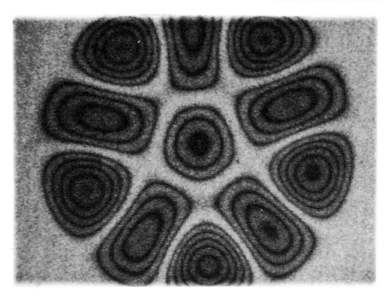

Fig. 10.15 Phase-stepped speckle fringes showing a vibration mode of a diaphragm (courtesy of National Physical Laboratory).

this way. Because such fringes are affected by speckle noise, the picture represents the average of sixteen results from randomly different speckle patterns, produced by altering the position of the source of laser illumination. A picture of the vibration mode can also be generated by performing the phase difference calculation used for deformation, with the object stationary and then vibrating. It can be seen in Fig. 10.16(*a*), where the object has also been tilted, that there is an apparent half-cycle phase shift for certain vibration amplitudes, which are those for which the Bessel function J_o takes negative values. In Fig. 10.16(*b*) there is no background movement; the binary nature of this type of display enables strikingly clear pictures to be obtained.

10.2.3 Spatial phase measurement

We have seen that by measuring the phase rather than the intensity of an interferogram, some analysis techniques can be developed. Perhaps the most serious drawback of phase-stepping methods is their requirement for a series of interferograms to be captured at different moments of time. This feature of the technique ultimately imposes a limit on the speed of data capture and analysis. For those applications which require the raw information to be acquired in a single image, there exist a number of fringe-analysis methods which rely on a spatial version of the temporal phase-measuring techniques which have been already covered in this

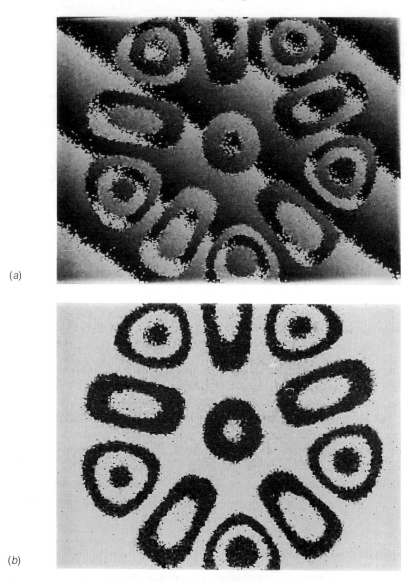

Fig. 10.16 Phase-stepped deformation speckle fringes showing diaphragm vibration: (*a*) surface tilted, (*b*) surface immobile (courtesy of National Physical Laboratory).

chapter (Macy, 1983; Mertz, 1983; Nugent, 1985). These spatial phase-measuring methods retain many of the advantages of phase stepping while removing the need to capture several images.

Womack (1984) describes several methods of extracting phase values from a single interferogram. In each of his methods, he relies on substantial tilt being present in the interferogram so that the measurement parameter is encoded as a deviation from straightness in the fringes and no closed-loop fringes are present (Hatsuzawa, 1985). In his **quadrature multiplicative moiré** method, for example, Womack considers an interferogram of the form

$$g(x, y) = a(x, y) + b(x, y) \cos [2\pi f_0 x + \phi(x, y)], \quad (10.8)$$

where $a(x, y)$, $b(x, y)$, $\phi(x, y)$ and f_0, respectively, represent the background intensity, amplitude, phase and spatial frequency of the fringes. A reference pattern $r(x, y)$ is then generated by and stored within a computer, being given by

$$r(x, y) = \cos (2\pi f_0 x). \quad (10.9)$$

Applying a low-pass filter to the product of equations (10.8) and (10.9) generates a function of the form

$$M_1(x, y) = \frac{1}{2} b(x, y) \cos [\phi(x, y)]. \quad (10.10)$$

In effect, $M_1(x, y)$ is a moiré fringe pattern which describes contours of equal phase in the original interferogram.

If the phase of the reference pattern is then shifted through $\pi/2$ and the above process is repeated, we generate a second function of the form

$$M_2(x, y) = \frac{1}{2} b(x, y) \sin [\phi(x, y)]. \quad (10.11)$$

The phase at a point (x, y) can then be calculated from

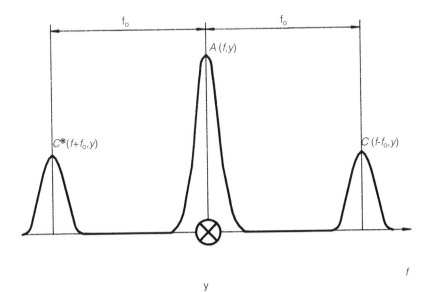

Fig. 10.17 Separated Fourier spectra of a tilted fringe pattern.

$$\phi(x, y) = \arctan [M_2(x, y)/M_1(x, y)]. \qquad (10.12)$$

Asundi and Yung (1991) discuss this technique using binary patterns.

Also using a tilted fringe pattern, Takeda and his colleagues (Takeda *et al.*, 1982) employed fast Fourier transform techniques to extract phase information. Following Takeda's method, we write equation (10.8) in the form

$$g(x, y) = a(x, y) + c(x, y) \exp(2\pi i f_0 x) + c^*(x, y) \exp(-2\pi i f_0 x), \qquad (10.13)$$

where

$$c(x, y) = \frac{1}{2} b(x, y) \exp[i\phi(x, y)] \qquad (10.14)$$

and $c^*(x, y)$ is the complex conjugate of $c(x, y)$.

Taking the Fourier transform of $g(x, y)$ with respect to x we get

$$G(f, y) = A(f, y) + C(f + f_0, y) + C^*(f + f_0, y), \qquad (10.15)$$

where the capital letters represent Fourier spectra and f represents spatial frequency in the x-direction.

If we assume that $a(x, y)$, $b(x, y)$ and $\phi(x, y)$ have frequencies which are much lower than f_0 then equation (10.15) will take the form of Fig. 10.17. The function $C(f - f_0, y)$

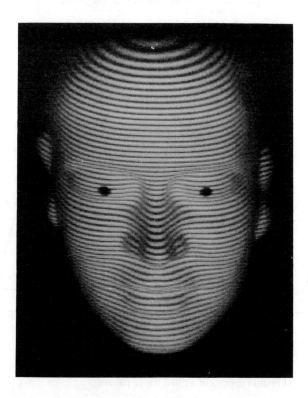

Fig. 10.18 (a) Grating pattern projected obliquely on to a face mask.(Courtesy of National Physical Laboratory.)

can then be isolated by a filter which is centred on f_0 and the function can be translated to the origin of Fig. 10.17 to give $C(f, y)$. Taking the inverse Fourier transform of $C(f, y)$ with respect to x, we get $c(x, y)$. The phase can then be calculated from

$$\phi(x, y) = \arctan \frac{\mathrm{Im}[c(x, y)]}{\mathrm{Re}[c(x, y)]}, \qquad (10.16)$$

where $\mathrm{Re}[c(x, y)]$ and $\mathrm{Im}[c(x, y)]$ represent the real and imaginary parts of $c(x, y)$ respectively.

Spatial phase calculations can be carried out in a particularly simple way (Freischlad *et al.*, 1990; Williams *et al.*, 1991) if the tilt in the interferogram is adjusted so that one fringe cycle nominally spans a particular small number of pixels. The intensity values from any set of successive pixels sampling one fringe cycle are then equivalent to a set of phase step values, so they can be used to calculate the fringe phase according to equation (10.7). By way of example, Fig. 10.18(*a*) shows a grating projected obliquely on to a face mask and Fig. 10.18(*b*) shows the corresponding phase map, produced by the same system as that used to generate Figs. 10.14 to 10.16.

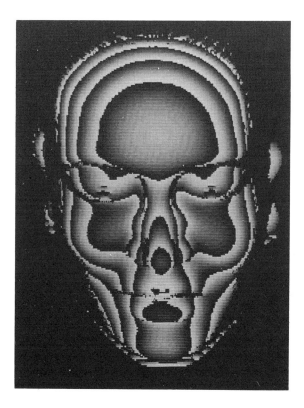

Fig. 10.18 (b) Contour map produced by spatial phase stepping (courtesy of National Physical Laboratory).

Automatic fringe-pattern analysis has had a major impact on interferometry, both by extending the breadth of interferometric applications and by improving the accuracy of interferometric measurement in well-established areas. Looking to the future, it seems reasonable to expect an increasingly wide use of automatic fringe analysis together with increased use of interferometric equipment in conjunction with other computer-based techniques such as finite-element analysis and computer-aided design.

Adapted by the editor from the author's review article in *Optics and Lasers in Engineering* (1986/7), **7**(1), pp.37–68.

REFERENCES

Asundi, A. and Yung, K. H. (1991) Phase-shifting and logical moiré. *J. Opt. Soc. Am. A*, **8**(10), 1591–600.
Augustyn, W. (1979) Versatility of microprocessor-based interferometric data reduction system. *Proc. SPIE*, **192**, 128–33.
Becker, F. and Yu, Y. (1985) Digital fringe reduction techniques applied to the measurement of three-dimensional transonic flow fields. *Opt. Engng*, **24**, 429–34.
Becker, F., Meier, G. E. A. and Wegner, H.(1982) Automatic evaluation of interferograms. *Proc. SPIE*, **359**, 386–93.
Bergquist, B. D. (1982) Coherent optical measurement of surface finish and form. *Proc. NELEX '82 Metrology Conference*. Paper 4.2. National Engineering Laboratory, East Kilbride, Glasgow.
Bruning, J. H. (1978) Fringe scanning interferometers, in *Optical Shop Testing* (ed. D. Malacara), Wiley, New York, pp.409–37.
Bruning, J. H., Herriott, D. R., Gallagher, J. E., Rosenfield, D. P., White, A. D. and Brangaccio, D. J. (1974) Digital wavefront measuring interferometer for testing optical surfaces and lenses. *Appl. Opt.*, **13**, 2693–703.
Button, B. L., Cutts, J., Dobbins, B. N., Moxon, C. J. and Wykes, C. (1985) The identification of fringe positions in speckle patterns. *Opt. and Laser Technol.*, **17**, 189–92.
Chambless, D. A. and Broadway, J. A. (1979) Digital filtering of speckle photography data. *Exp. Mech.*, **19**, 286–9.
Cheng, Y. Y. and Wyant, J. C. (1985a) Multiple wavelength phase-shifting interferometry. *Appl. Opt.*, **24**, 804–7
Cheng, Y. Y. and Wyant, J.C. (1985b) Phase shifter calibration in phase shifting interferometry. *Appl. Opt.*, **24**, 3049–52.
Choudry, A. (1981) Digital holographic interferometry of convective heat transport. *Appl. Opt.*, **20**, 1240–4.
Cline, H. E., Holik, A. S. and Lorensen, W. E. (1982) Computer-aided surface reconstruction of interference contours. *Appl. Opt.*, **21**, 4481–8.
Cline, H. E., Lorensen, W. E. and Holik, A. S. (1984) Automatic moiré contouring. *Appl. Opt.*, **23**, 1454–9.
Creath, K. (1985) Phase-shifting speckle interferometry. *Appl. Opt.*, **24**, 3053–8.
Creath, K. (1988) Phase-measurement interferometry techniques, in *Progress in Optics, XXVI* (ed.E. Wolf), North-Holland, Amsterdam, pp.349–93.
Dändliker, R. (1980) Heterodyne holographic interferometry, in *Progress in Optics, XVII*, (ed.E Wolf), North-Holland, Amsterdam , pp.1–84.
Dändliker, R. and Thalmann, R. (1983) Determination of 3-D displacement and strain by holographic interferometry for non-plane objects. *Proc. SPIE*, **398**, 11–16.
Dändliker, R. and Thalmann, R. (1985) Heterodyne and quasi-heterodyne holographic interferometry. *Opt. Engng*, **24**, 824–31.

Dändliker, R. and Thalmann, R. (1986) Electronic processing for holographic interferometry, in *Optical Metrology*, (ed. O.D.D. Soares), NATO ASI Series, Martinus Nijhoff, The Hague, pp.426–40.
Dändliker, R., Marom, E. and Mottier, F. M. (1976) Two reference beam holographic interferometry. *J. Opt. Soc. Am.*, **66**, 23–30.
Dändliker, R., Thalmann, R. and Willemin, J. F. (1982) Fringe interpolation by two reference beam holographic interferometry: reducing sensitivity to hologram misalignment. *Opt. Comm.*, **42**, 301–6.
der Hovanesian, J. and Hung, Y. Y. (1982) Three-dimensional shape recognition by moiré correlation. *Proc. SPIE*, **360**, 88–9.
Dew, G. D. (1964) A method for the precise evaluation of interferograms. *J.Sci. Instrum.*, **41**, 160–2.
Drain, L. E. (1980) *The Laser Doppler Technique*, Wiley, Chichester.
Dyson, J. (1963) The rapid measurement of photographic records of interference fringes. *Appl. Opt.*, **2**, 487–9.
Ennos A. E., Robinson, D. W. and Williams, D. C. (1985) Automatic fringe analysis in holographic interferometry. *Opt. Acta.*, **32**, 135–45.
Erf, R. K. ed. (1974) *Holographic Non-destructive Testing*, Academic Press, New York.
Freischlad, K., Küchel, M., Schuster, K. H., Wegmann, U. and Kaiser, W. (1990) Real-time wavefront measurement with lambda/10 fringe spacing for the optical shop. *Proc. SPIE*, **1332** (1), 18–24.
Funnell, W. R. J. (1981) Image processing applied to the interactive analysis of interferometric fringes. *Appl. Opt.* **20**, 3245–9.
Greivenkamp, J. E. (1984) Generalised data reduction for heterodyne interferometry. *Opt. Engng*, **23**, 350–2.
Halioua, M., Krishnamurthy, R. S., Liu, H. C. and Chiang, F. P. (1985) Automated 360° profilometry of 3-D diffuse objects. *Appl. Opt.*, **24**, 2193–6.
Hariharan, P. (1985) Quasi-heterodyne hologram interferometry. *Opt. Engng*, **24**, 632–8.
Hariharan, P. and Oreb, B. F. (1984) Two-index holographic contouring: application of digital techniques. *Opt. Comm.*, **51**, 142–4.
Hariharan, P. and Oreb, B. F. (1986) Stroboscopic holographic interferometry: application of digital techniques. *Opt. Comm.*, **59**, 83–6.
Hariharan, P., Oreb, B. F. and Brown, N. (1983) Real-time holographic interferometry: a microcomputer system for the measurement of vector displacements. *Appl. Opt.*, **22**, 876–80.
Hatsuzawa, T. (1985) Optimization of fringe spacing in a digital flatness test. *Appl. Opt.*, **24**, 2456–9.
Hot, J. P. and Durou, C. (1979) System for the automatic analysis of interferograms obtained by holographic interferometry. *Proc. SPIE*, **210**, 144–51.
Hunter, J. C. and Collins, M. W. (1990) The semi-automatic analysis of compressible flow interferograms. *Meas. Sci. Technol.*, **1**, 238–46.
Idesawa, M., Yatagai, T. and Soma, T. (1977) Scanning moiré method and automatic measurement of 3-D shapes. *Appl. Opt.*, **16**, 2152–62.
Indebetouw, G. (1979) Profile measurement using projection of running fringes. *Appl. Opt.*, **17**, 2930–3.
Jaerisch, W. and Makosch, G. (1973) Optical contour mapping of surfaces. *Appl. Opt.*, **12**, 1552–7.
Jones, R. and Wykes, C. (1988) *Holographic and Speckle Interferometry*, 2nd edn, Cambridge University Press, London.
Kaufmann, G. H., Ennos, A. E., Gale, B. and Pugh, D. J. (1980). An electro-optical read-out system for analysis of speckle photographs. *J. Phys. E: Sci. Instrum.*, **13** (5), 579–84.
Kreis, T. M. and Kreitlow, H. (1979) Quantitative evaluation of holographic interferograms under image processing aspects. *Proc. SPIE*, **210**, 196–202.
Kreitlow, H. and Kreis, T. M. (1979) Automatic evaluation of Young's fringes related to the study of in-plane deformations by speckle techniques. *Proc. SPIE*, **210**, 18–24.
Lavan, M. J., Cadwallender, W. K. and Deyoung, T. F. (1975) Heterodyne interferometer to determine relative optical phase changes. *Rev. Sci. Instrum.*, **46**, 525–7.
Livnat, A., Kafri, O. and Erez, G. (1980) Hills and valleys analysis in optical mapping and its application to moiré contouring. *Appl. Opt.*, **19**, 3396–400.

McKelvie, J., Pritty, D. and Walker, C.A. (1978) An automatic fringe analysis interferometer for rapid moiré stress analysis. *Proc. SPIE*, **164**, 175–88.

Macy, W.W. (1983) Two-dimensional fringe-pattern analysis. *Appl. Opt.*, **22**, 3898–901.

Malacara D. ed. (1978), *Optical Shop Testing*, Wiley, New York.

Marshall, S. J., Rixon, R. C., Caulfield, M. M. and Mackenzie, P. M.(1987) The application of automatic fringe analysis in fracture mechanics. *Opt. and Lasers in Engng*, **7**, 175–93.

Massie, N. A. (1980) Real-time digital heterodyne interferometry. *Appl. Opt.*, **19**, 154–60.

Massie, N. A., Nelson, R. D. and Holly, S. (1979) High performance real-time heterodyne interferometry. *Appl. Opt.*, **18**, 1797–803.

Mertz, L. (1983) Real-time fringe-pattern analysis. *Appl. Opt.*, **22**, 1535–9.

Morgan, C. J. (1982) Least-squares estimation in phase-measurement interferometry. *Opt. Lett.*, **7**, 368–70.

Nakadate, S. (1985) Shearing heterodyne interferometry using acousto-optic modulators. *Appl. Opt.*, **24**, 3079–87.

Nakadate, S. and Saito, H. (1985) Fringe scanning speckle-pattern interferometry. *Appl. Opt.*, **24**, 2172–80.

Nakadate, S., Yatagai, T. and Saito, H. (1983) Computer-aided speckle pattern interferometry. *Appl. Opt.*, **22**, 237–43.

Nugent, K. A. (1985) Interferogram analysis using an accurate fully automatic algorithm. *Appl. Opt.*, **24**, 3101–5.

Peacock, G., ed. (1855) *Miscellaneous Works of the Late Thomas Young*, John Murray, London.

Perrin, J. C. and Thomas, A. (1979) Electronic processing of moiré fringes: application to moiré topography and comparison with photogrammetry. *Appl. Opt.*, **18**, 563–74.

Pratt, W. K. (1978) *Digital Image Processing*, Wiley, New York.

Prettyjohns, K. N. (1984) Charge-coupled device image acquisition for digital phase measurement interferometry. *Opt. Engng.* **23**, 371–8.

Prettyjohns, K. N., Devore, S., Deraniak, E. and Wyant, J. C. (1985) Direct phase measurement interferometer working at 3.8 μm. *Appl. Opt*, **24**, 2211–16.

Pugh, D. J. and Jackson, K. (1982) Automatic gauge block measurement. *Proc. NELEX '82 Metrology Conference*, paper 1.2. National Engineering Laboratory, East Kilbride, Glasgow.

Reid, G. T. (1983) Surface form measurement by heterodyne moiré contouring. *Proc. SPIE*, **376**, 8–14.

Reid, G. T. (1986) Phase-measuring moiré topography, in *Optical Metrology* (ed. O.D.D. Soares) Martinus Nijhoff, The Hague, pp.256–68.

Reid, G. T., Rixon, R. C. and Marshall, S. J. (1985) 3-D machine vision for automatic measurements of complex shapes. *Proc. 7th Int. Conf. on Automated Inspection and Product Control*. Birmingham, UK.

Reid, G. T., Rixon, R.C. and Messer, H. I. (1984) Absolute and comparative measurements of three-dimensional shape by phase-measuring moiré topography. *Opt. and Laser Technol.*, **16**, 315–19.

Reid, G. T., Rixon, R. C., Marshall, S. J. and Stewart, H. (1986) Automatic on-line measurements of 3-D shape by shadow casting moiré topography. *Wear*, **109**, 297–304.

Robinson, D. W. (1983a) Role for automatic fringe analysis in optical metrology. *Proc. SPIE*, **376**, 20–5.

Robinson, D. W. (1983b) Automatic fringe analysis with a computer image processing system. *Appl. Opt.*, **22**, 2169–76.

Robinson, D. W. and Williams, D. C. (1986) Digital phase-stepping speckle interferometry. *Opt. Comm.*, **57**, 26–30.

Rodriguez-Vera, R., Kerr, D. and Mendoza-Santoyo, F. (1991) 3-D contouring of diffuse objects by Talbot-projected fringes. *J. Mod. Opt.*, **38**(10), 1935–45.

Schemm, J. B. and Vest, C. M. (1983) Fringe pattern recognition and interpolation using non-linear regression analysis. *Appl. Opt.*, **22**, 2850–3.

Schwider, J., Burow, R., Elssner, K. E., Gryanna, J., Spolaczyk, R. and Merkel, K. (1983) Digital wave-front measuring interferometry: some systematic error sources. *Appl. Opt.*, **22**, 3421–32.

Seemuller, W. W. (1982) Coherent image fringe contrast measurement with sensing arrays. *IEEE Trans. Instrum. Meas.*, **IM-31**, 180–4.

Smythe, R. and Moore, R. (1984) Instantaneous phase measuring interferometry. *Opt. Engng*, **23**, 361–4.
Snyder, J. J. (1980) Algorithm for fast digital analysis of interference fringes. *Appl. Opt.*, **19**, 1223–5.
Srinivasan, V., Liu, H. C. and Halioua, M. (1984) Automated phase-measuring profilometry of 3-D diffuse objects. *Appl. Opt.*, **23**, 3105–8.
Stricker, J. (1985) Electronic heterodyne readout of fringes in moiré deflectometry. *Opt. Lett.*, **10**, 247–9.
Takeda, M., Ina, H. and Kobayashi, S. (1982) Fourier-transform method of fringe-pattern analysis for computer-based topography and interferometry. *J. Opt. Soc. Am.*, **72**, 156–60.
Thalmann, R. and Dändliker, R. (1985a) Holographic contouring using electronic phase measurement. *Opt. Engng*, **24**, 930–5.
Thalmann, R. and Dändliker, R. (1985b) High resolution video processing for holographic interferometry applied to contouring and measuring deformations. *Proc. SPIE*, **492**, 299–306.
Theocaris, P. S. (1969) *Moiré Fringes in Strain Analysis*, Pergamon Press, Oxford.
Tichenor, D. A. and Madsen, V.P. (1978) Computer analysis of holographic interferograms for non-destructive testing. *Proc. SPIE*, **155**, 222–7.
Trolinger, J. D. (1985) Automated data reduction in holographic interferometry. *Opt. Engng*, **24**, 840–2.
van Bally, G., ed. (1979) *Holography in Medicine and Biology*, Springer Verlag, Berlin.
Varman, P. O. (1984) A moiré system for producing numerical data of the profile of a turbine blade using a computer and video store. *Opt. and Lasers in Engng*, **5**, 41–58.
Virdee, M. S., Williams, D. C., Banyard, J. E. and Nassar, N.S. (1990) A simplified system for digital speckle interferometry. *Opt. and Laser Technol.*, **22**(5), 311–16.
Williams, D. C., Nassar, N. S., Banyard, J. E. and Virdee, M.S. (1991) Digital phase-step interferometry: a simplified approach. *Opt. and Laser Technol.*, **23**(3), 147–50.
Womack, K. H. (1984) Interferometric phase measurement using spatial synchronous detection. *Opt. Engng*, **23**, 391–5.
Wyant, J. C. (1982) Interferometric optical metrology: basic principles and new systems. *Laser Focus*, May, 65–71.
Wyant, J. C., Oreb, B. F. and Hariharan, P. (1984) Testing aspherics using two-wavelength holography: use of digital electronic techniques. *Appl. Opt.*, **23**, 4020–3.
Wyant, J. C., Koliopoulos, C. L., Bhushan, B. and George, O. E. (1983) An optical profilometer for surface characterisation of magnetic media. *Trans. ASLE*, **27**, 101–13.
Yatagai, T. (1984) Fringe scanning Ronchi test for aspherical surfaces. *Appl. Opt.*, **23**, 3676–9.
Yatagai, T. (1991) Automated fringe analysis techniques in Japan. *Opt. and Lasers in Engng*, **15**(2), 79–91.
Yatagai, T. and Idesawa, M. (1982) Automatic fringe analysis for moiré topography. *Opt. and Lasers in Engng*, **3**, 73–83.
Yatagai, T. and Kanou, T. (1984) Aspherical surface testing with shearing interferometer using fringe scanning detection method. *Opt. Engng*, **23**, 357–60.

11
Monomode fibre optic sensors

J. D. C. JONES

11.1 MEASUREMENT USING OPTICAL FIBRES

An optical sensor may be defined formally as a device in which an optical signal is changed in some reproducible way by an external stimulus, such as temperature or strain. This definition covers a very wide range of devices, because an optical beam is characterized by a number of independent variables such as intensity, wavelength spectrum, phase and state of polarization. In an optical sensor, any one or a combination of these may be modulated by the measurand (parameter which is to be measured).

A fibre-optic sensor is, of course, an optical sensor which makes use of optical fibres. The fibre may actually be used as the sensing element, or simply as a flexible waveguide which conveys light to and from the region of measurement. As explained in section 6.4, those devices in which the beam is guided by a fibre at the measurement region are called **intrinsic**, the other type is called **extrinsic**. As will be described further below, fibre components may be used to control or modulate the guided beam, thus facilitating signal processing.

This chapter is solely concerned with the use of monomode optical fibre, which has the very important property of preserving the spatial coherence of the guided beam (Adams, 1981). Such fibre is therefore most suitable for sensor applications which exploit coherent optical techniques (Jackson and Jones, 1986a) – that is, those sensors in which the transduction mechanism is the modulation of optical phase, measured by interferometry (interferometric sensors), or those based on the modulation of the state of polarization (polarimetric sensors).

Interferometry is a well-established classical optical technique, and has been used for making very high-resolution measurements (Steel, 1983). However, classical interferometry demands the use of precision optical components mounted with great stability, and is thus often impractical for applications outside the research laboratory. The use of fibre optic techniques has greatly extended the range of applications of interferometry. It should also be pointed out that monomode fibre sensors share the advantages of other optical sensors, such as

Optical Methods in Engineering Metrology. Edited by D.C. Williams. Published in 1993 by Chapman & Hall, London. ISBN 0 412 39640 8

intrinsic safety, freedom from electromagnetic interference, and potential compatibility with optical communication systems.

In the next section, the basic transduction mechanisms exploited in these monomode fibre sensors are described. Fibre optic sensors have been demonstrated in which the transduction mechanism involves modulation of almost all of the characteristics of the optical signal. But, as noted above, the objectives of this chapter are more limited, and are restricted to sensors which involve modulation of either the phase or the state of polarization of the beam and for these particular characteristics, the use of monomode fibres is indicated.

Multimode fibres appeared long before monomode ones. They have core diameters much larger than the wavelength of light – 50 μm is a typical value. Consequently, many of their properties can be described using a ray optics model, and it is evident that a large number of possible ray paths exist. Hence, the output from such a fibre comprises a summation of the wavefronts from all of the possible ray paths, or modes. It is therefore not possible to ascribe a unique value to the phase of the emerging beam, and thus sensors based on phase modulation are not practical. Similarly, different modes experience different changes in their states of polarization as they propagate, so that the guided beam becomes depolarized, rendering polarization based sensors impractical also. Nevertheless, it is possible to use multimode fibres in phase- or polarization-modulated systems, but only provided that the phase or polarization state is transduced by an optical system which converts them to an intensity change, for example by using an interferometer or some polarizing optical components, for transmission by the multimode fibre. Such an approach, however, may lead to unwarranted complexity.

Multimode optical fibre sensors have been used with considerable success in intensity- and wavelength-modulated systems (Jones *et al.*, 1989). Nevertheless, they have their limitations. Intensity-modulated sensors are limited in their performance, because there are so many interfering effects other than the measurand which can alter the intensity within an optical system. Although various compensation schemes exist, they add complexity to the system. Wavelength-modulated sensors are very successful in certain specific applications, but offer direct transduction mechanisms for only a few measurands. Monomode fibres can be used in intensity and wavelength modulated sensors, but they offer no significant performance advantage and are more difficult to use in practice.

We shall see in section 11.2 that the mechanisms of phase and polarization modulation in monomode fibres provide access to a wide range of transduction principles and are the basis for a generic sensing technique. However, the phase modulation of an optical signal is only the first step in determining and displaying the value of a measurand. In all practical cases, some form of electrical output is required from the sensor, but optical detectors give an electrical output dependent on optical power. Thus optical processing is required to convert phase to power, and this is achieved by interferometry, the process being like that described in section 5.1.1. There are many available designs of optical fibre interferometers, and a selection of these are described in section 11.3, which also explains that the measurement of polarization state is a process closely related to interferometry.

A fibre optic interferometer requires many components in addition to the fibre itself. These are the subject of section 11.4, which includes both passive components such as directional couplers, which are the fibre equivalents of beamsplitters, and active components which are used to modulate the guided beam as an aid to signal processing.

A simple measurement of power at the output of an interferometer is not sufficient to determine unambiguously the value of the measurand. Successful demodulation demands a combination of optical and electronic signal processing. This involves interferometer design, electronic processing, and an interconnection between the two, often by using active components within the interferometer. These problems are discussed in section 11.5, which also describes the various noise sources present that control the resolution of the sensor system.

Section 11.6 is intended to give a representative view of the range of applications for which monomode fibre sensors are suited. It is divided into two parts: the first is concerned with intrinsic sensors, where the light is guided within the sensing element; the second is concerned with extrinsic sensors, which are effectively adaptations of classical coherent optical techniques in which fibres are used to enhance versatility or performance.

11.2 TRANSDUCTION MECHANISMS

We first consider in general terms how an optical beam will be modified by the response of a sensing element to a measurand.

11.2.1 Sensor transfer function

It is possible completely to describe an optical beam by its electric field vector, E. It is therefore possible to describe the propagation of a beam through a sensing element by the equation

$$E_{out} = T(X) E_{in} \qquad (11.1)$$

as shown in Fig. 11.1, where E_{in} and E_{out} are the electric field vectors of the beam before and after passing through the sensing element. The matrix T describes the optical properties of the sensing element, which are a function of the measurand field X. The measurand may be a scalar quantity, such as temperature, or a vector quantity, such as an electromagnetic field.

The purpose of the sensor system is to recover the value of the measurand. This is achieved by first measuring the fields E_{in} and E_{out}, and hence deriving the matrix T. From a knowledge of the manner in which T depends on X the measurand may be determined.

418 Monomode fibre optic sensors

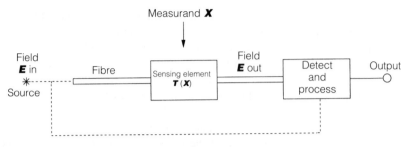

Fig. 11.1 The optical fibre sensor: basic principles.

For simplicity, we shall consider that the optical source is perfectly monochromatic, and that it is fully polarized. We therefore describe the electric field using a two element vector, such that

$$\boldsymbol{E} = (E_x, E_y)\, e^{i\omega t}, \qquad (11.2)$$

where E_x and E_y are the electric field amplitudes in the orthogonal x-and y-directions, transverse to the directions of propagation of the beam, ω is the optical angular frequency, and t is time. The state of polarization is described by the relative magnitudes of E_x and E_y, and the phase difference between them; E_x and E_y are thus complex scalars. The polarization properties of the sensing element may then be described by a 2×2 complex matrix known as the Jones matrix. With these simplifications, equation (11.1) becomes

$$\boldsymbol{E}_{\text{out}} = a\, \boldsymbol{B}\,(X)\, \exp\,[i\varphi_1\,(X)]\, \boldsymbol{E}_{\text{in}}, \qquad (11.3)$$

where $\boldsymbol{B}$ is the Jones matrix of the sensing element, and φ_1 is the mean phase retardance experienced by the beam as a result of propagating through the sensing element; a is a scalar describing the transmittance of the system, which we shall assume to be constant.

11.2.2 Phase-modulated sensors

For simplicity, we shall first consider non-birefringent systems, in which the phase velocity of the light is independent of the state of polarization. In this case $\boldsymbol{B} = \boldsymbol{I}$, the identity matrix, and equation (11.3) becomes

$$\boldsymbol{E}_{\text{out}} = a\, \boldsymbol{E}_{\text{in}}\, \exp\,[i\varphi_1\,(X)]. \qquad (11.4)$$

We shall now examine the way in which φ_1 depends on X. For any optical system, the phase retardance may be written

$$\varphi_1 = 2\pi n l / \lambda, \qquad (11.5)$$

where n is the refractive index, l is the physical path length and λ is the vacuum wavelength of the light. Generally speaking, when the measurand is applied to the sensing element, such as a length of monomode optical fibre, it will act to change both the physical length and the refractive index. For example, we may find the sensitivity of φ_1 to temperature T by differentiating equation (11.5) to give

$$\frac{\delta\varphi_1}{\delta T} = \frac{2\pi}{\lambda}\left(n\frac{\delta l}{\delta T} + l\frac{\delta n}{\delta T}\right), \tag{11.6}$$

where the first term in the bracket corresponds to the thermal expansion of the fibre and the second to the dependence of refractive index on temperature (the thermo-optic effect). We may determine the sensitivity of the fibre to other measurands by analogy. For example, for strain measurement the sensitivity arises from contributions due to physical extension and also to the strain-optic effect. Some practical values for fibre sensitivities are given in Table 11.1 (Jones et al., 1986; Rashleigh, 1983).

Table 11.1 Temperature and strain sensitivity coefficients measured using 0.1 m of highly birefringent monomode optical fibre (York Technology 'bow-tie' fibre, 3 mm beat length) at a wavelength of 633 nm.

X	$(1/l)(\delta\varphi_1/\delta X)$	$(1/l)(\delta\varphi_2/\delta X)$
T	100	5 rad K^{-1} m^{-1}
$\Delta l/l$	6.5×10^6	6.5×10^4 rad m^{-1}

11.2.3 Polarization-modulated sensors

Birefringence is induced by anisotropies in the sensing element. For example, if it is stressed then the phase velocities for linear states of polarization in the directions parallel and perpendicular to the direction of stress become unequal. A good example of this is those fibres which are designed to have a high degree of intrinsic linear birefringence. This is achieved by manufacturing the fibres with a high degree of internal stress (Payne et al., 1982). In common with all linearly birefringent materials (Fowles, 1975), there exist two specific linear states of polarization (the **eigenmodes**) which propagate without change and without coupling; however, the eigenmodes propagate at different phase velocities, and hence experience different phase retardances. For such fibres, equation 11.3 becomes

$$\boldsymbol{E}_{\text{out}} = a\boldsymbol{E}_{\text{in}}\begin{bmatrix} e^{i\varphi_2/2} & 0 \\ 0 & e^{-i\varphi_2/2} \end{bmatrix}\exp[i\varphi_1], \tag{11.7}$$

where the co-ordinate axes have been chosen so that they are aligned with the eigenmodes.

These special linearly birefringent fibres are very suitable for use in sensor applications, because their modal retardance φ_2 is a function of such parameters as temperature and strain (Jones et al., 1986; Rashleigh, 1983). We can calculate the sensitivity by analogy with the phase-modulated sensor. The modal retardance may be written

$$\varphi_2 = 2\pi (n_s - n_f) l/\lambda \tag{11.8}$$

where n_s and n_f are the effective refractive indices corresponding to the polarization eigenmodes. To consider the effect of a temperature change, for example, we differentiate as before to find

$$\frac{\delta \varphi_2}{\delta T} = \frac{2\pi}{\lambda} \left(\Delta n \frac{\delta l}{\delta T} + l \frac{\delta}{\delta T} \Delta n \right), \tag{11.9}$$

where $\Delta n = n_s - n_f$. The first term in the brackets corresponds to the thermal expansion of the fibre, and the second to the change in relative refractive index with temperature – that is, to the temperature dependence of the birefringence. Sensitivities to other measurands, such as strain and pressure, may be determined by analogy.

The measurand dependence of linear birefringence is a strong function of the structure of the particular fibre used. For example, a common technique by which fibres can be made birefringent is to build in some thermal stress during manufacture. It is not surprising that such fibres show an enhanced temperature sensitivity. An example of typical sensitivities for a particular fibre is shown in Table 11.1.

It is also possible to construct sensors based on measurand-induced changes in polarization azimuth (Langeac, 1982). A very important example is the measurement of magnetic field based on the Faraday effect (Fowles, 1975; Smith, 1978; Berwick et al., 1987). This effect may be considered as magnetically induced circular birefringence: when a beam of polarized light propagates through a medium in a magnetic field then the polarization azimuth is rotated. This technique may be employed in a low birefringence fibre, as is usual, or in a special fibre with high circular birefringence.

11.3 OPTICAL PROCESSING

We saw in section 11.2 that an important transduction mechanism is the modulation of the phase φ_1 of an optical signal. However, it is clearly impossible to measure this phase directly, because the optical frequency is extremely high (~500 THz for visible light) and thus beyond the bandwidth of practical photodetectors. It is therefore necessary to mix the phase-modulated optical signal coherently with one or more reference beams of closely similar frequency to produce a low-frequency difference signal within the detector bandwidth. This process is, of course, optical interferometry.

Many different types of fibre interferometer have been exploited for sensor applications, and a representative range of these are described below. We shall also describe how the concepts of interferometry can be extended for use in polarization state measurement.

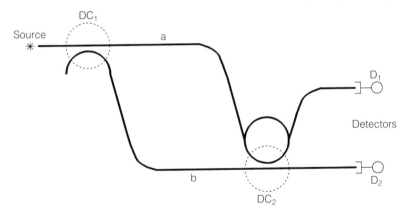

Fig. 11.2 The fibre-optic Mach – Zehnder interferometer: DC are directional couplers.

11.3.1 Two-beam interferometers

(a) The Mach–Zehnder interferometer

The simplest interferometers are those based on the mixing of two optical beams. As an example, consider the fibre optic Mach–Zehnder interferometer shown in Fig. 11.2. Light is coupled from a coherent laser source into a monomode optical fibre, and is amplitude divided into two paths, denoted *a* and *b*, by means of a directional coupler DC_1. A directional coupler is the fibre optic analogue of the conventional beamsplitter. In a sensor, we could consider path *a* to represent the signal beam and *b* to represent the reference beam. The signal beam is phase modulated by the measurand, whereas the phase of the reference beam remains constant. The signal and reference beams are recombined at a second directional coupler DC_2, giving two optical outputs at the detectors D_1 and D_2.

To analyse the behaviour of the interferometer, we shall make several simplifying assumptions.

1. The laser has a very long coherence length.
2. The directional couplers divide the optical power equally between their two outputs.
3. Polarization effects are small enough to be neglected.
4. There are no power losses in the system.

Let us denote the phase retardance of the signal and reference arms by φ_a and φ_b respectively as calculated from equation (11.5), and hence determine the electric field at detector D_1. Thus

$$E_1 = \tfrac{1}{2} E_0 [\exp(i\varphi_a) + \exp(i\varphi_b + \pi)], \tag{11.10}$$

where E_0 is the source electric field. This equation incorporates the fact that a beam coupled across a directional coupler experiences a $\pi/2$ phase shift in comparison with the transmitted beam (McIntyre and Snyder, 1973). Similarly, we may show that the field at detector D_2 is

$$E_2 = \tfrac{1}{2} E_0 [\exp(i\varphi_a + \pi/2) + \exp(i\varphi_b + \pi/2)]. \qquad (11.11)$$

We can now calculate the intensities I at the detectors using

$$I \propto \langle E \cdot E^* \rangle, \qquad (11.12)$$

where the brackets indicate a time average and the asterisk denotes the complex conjugate, so that

$$I_1 = \tfrac{1}{2} I_0 [1 - \cos(\varphi_a - \varphi_b)] \qquad (11.13)$$

and

$$I_2 = \tfrac{1}{2} I_0 [1 + \cos(\varphi_a - \varphi_b)], \qquad (11.14)$$

where $I_0 = E_0 \cdot E_0^*$. We have thus obtained two outputs, each of which depends on the phase difference between the arms of the interferometer. It may be seen that the two outputs vary in antiphase, so that $I_1 + I_2 = I_0$ for all values of $\varphi_a - \varphi_b$, thus ensuring that energy is conserved.

Equations (11.13) and (11.14) suggest that at either output the intensity will range from a maximum value $I_{max} = I_0$ to a minimum $I_{min} = 0$, depending on the value of $\varphi_a - \varphi_b$. In practice, this complete modulation will not be observed, and equations (11.13) and (11.14) become

$$I_{2,1} = I_0 [1 \pm V\cos(\varphi_a - \varphi_b)], \qquad (11.15)$$

where V is the **visibility** of the interference, defined by

$$V = \frac{I_{max} - I_{min}}{I_{max} + I_{min}}, \qquad (11.16)$$

which can take any value between zero and unity.

The visibility will be less than unity for one or a combination of the following reasons.

1. The coherence length of the source is not large enough in comparison with the optical path imbalance between the interferometer arms;
2. The components of signal and reference beams reaching a specific detector are unequal in intensity;
3. The states of polarization of the signal and reference beams are unequal.

Source coherence effects are discussed further below. In practice, single-longitudinal-mode diode lasers (Wilson and Hawkes, 1983) are often used as sources and have a coherence length of several metres (section 5.7.1), which is generally much greater than the path imbalance of the interferometer. Equality of signal and reference beam intensities can be ensured by selecting directional couplers with suitable power splitting ratios.

It is difficult to ensure that the signal and reference beams have the same state of polarization. This is because normal circular core optical fibre is slightly birefringent, with much of this birefringence arising from extrinsic sources such as bends and twists in the fibre (Kaminow, 1981). The resulting output polarization is therefore variable and unpredictable. The directional couplers may also introduce some birefringence. One solution is to construct the entire interferometer using highly birefringent fibre (Payne *et al.*, 1982), and to illuminate only one polarization eigenmode. Because the eigenmodes do not couple, the state of polarization of the guided beam is preserved. However, this approach also demands that polarization preserving couplers are used. A more straightforward solution is therefore to use normal circular core fibre, and to incorporate polarization controllers, described in section 11.4, into the interferometer. It is then possible to adjust the states of polarization in the two beams until they are equal, indicated by maximum observed visibility. Such polarization controllers may operate by introducing an adjustable amount of bending or twisting into the fibre (Lefevre, 1980).

(b)   The Michelson interferometer

Another commonly used form of two-beam interferometer is the Michelson, shown in Fig. 11.3. This is the fibre analogue of Fig. 5.1. It is similar to the Mach–Zehnder, except that it is a reflective configuration, in that the far ends of fibres a and b behave as mirrors, so that the guided beam returns and recombines at the same coupler as was used for beam division. These mirrors may be formed by depositing a reflective coating directly on to the fibre ends, which may be either metallic or a dielectric multi-layer. Alternatively, conventional mirrors may be used, either butted against the fibres or with suitable lenses interposed. However, the simplest approach is to leave the fibre ends uncoated, and to rely on the small Fresnel reflectivity of about 4% occurring at the quartz–air interface. In this case, the low reflectivity leads to a much reduced detected power. A special type of fibre reflector is described in (c) below.

The transfer function may be derived for the Michelson by analogy with the Mach–Zehnder, and it may be shown that

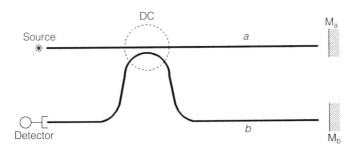

Fig. 11.3 The fibre-optic Michelson interferometer: M_a and M_b are mirrors.

$$I_1 = \tfrac{1}{2} I_0 \left[1 + V \cos (\varphi_a - \varphi_b) \right] \tag{11.17}$$

and

$$I_2 = \tfrac{1}{2} I_0 \left[1 - V \cos (\varphi_a - \varphi_b) \right]. \tag{11.18}$$

In calculating φ_a and φ_b, allowance must be made for the fact that in this reflective configuration the beams traverse each fibre twice, once in the outward and once in the return direction. For the visibility to be unity, assumptions (1), (3) and (4) in (*a*) above must be satisfied, and the reflectivities of the two mirrors must be equal. However, it may be shown that the visibility is independent of the coupler splitting ratio.

In some applications the Michelson is preferable to the Mach–Zehnder because the sensing fibre is connected only at one end, so that it may be used as a probe. Also, only one coupler is required, and its splitting ratio is unimportant. Conversely, although the interferometer has two complementary outputs, one of these is directed towards the source, and so it is not readily accessible. The availability of complementary outputs facilitates signal processing, as will be discussed further in section 11.5. It is clear that considerable optical power is fed back to the source, which may therefore become unstable. Diode lasers are particularly susceptible to optical feedback (Goldberg *et al.*, 1982) and an isolator such as those based on the Faraday effect (Stolen and Turner, 1982) should be used.

(c) The Sagnac interferometer

The Sagnac interferometer is a specialized form of two-beam interferometer, shown in Fig. 11.4, which is said to be reciprocal. We see that the two beams each propagate around the same closed loop, but in opposite directions. At first sight, it would appear that the two beams have identical optical path lengths, so that no phase

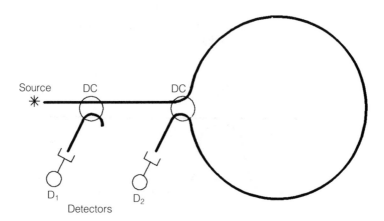

Fig. 11.4 The fibre-optic Sagnac interferometer.

shift would ever occur. However, certain stimuli do produce unequal, non-reciprocal phase shifts. The most important of these is angular velocity. If the closed loop of fibre is rotated in its own plane with an angular velocity Ω, then the effective optical paths of the two beams become unequal, and a phase shift

$$\varphi_\Omega = 4 \, \Omega \, \omega \, A/c^2 \tag{11.19}$$

is produced, where ω is the optical angular frequency, A is the enclosed area of the fibre loop and c is the velocity of light in free space. This is the operating principle of the fibre optic gyroscope (Culshaw and Giles, 1983).

The Sagnac interferometer may also be used as a loop reflector (Miller *et al.*, 1987; Millar *et al.*, 1988). That is, it can be made to act as a mirror. For example, when using a coupler with equal splitting ratio, in the absence of non-reciprocal phase shifts, and neglecting polarization effects and attenuation, all the optical power is coupled back into the input lead so that the interferometer acts as a fully reflective mirror. By varying the power splitting ratio of the coupler, any desired power fraction can be reflected.

11.3.2 Multiple-beam interferometers

(a) Fabry–Perot

Multiple-beam interferometers are used in situations where very high resolution is required. That is, it is possible to construct multiple-beam interferometers in which the intensity varies with phase more rapidly than is possible with two-beam designs. Probably the best-known multiple-beam interferometer is the Fabry–Perot (Fowles, 1975), and a fibre optic version is shown in Fig. 11.5 (Kersey *et al.*, 1983a; Stone, 1985). The Fabry–Perot may be considered as an optically resonant cavity, in which the beam is multiply reflected between the two highly reflective mirrors. In a fibre optic version, the mirrors may be formed using the same techniques as for the Michelson, described in section 11.3.1.

The derivation of the transfer function for a Fabry–Perot interferometer is given in many standard text books (Fowles, 1975). It may be shown that the transmitted intensity is

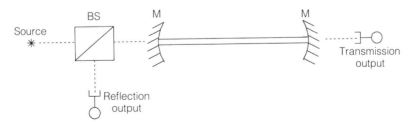

Fig. 11.5 The fibre-optic Fabry – Perot interferometer: BS is a beamsplitter

$$I = I_0/[1 + F \sin^2 (\varphi/2)] \tag{11.20}$$

where φ is the round-trip phase retardance, and the **finesse** F is

$$F = 4R/(1-R)^2, \tag{11.21}$$

where R is the mirror reflectivity, neglecting attenuation. The form of this transfer function comprises a set of sharp peaks, periodically spaced. As the finesse is increased by increasing the mirror reflectivity, the peaks become sharper. It may be seen that provided the interferometer is operated on the side of one of the peaks, then the intensity varies more quickly with phase than for the two-beam interferometer (see, for example, equation (11.13)). In the absence of attenuation, the reflection output is complementary, the sum of the transmitted and reflected intensities being constant.

Because of the requirement for high reflectivity mirrors, it is fairly difficult to fabricate a fibre Fabry–Perot having a large finesse. Conversely, it is particularly easy to make a low-finesse Fabry–Perot, in which the mirrors are simply the cleaved end faces of the fibre, which give a small reflection (Kersey et al., 1983b). This very simple interferometer is useful for many applications where a small sensing element is required. It may be deployed remotely by taking a short length of cleaved fibre, which acts as the Fabry–Perot sensing element, and splicing it to a connecting fibre of arbitrary length in such a way that a reflection occurs at the splice (Akhavan Leilabady et al., 1986). The Fabry–Perot may then be illuminated and interrogated via the connecting fibre.

(b) Ring resonator

A special form of multiple interferometer can be conveniently implemented in fibre form. This is the ring resonator, shown in Fig. 11.6, which has the considerable advantage that it requires no mirrors (Tai et al., 1986). Light from the source is launched into the input fibre, and some light is transmitted by the coupler to the detector. However, most of the light is coupled into the ring where it circulates. On each circulation, some light is coupled out of the ring and is transmitted to the detector. Multiple beams are therefore incident on the detector, where they interfere, yielding the transfer function

$$\frac{I}{I_0} = (1-\gamma_0)\left[\frac{(1-k)(1-k_r)}{(1+\sqrt{[kk_r]})^2 - 4\sqrt{[kk_r]}\sin^2\frac{1}{2}\left(\frac{\varphi}{2}-\pi\right)}\right], \tag{11.22}$$

where k and γ_0 are the power coupling coefficient and excess loss of the directional coupler, respectively, and φ is the phase retardance for one circulation of the ring. k_r is the resonant coupling ratio and is given by

$$k_r = (1-\gamma_0)\,e^{-2\alpha_0 L}, \tag{11.23}$$

where α_0 is the attenuation coefficient and L is the length of the fibre ring. This analysis does not include coherence or polarization effects.

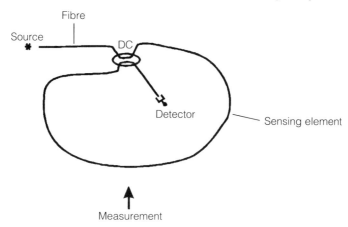

Fig. 11.6 The fibre optic ring resonator.

Equation (11.22) shows that the transfer function of the ring resonator is very similar to that of a Fabry–Perot interferometer operated in reflection. It is also possible to produce a complementary transmission-like output by inserting a second coupler in the ring. The ring resonator finds applications not only in sensing, but also as a resonant cavity for fibre lasers.

11.3.3 Polarimetric techniques

In section 11.2, it was explained that an important class of monomode fibre optic sensors is formed by those in which the transduction mechanism is the modulation of the state of polarization of the guided beam by the measurand, that is, those sensors based on changes in birefringence. In this section we discuss optical techniques by which this change in state of polarization may be converted to a change of intensity. Polarimetry may be considered as analogous to two-beam interferometry. In an interferometer, the two beams, ideally of equivalent state of polarization, follow spatially separate paths and mix coherently on a detector. In polarimetry, the two beams may follow physically identical paths but have orthogonal states of polarization. The orthogonal states are then forced to interfere by resolving them both in the same direction using a polarization analyser.

An arrangement for the measurement of changes in linear birefringence is shown in Fig. 11.7. The sensing element could conveniently be a length of monomode fibre with a high degree of intrinsic linear birefringence, of the type described in section 11.2. The measurement principle is interference between the polarization eigenmodes. It is therefore necessary to use a polarized source which couples to both of the fibre eigenmodes. Some form of polarization analyser is also required to resolve the eigenmodes along a common direction. For example, as shown in Fig. 11.7, a polarizing beamsplitter may be used. A half-wave plate is useful for rotating the

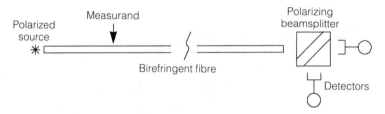

Fig. 11.7 The basic polarimetric sensor.

polarization azimuth of the beam emerging from the fibre, so that the angle α between the eigenaxes of the fibre and of the polarizing beamsplitter can be adjusted.

To derive the transfer function of the system, we note that the electric field at either of the detectors may be written

$$E_1 = P_i(\alpha) B_L E_{in} e^{i\varphi_1}; \quad i = 1, 2 \tag{11.24}$$

in the absence of attenuation and coherence effects, where E_{in} is the electric field at the source and φ_1 is the mean phase retardance for the optical system; $P_i(\alpha)$ and B_L are the Jones matrices for the polarizing beamsplitter and the fibre respectively. We see from equation (11.7) that

$$B_L = \begin{bmatrix} e^{i\varphi_2/2} & 0 \\ 0 & e^{-i\varphi_2/2} \end{bmatrix}, \tag{11.25}$$

where φ_2 is the modal retardance. For the beamsplitter

$$P_1(\alpha) = R(-\alpha) P_i(0) R(+\alpha), \tag{11.26}$$

where $R(\alpha)$ is a rotation matrix given by

$$R(\alpha) = \begin{pmatrix} \cos\alpha & -\sin\alpha \\ \sin\alpha & \cos\alpha \end{pmatrix} \tag{11.27}$$

and $P_i(0)$ are the Jones matrices for the beamsplitter with $\alpha = 0$, such that

$$P_1(0) = \begin{pmatrix} 1 & 0 \\ 0 & 0 \end{pmatrix}; \quad P_2(0) = \begin{pmatrix} 0 & 0 \\ 0 & 1 \end{pmatrix}. \tag{11.28}$$

For simplicity, we shall consider the optimum case, which gives unit visibility polarization interference with maximum intensity. To achieve this, it is necessary to populate equally the polarization modes of the fibre, and to resolve them equally with the polarizing beamsplitter. For example, we may choose as an input a linear state with an azimuth of $\pi/4$ rad relative to the fibre eigenaxes, so that

$$E_{in} = \frac{1}{\sqrt{2}} \begin{pmatrix} 1 \\ 1 \end{pmatrix} E_0 \tag{11.29}$$

where E_0 is the magnitude of the source electric field. To resolve the two modes equally, we require that $\alpha = \pi/4$, giving

$$P_1(\pi/4) = \tfrac{1}{2}\begin{pmatrix} 1 & 1 \\ 1 & 1 \end{pmatrix}; \; P_2(\pi/4) = \tfrac{1}{2}\begin{pmatrix} 1 & -1 \\ -1 & 1 \end{pmatrix}. \quad (11.30)$$

We may thus substitute equations (11.29) and (11.30) into equation (11.24), and calculate the output intensities

$$I_1 = \tfrac{1}{2} I_0 (1 + \cos \varphi_2) \quad (11.31)$$
$$I_2 = \tfrac{1}{2} I_0 (1 - \cos \varphi_2), \quad (11.32)$$

where I_0 is the source intensity. We thus see that the two outputs are complementary.

11.3.4 Spectral techniques

(a) Two wavelengths

A technique for extending the measurement range of interferometers which has a long history in classical interferometry is to use two different sources of slightly different wavelength, as described in section 5.7.3.

A suitable experimental arrangement is shown in Fig. 11.8. Two sources are used, with wavelengths λ_1 and λ_2. Let us consider the situation when only the first source is illuminated; then, with the simplifying assumptions 1 to 4 of section 11.3.1(*a*), the transfer function is

$$I(\lambda_1) = I_0(\lambda_1)[1 + \cos \varphi(\lambda_1)]; \; \varphi(\lambda_1) = 2\pi n l/\lambda_1, \quad (11.33)$$

whereas, with only the second source illuminated,

$$I(\lambda_2) = I_0(\lambda_2)[1 + \cos \varphi(\lambda_2)]; \; \varphi(\lambda_2) = 2\pi n l/\lambda_2, \quad (11.34)$$

with the assumption that λ_1 is close to λ_2, so that we can neglect dispersion by treating n as a constant with respect to wavelength. With both sources illuminated, and taking $I_0(\lambda_1) = I_0(\lambda_2) = I_0$, we have

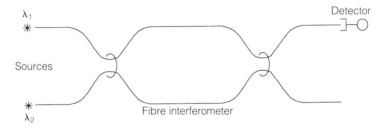

Fig. 11.8 Extended measurement range interferometer using two source wavelengths.

$$I = I(\lambda_1) + I(\lambda_2) = 2I_0 \left\{ 1 + V \cos\left[\pi n l \left(\frac{\lambda_1 + \lambda_2}{\lambda_1 \lambda_2}\right)\right] \right\} \quad (11.35)$$

with

$$V = \cos\left[\pi n l \left(\frac{\lambda_2 - \lambda_1}{\lambda_1 \lambda_2}\right)\right]. \quad (11.36)$$

Therefore, by making a simultaneous measurement of both the visibility and the phase of the interference, the unambiguous measurement range has been extended by a factor of $\lambda_2/(\lambda_2 - \lambda_1)$ in comparison with the interferometer illuminated by λ_1 alone.

(b) Low-coherence source

It should be clear that it would be possible to extend the measurement range further by using three or even more wavelengths. One may generalize this argument to an infinite number of wavelengths, corresponding to a single source of finite spectral width (Al–Chalabi *et al.*, 1983). Such a source is obviously not monochromatic, and the spectral width $\Delta\lambda$ may be characterized in terms of the coherence length l_c of the source, where

$$l_c \sim \lambda^2 / \Delta\lambda. \quad (11.37)$$

If the path imbalance in the interferometer Δl exceeds the coherence length, then the visibility of the resulting interference will be small. For example, for a two-beam interferometer in which conditions 2 to 4 of section 11.3.1 (*a*) are satisfied, the transfer function is given by equation (11.16) and the visibility becomes

$$V = \exp - (|\Delta l| / l_c). \quad (11.38)$$

We therefore see that the position of maximum visibility corresponds to $\Delta l = 0$, thus defining a unique point in the interferometer transfer function.

A useful practical arrangement is shown in Fig. 11.9. This comprises a remote sensing interferometer of nominal path imbalance Δl_s ($\neq 0$) connected by a fibre to a local receiving interferometer of adjustable path imbalance Δl_R ($\neq 0$). In operation, Δl_R is adjusted until maximum visibility is obtained, at which point

$$|\Delta l_R| - |\Delta l_s| = 0 \quad (11.39)$$

and thus Δl_s may be determined.

11.4 MODULATORS AND COMPONENTS

Signal processing in fibre interferometers is concerned largely with controlling the guided beam. For example, the optical arrangements of section 11.3 have involved the use of passive components for amplitude division of the beam, and for various polarization operations. We shall see in section 11.5 that it is often also necessary to

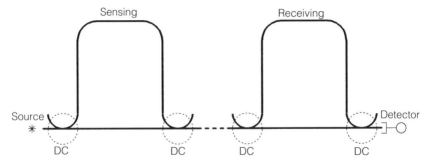

Fig. 11.9 Dual interferometers.

employ various active components for the modulation of phase, polarization or optical frequency. Devices which carry out these operations are described below.

11.4.1 Phase modulation

A direct means of controlling the phase retardance of the guided beam is to vary the optical path length. The first technique reported, and one which is in general use, is to employ a piezoelectric transducer to strain the fibre in its axial direction (Jackson *et al.*, 1980). The amplitude of the induced phase change can be enhanced by wrapping a length of the fibre round a piezoelectric cylinder.

An alternative approach involves heating the fibre to induce a phase change. This is achieved conveniently by applying a metallic coating to the fibre and passing an electric current through it. In this case only a low-voltage but high-current power supply is required. This approach is advantageous because the thermo-optic coefficient of the fibre is fairly high, and because the coating can be applied over a substantial length, allowing phase modulations of large amplitude (White *et al.*, 1987).

It is not always desirable to use an electrically active phase-shifting element, perhaps for reasons of safety. An alternative technique is to tune the wavelength of the source, in conjunction with an unbalanced interferometer. This approach may conveniently be adopted when diode laser sources are used, because their optical frequency shows some dependence on the injection or supply current. For a typical diode laser, a frequency shift of about 3 GHz/mA is produced. The maximum amplitude of frequency modulation that can be applied is limited by the onset of mode hopping; that is, by the abrupt switching of the laser output to an adjacent longitudinal cavity mode. For most lasers the maximum practical frequency deviation is about 30 GHz. This topic is discussed in more detail in section 5.7.

To calculate the phase shift, we recall from equation (11.5) that

$$\varphi_1 = 2\pi n \frac{l}{\lambda} = 2\pi n l \frac{v}{c}, \qquad (11.40)$$

where v is the optical frequency of the source. Hence an optical frequency shift Δv will produce a phase change

$$\Delta\varphi_1 = 2\pi n l \frac{\Delta v}{c} \tag{11.41}$$

and we note that the phase change is thus proportional to the path imbalance. However, for reasons that will be explained in section 11.5.3, it is desirable to minimize this path imbalance in a practical design. An obvious disadvantage of the technique is that the current modulation inevitably introduces an unwanted intensity modulation of the laser.

11.4.2 Polarization modulation

(a) Polarization ellipticity

State of polarization can be controlled either by inducing birefringence into fibres which naturally have a low birefringence (that is, fibres with nominal cylindrical symmetry), or by exploiting the properties of highly birefringent fibre. For example, many polarization controllers are based on introducing controlled bends or twists into fibres using various types of transducer; for a summary of such techniques, see Tatam *et al.* (1987a). It was shown in section 11.2 that the modal retardance φ_2 of a fibre with high linear birefringence controls the ellipticity of the state of polarization. It is therefore possible to make a polarization controller by means such as straining or heating a birefringent fibre. For example, a piezoelectric transducer can be used, in a manner analogous to that described in section 11.4.1. However, it is generally undesirable to coil a birefringent fibre on to a piezoelectric cylinder, because excessive bending will cause power coupling between the eigenmodes.

An alternative approach is to use a frequency-modulated source. The optical path lengths corresponding to the eigenmodes are unequal, so that a change in optical frequency will modulate φ_2. However, the difference in the path length is small, being one wavelength per beat length of fibre, typically a few millimetres, so that long fibres are needed to produce useful modulations. To overcome this, a two-beam interferometer (fibre or otherwise) may be constructed in which polarization components are used to produce orthogonally polarized beams in the two arms. The phase difference between the two arms is then equivalent to φ_2. The path imbalance can be designed as any convenient value, so that any desired polarization control can be achieved by laser frequency modulation (Jackson and Jones, 1986b). Alternatively, any phase-modulation technique can be used in one of the interferometer arms to produce polarization modulation. It is difficult to construct such a polarization interferometer in all-fibre form, because the polarization components used must be of high quality and stable in their properties (Tatam *et al.*, 1988). It is straightforward to convert a polarization state of controlled ellipticity to one of controlled azimuth, simply by using a quarter wave plate whose eigenaxis is at $\pi/4$ rad to the polarization eigenmodes of the incident beam (Jackson and Jones, 1986b).

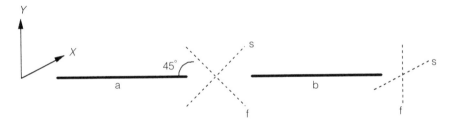

Fig. 11.10 Full polarization controller: a and b are highly birefringent fibres with eigenmodes shown as f(fast) and s(slow). The coordinate system is shown by x and y.

(b) Complete polarization control

In many situations, it is important to have complete control over the state of polarization, both in ellipticity and in azimuth. This basically demands two separate modulators which are concatenated. An example is shown in Fig. 11.10, which comprises two sections of linearly birefringent fibre, each with separately controllable modal retardance φ_2 perhaps by using piezoelectric stretchers (Tatam *et al.*, 1987b). In one implementation, the input state of polarization is linear, and is of azimuth $\pi/4$ with respect to the eigenaxes of the first fibre section. In this case, through control of the modal retardances, say α and β for the first and second fibre sections, respectively, it is possible to produce any desired output state. To see this, we may calculate the output electric field from

$$E_{\text{out}} = B_\beta R\left(-\frac{\pi}{4}\right) B_\alpha R\left(+\frac{\pi}{4}\right) E_{\text{in}} ; E_{\text{in}} = \begin{pmatrix} 1 \\ 0 \end{pmatrix}, \quad (11.42)$$

where B_α and B_β are the Jones matrices for the first and second fibre sections, respectively. We thus find that

$$E_{\text{out}} \propto \begin{pmatrix} \cos(\alpha/2) \\ \sin(\alpha/2) \exp - i(\beta + \pi/2) \end{pmatrix} \quad (11.43)$$

which is an arbitrary state of polarization that may take any value depending on the values of α and β. It is obviously possible to use this controller to perform the inverse function; that is, to convert an arbitrary state of polarization to a known linear one. To carry out the more general task of converting an arbitrary state to any other arbitrary state requires a controller incorporating additional sections of fibre.

(c) Passive components

The most important passive component is the polarization analyser, and devices have been reported based on both normal and highly birefringent fibre. The evanescent field polarizer exploits the fact that a metal-clad waveguide shows a strong polarization dependence for attenuation. A device based on this principle may be fabricated by locally polishing the cladding of a fibre to expose the evanescent field and then depositing a metal coating. Extinction ratios of ≥ 45 dB have been reported for this type of component (Gruchmann et al., 1983). An alternative approach involves depositing some birefringent material on to the polished fibre (Liu et al., 1986). If the refractive indices for the eigenstates of the birefringent material are greater and less than the effective refractive index for the guided wave, respectively, then one of the states will be very strongly attenuated with respect to the other. By using a birefringent crystal, extinction ratios of > 60 dB have been obtained (Bergh et al., 1980a).

Polarizers may readily be fabricated using birefringent fibre. For a sufficiently long wavelength, the attenuation coefficients for the eigenmodes differ significantly. Therefore, at this wavelength, if the beam propagates through a sufficient length of the fibre, it will become polarized. The effect may be enhanced by exploiting the fact that, for a given wavelength, the increased attenuation coefficients produced by bending the fibre are polarization dependent (Varnham et al., 1983). These coiled fibre polarizers have been demonstrated with extinction ratios ~ 60 dB (Varnham et al., 1984).

We noted above that polarization controllers could be made by bending or squeezing the fibre, thus inducing birefringence to form fibre equivalents of wave plates or rotators. For example, a coiled fibre exhibits linear birefringence, where the retardance is adjustable via the bend radius and length, and the tension. A full polarization state controller may be realized by forming two coils in series from a single fibre, each with a linear retardance of $\lambda/4$ (Lefevre, 1980). By rotating the coils relative to the input state and to each other, any output state may be generated.

11.4.3 Frequency modulation

One of the most important methods of signal processing in fibre interferometers is the heterodyne technique, which is described in section 11.5.3. Implementation of this technique requires that a shift in optical frequency is produced in one of the arms of the interferometer (section 5.3). In a classical optical system, a very useful frequency-shifting component is the Bragg cell. This device, which is also described in section 6.1.2, operates by coupling ultrasonic travelling waves into an optical medium such as a quartz block. The acoustic pressure waves effectively produce a moving diffraction grating, which is formed by the changes in refractive index corresponding to the pressure differences caused by the acoustic wave. Light which is diffracted by this moving grating is consequently frequency shifted. Commercially available Bragg cells typically provide a frequency shift of 40 or 80 MHz.

(a) Fibre optic frequency shifters

The bulk-optic Bragg cell is clearly not easily compatible with fibre optic systems. For this reason, research into fibre optic frequency shifters is very active at the time

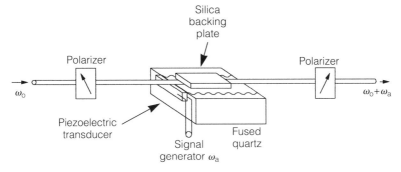

Fig. 11.11 Frequency shifter using travelling acoustic waves in birefringent fibre.

of writing. One development, which has strong analogies with the Bragg cell, is shown in Fig.11.11 (Risk *et al.*, 1984). It uses a birefringent optical fibre which is clamped between a quartz block and a backing plate. Travelling acoustic waves, propagating with a component in the direction of the fibre axis, are coupled into the quartz block. The acoustic wavelength and the direction of the fibre are chosen so that the component of the wavelength in the direction of the fibre axis is resonant with the polarization beat length of the fibre. These travelling waves periodically squeeze the fibre, which is orientated so that the axis of the squeezing bisects the fibre eigenaxes. Light is launched initially into only one of the fibre eigenmodes. However, the squeezing causes the eigenmodes to be coupled, thus transferring optical power to the other mode. Because this coupling is caused by a moving disturbance, the coupled power, which is all in the previously unilluminated mode, is frequency shifted. The frequency-shifted beam may be selected at the output using a polarizer.

In an alternative design, the fibre is not supported by a quartz block, but is held at its ends in free space (Pannell *et al.*, 1988). Near to one end of the fibre, ultrasonic flexure waves are coupled into it. That is, one end of the fibre is made to vibrate transversely. This causes bends to travel along the fibre, with the supports designed so that bends move unidirectionally. The bending causes mode coupling, and therefore a frequency shift is produced as before. Another possibility is to use two-moded low birefringence fibre (Kim *et al.*, 1986); that is to say, a fibre illuminated with an optical wavelength sufficiently short to allow the two lowest-order transverse modes to propagate, rather than just the fundamental mode. Travelling flexure waves may then be used to produce mode coupling and frequency shifting, as before.

(b) Pseudo-frequency shifting by phase modulation

The techniques described above are progressing rapidly, but could not yet be described as mature. Therefore, many currently used heterodyne processing techniques are based on using periodic phase modulation in order to produce a pseudo-frequency shift. The simplest example is the use of serrodyne or saw-tooth ramp phase modulation (Jackson *et al.*, 1982). During the linearly rising part of the ramp, a constant rate of change of phase is produced, which is equivalent to a frequency shift. During the flyback, the phase returns to its original value, and the process repeats. Because of the flyback, the frequency shift is impure; it is not a true single-sideband frequency shift in the sense that the flyback corresponds to a frequency shift of opposite sign. However, a satisfactory heterodyne carrier (see section 11.5.3) can be produced by band-pass filtering, with appropriate adjustment of the phase-modulation amplitude.

In principle, any asymmetric modulating waveform can be used to synthesize a heterodyne carrier. Alternatively, even symmetrical modulating waveforms like sinusoidal can be made asymmetric by gating on and off the signal from the photodiode at the output of the interferometer (Cole *et al.*, 1982). Any of the phase-modulation techniques of section 11.4.1 can be used in these pseudo-heterodyne techniques. Simple-harmonic sinusoidal phase modulation is especially desirable for use with piezoelectric phase shifters, because they can then be driven at their mechanical resonance frequency.

11.4.4 Directional couplers

The directional coupler is the fibre optic equivalent of the beamsplitter, and thus lies at the heart of most interferometer designs. The basic arrangement is shown in Fig. 11.12; it is a 2 × 2 coupler, which is the simplest possible kind, having two input fibres and two outputs, and is bi-directional. There are two fabrication techniques which are commonly used, namely polished and fused. These are described below.

(a) Fabrication techniques

A polished coupler (Bergh *et al.*, 1980b; Parriaux *et al.*, 1981) is fabricated by setting the fibre in a suitable substrate, such as a groove in a quartz block. The block is then polished to remove locally much of the cladding material, so that the evanescent field of the guided wave becomes exposed. The polished region is then

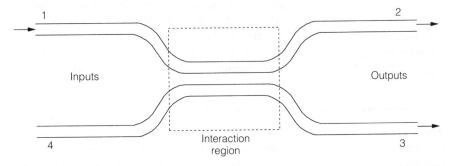

Fig. 11.12 Schematic representation of a directional coupler.

placed in contact with the polished region of a second similar fibre-block assembly, and the two are coupled by a thin film of index-matching fluid. Micropositioners may be used to adjust laterally the relative positions of the two blocks. In this way, the power coupling ratio (the fraction of the power coupled from one fibre core to the other) can be adjusted. This facility can be very useful in laboratory applications, although polished couplers are rather expensive for general use.

In the fused coupler (Kawasaki *et al.*, 1981; Villaruel and Moeller, 1981), the two fibres are slightly twisted together, and heated until they soften. They are then drawn axially, so that the diameter in the coupling region is greatly reduced. In the tapered region, the core diameters are vanishingly small, and the beam is guided in the cladding-core composite structure. The power is carried by the two lowest-order modes, symmetric and antisymmetric. As described below, power exchange between the two fibres then occurs as a result of beating between the modes. The power-splitting ratio may be controlled during the fabrication process, whilst the fibres are begin drawn. Following manufacture, the splitting ratio is fixed. Fused couplers are now mass-produced, and their use is widespread. They are robust, and reasonably inexpensive.

(b) Specifications

Couplers can be characterized in terms of the following parameters: insertion loss, backscatter and extinction ratio. The insertion loss is defined by

$$\Lambda = P_1/(P_2 + P_3) \qquad (11.44)$$

and the backscatter by

$$b = P_1/P_4, \qquad (11.45)$$

where P_i is the optical power in the *i*th arm of the coupler, as shown in Fig. 11.12. The extinction ratio relates to a coupler designed to couple as much power as possible from the first fibre into the second, and in this case is given by

Table 11.2 Demonstrated performance parameters for polished and fused type directional couplers, expressed in db.

Parameter	Polished	Fused
Insertion loss Λ	0.005[1]	0.05[3]
Backscatter b	70[2]	60
Extinction ratio η	50[2]	30[4]

1. Yu and Hall (1984)
2. Digonnet and Shaw (1982)
3. Beasley et al. (1983)
4. Bilodeau et al. (1987)

$$\eta = P_3/P_2. \tag{11.46}$$

It is usual to express Λ, b and η in decibels, and Table 11.2 shows typical values for the best that can currently be achieved for both the polished and fused types.

(c) Polarization effects

All couplers show polarization effects. The coupling ratio depends on polarization state, and the coupler itself is birefringent so that the guided beam changes its state of polarization. The polishing process makes a fibre asymmetrical, and therefore birefringent (Digonnet and Shaw, 1982; Stolen, 1986). The polarization eigenmodes are parallel and perpendicular to the plane of polishing. Therefore the state of polarization can be preserved provided that only a single eigenmode is excited. Fused couplers show considerably greater birefringence due to anisotropic stresses in the fused region (Bricheno and Fielding, 1984; Villaruel et al., 1983).

(d) Theory

It is possible to physically represent the coupling between fibres by considering the two adjacent cores as a single waveguide which supports the two lowest-order modes, being the symmetrical and anti-symmetrical modes referred to above (Eyges and Wintersteiner, 1981; Feit and Fleck, 1981); these two modes have propagation constants β_s and β_a respectively. The two modes are excited approximately equally by the incoming beam and commence in phase, so that all the power is in the core of the first fibre. As the modes propagate, they slip in relative phase, given that $\beta_s \neq \beta_a$. Eventually the condition

$$|\beta_s - \beta_a| l_0 = \pi \tag{11.47}$$

will be satisfied, where l_0 is the coupling length. At this point the modes are in antiphase, and all the power will be in the second fibre. Clearly, any desired coupling ratio can be produced by controlling the interaction length to produce the desired modal retardance (the phase difference between the modes). When the modal retardance is $\pi/2$, power is divided equally between the two output ports. It is then easily shown that there is a phase difference of $\pi/2$ between the two outputs, as noted previously in section 11.3.1.

As the interaction length is increased, power will be coupled alternately between the two output ports. For a typical fused coupler, the interaction length can be as short as 1 mm (Digonnet and Shaw, 1982).

(e) Special couplers

The propagation constants, β_s and β_a, depend both on state of polarization and wavelength. In this way it is possible to make couplers which are the equivalent of polarizing beamsplitters (Dyott and Bello, 1983; Pleibel *et al.*, 1983; Yokohama *et al.*, 1986; Kawachi *et al.*, 1982), or may be used for wavelength division (Digonnet and Shaw, 1983; Meltz *et al.*, 1983). For example, for some value of the interaction length the following conditions will hold:

$$|\beta_s^{(1)} - \beta_a^{(1)}| \, l = 2N\pi \qquad (11.48a)$$

$$|\beta_s^{(2)} - \beta_a^{(2)}| \, l = (2N+1)\pi \qquad (11.48b)$$

where the superscripts (1) and (2) refer to the two polarization eigenstates. In this case, state (1) will pass through the coupler to port 2, whereas state (2) will be completely coupled to port 3. Alternatively, (1) and (2) may represent two wavelengths λ_0 and $\lambda_0 + \Delta\lambda$, which would then be transmitted and coupled respectively. For wavelength division couplers, for fixed values of β_s and β_a, the channel spacing $\Delta\lambda$ decreases with interaction length l.

The construction of true polarization preserving couplers demands the use of highly birefringent fibre. The technical difficulties are then increased considerably. One serious problem is that the eigenaxes of the two fibres must be made accurately parallel.

It is also possible to construct couplers with a large number of ports. One approach is to splice together a number of 2×2 devices, or to fabricate a set of 2×2 couplers simultaneously to form a network which is sometimes called a **tree coupler.** One type of coupler which has special signal processing applications, noted in section 11.5.2, is the 3×3 (Koo *et al.*, 1982). In such a device, the waveguiding properties of the composite interaction region are complicated, and lead to phase differences between the outputs; it is these phase differences which are exploited in the signal processing applications.

11.5 ELECTRONIC PROCESSING

A fundamental problem with all monomode fibre optic sensors, whether based on interferometry or polarimetry, is that their transfer functions are periodic. We have shown that the complementary outputs from a two-beam system take the form

$$I_1 = I_0(1 - V\cos\varphi), \qquad (11.49a)$$

$$I_2 = I_0(1 + V\cos\varphi), \qquad (11.49b)$$

from equation (11.15) for example. The effect of this periodicity is that the value of the recovered measurand is ambiguous and the sensitivity is variable.

We may write

$$\varphi = \varphi_m + \varphi_d, \qquad (11.50)$$

where φ is the phase difference between the two beams, which is made up of the measurand-induced phase φ_m and an unwanted term φ_d arising from environmental drift and noise. We may calculate the sensitivity from

$$\frac{\delta I_1}{\delta\varphi_m} = VI_0 \left[\sin\varphi_m \cos\varphi_d + \cos\varphi_m \sin\varphi_d\right]. \tag{11.51}$$

It is instructive to consider the case where the measurand-induced phase φ_m is small, so that we can approximate equation (11.51) to

$$\frac{\delta I_1}{\delta\varphi_m} \simeq VI_0 \sin\varphi_d. \tag{11.52}$$

We therefore see that the sensitivity is dependent on the drift term φ_d, and therefore is certainly not constant. Furthermore, when φ_d is a multiple of π the sensitivity is zero and the signal fades completely. Ideally, we would wish to operate with $\varphi_d = \pi/2$ (modulo π), where the sensitivity is a maximum; this is called the **quadrature** condition.

The objective of any signal processing scheme is to produce an output of constant sensitivity, which is free from fading and is linear with respect to the measurand. There exist two basic classifications of processing scheme: homodyne and heterodyne. In the homodyne technique, phase control is used within the interferometer (or polarization control in a polarimeter) in order to hold it at the quadrature point, thus maintaining constant sensitivity. In the heterodyne technique, a frequency shift is imposed on one of the beams to generate a heterodyne carrier signal which is modulated by the measurand. A number of electronic techniques exist for the demodulation of such a carrier. We shall describe a representative range of both types of techniques below and conclude with a discussion of the noise present in the output signal.

11.5.1 Active homodyne processing

The basic principle of active homodyne signal processing is as follows. The intensity output from the interferometer (or polarimeter) is measured relative to some reference level which corresponds to the output at quadrature. The relative measurement is therefore zero when the interferometer is at quadrature. Otherwise, the relative measurement serves as an error signal in a servo loop feeding back to a phase control element in the interferometer, which is hence driven to quadrature.

As an example, Fig. 11.13 shows an arrangement based on a Mach–Zehnder interferometer with a piezoelectric element for phase modulation (Jackson et al., 1980). However, the method is general, and can be adapted to cover the various optical arrangements of section 11.3, using the modulators described in section 11.4. An advantage of using the Mach–Zehnder interferometer is that the complementary outputs are readily available, so that when they are combined differentially a signal of the form

$$i \propto I_2 - I_1 = 2I_0 V \cos\varphi \tag{11.53}$$

is produced, following from equations (11.49a,b). We see that the output i is thus zero at the quadrature condition. Therefore under these circumstances no separate reference level is required.

Electronic processing 441

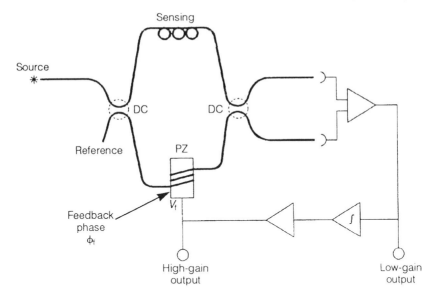

Fig. 11.13 Active homodyne signal processing: PZ is a piezoelectric phase modulator.

Let us suppose that the servo loop is initially in equilibrium, and that the equilibrium is disturbed by a phase change in the signal arm produced by the measurand. The output i will then no longer be zero and is fed back via an integrator, applying a voltage change to the piezoelectric element and so producing a change in reference phase exactly equal to the change in signal phase, thus restoring the interferometer to quadrature. The piezoelectric modulator is a linear device, so that the change in phase is directly proportional to the applied voltage. Therefore the change in measurand is linear with the modulator voltage, and the coefficient of proportionality may be found by calibration.

Whilst only d.c. methods of locking the interferometer phase have been described, a.c. techniques exploiting lock-in amplification are available. For example, a small a.c. dither may be applied to the path difference, so that by locking to zero modulation at the frequency of the dither signal using a phase sensitive detector as in section 2.4.2, the interferometer phase is stabilized. The phase will then be locked to $N\pi$ radians, where N is an integer, rather than the quadrature point. Alternatively, by locking to a null in the second harmonic of the dither frequency, the interferometer is stabilized at a quadrature point as required. Despite the availability of such techniques, the simplicity of d.c. phase stabilization is often compelling.

For many applications the use of a piezoelectric phase modulator is undesirable, because it is a high-voltage component which stores energy and may therefore be a risk to safety. A suitable alternative approach is to control the phase by using a frequency modulated laser diode in conjunction with an unbalanced interferometer.

442 Monomode fibre optic sensors

Active homodyne processing is also appropriate for use in polarimetric sensors. However, in this case the phase modulator must be replaced with a polarization modulator, using the techniques described in section 11.4.2.

Active homodyne processing has the considerable advantage of simplicity. However, the measurement range is always restricted by the finite range of phase or polarization compensation which can be provided by the modulator. For example, consider the use of the piezoelectric phase modulator in a situation where the measurand is increasing steadily. The phase compensation fed back must increase concomitantly. Eventually the modulator will reach the limit of its range, set either by the maximum available voltage or the maximum safe fibre strain. At this point the zero must be reset, and the value of the measurand will be lost. Similar arguments apply to all modulator types. For example, if one attempts to modulate the frequency of a laser diode too far, then it will discontinuously switch operation to an adjacent longitudinal mode; the maximum practical amplitude of frequency modulation is ~30 GHz. Two approaches exist which allow phase measurement over a wide range; these are passive homodyne processing, discussed in section 11.5.2, and heterodyne processing in section 11.5.3.

11.5.2 Passive homodyne processing

The objective of any passive homodyne processing scheme is to derive two or more outputs from the interferometer or polarimeter which bear a constant phase relationship to each other; that is, we wish to produce outputs of the form

$$I_A = \tfrac{1}{2} I_0 (1 + V\cos \varphi) \tag{11.54a}$$

$$I_B = \tfrac{1}{2} I_0 [1 + V\cos (\varphi + \varphi_B)] \tag{11.54b}$$

where φ_B is a bias phase introduced by some passive component. From equations (11.54a,b) we see that the two outputs cannot fade simultaneously, provided that $\varphi_B \neq 0$ is not a multiple of π. Ideally, we require that $\varphi_B = \pi/2$ (modulo π), so that when one output has faded the other is at quadrature, and vice versa. Such outputs are required in conventional interferometry for fringe counting (section 5.1.4). We shall first describe some of the techniques that can be used to generate these quadrature outputs, and then explain methods by which the quadrature outputs can be processed to give a signal which is linear with phase.

(a) 3 × 3 directional coupler

One way in which quadrature outputs can be derived is to use a 3 × 3 directional coupler (Koo *et al.*, 1982) as the recombiner in a Mach–Zehnder interferometer, as shown in Fig. 11.14. We recall that in a 2 × 2 directional coupler, a phase shift of $\pi/2$ is produced between the transmitted and coupled beams. In the 3 × 3 coupler, phase

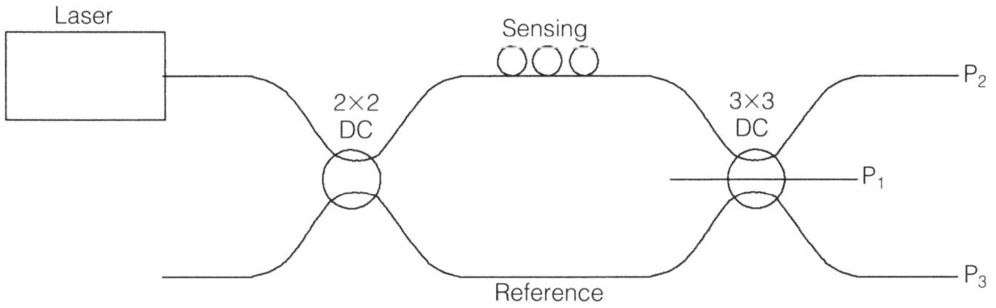

```
                              Laser                    Sensing
```

Fig. 11.14 Passive homodyne interferometer utilizing a 3 × 3 directional coupler.

shifts are also produced between the outputs, with the magnitude of the phase changes dependent on the properties of the coupler. Thus, for the three outputs of the interferometer in the general case we can write

$$I_1 = 2B_2 [1 + \cos\varphi] \tag{11.55}$$

$$I_2 = B_1 + B_2 \cos\varphi + B_3 \sin\varphi \tag{11.56}$$

$$I_3 = B_1 + B_2 \cos\varphi - B_3 \sin\varphi, \tag{11.57}$$

where for simplicity we have assumed that the visibility is unity. We see that any phase changes between the outputs can be represented by choosing appropriate values of B_1, B_2 and B_3. The outputs I_2 and I_3 may then be combined electronically in the following manner:

$$i_1 = I_2 + I_3 - 2B_1 = 2B_2 \cos\varphi \tag{11.58}$$

$$i_2 = I_2 - I_3 = 2B_3 \sin\varphi \tag{11.59}$$

thus giving the quadrature outputs required.

(b) Phase-switched technique

An alternative technique for providing the bias φ_B is to use a phase modulator. For example, we may drive the modulator with a square wave giving a phase modulation amplitude of $\pi/2$, so that when the square wave is low

$$I_A = \tfrac{1}{2} I_0 (1 + V\cos\varphi) \tag{11.60}$$

and when it is high

$$I_B = \tfrac{1}{2} I_0 [1 + V\cos(\varphi + \pi/2)] \tag{11.61}$$

thus yielding the desired quadrature outputs. Any of the phase-modulation techniques of section 11.4.1 can be used, but in a passive arrangement the frequency

444 Monomode fibre optic sensors

modulated diode laser in conjunction with an unbalanced interferometer is the most suitable approach (Kersey et al., 1983b).

(c) Electronic processing

Once the quadrature outputs have been produced, a number of electronic techniques exist to produce a signal linear with phase. We shall describe a method which is applicable to the demodulation of small-amplitude, high-frequency measurands. In this case, from equations (11.54a,b) with

$$\varphi_B = \pi/2, \; \varphi = \varphi_d + \varphi_{om} \sin\omega_m t \tag{11.62}$$

we see that

$$I_A = \tfrac{1}{2} I_0 (1 + V\cos\varphi_d - V\sin\varphi_d . \varphi_{om} \sin \omega_m t) \tag{11.63a}$$

and

$$I_B = \tfrac{1}{2} k I_0 (1 + V\sin\varphi_d - V\cos\varphi_d . \varphi_{om} \sin \omega_m t). \tag{11.63b}$$

We now high pass filter these signals to produce

$$I_A = -\tfrac{1}{2} k I_0 V \varphi_{om} \sin\varphi_d . \sin \omega_m t \tag{11.64a}$$

and

$$I_B = -\tfrac{1}{2} k I_0 V \varphi_{om} \cos\varphi_d . \sin \omega_m t \tag{11.64b}$$

so that by squaring and adding we obtain

$$\sqrt{(I_A^2 + I_B^2)} = \tfrac{1}{2} k I_0 V \varphi_{om} \sin \omega_m t \tag{11.65}$$

which is linear with the measurand phase φ_{om}, as required. It is evident that because equation (11.65) contains a square root, the sign of φ_{om} is ambiguous. However, in many practical applications involving small high-frequency measurands such as the fibre optic hydrophone (section 11.6.1(*b*)), only the amplitude of φ_{om} is required and not its sign. Furthermore, the square and add technique is applicable only when φ_{om} is sufficiently small that the small angle approximation applies. A more sophisticated technique which is applicable to measurands of larger magnitude is described by Koo et al. (1982).

(d) Conclusions

The advantages of the passive homodyne scheme over the active homodyne are that the measurement range is increased and that no electrically active elements are required in the region of measurement. However, it has the disadvantage of requiring special optical elements such as the 3 × 3 coupler, or using an unbalanced interferometer, thus increasing noise – see section 11.5.4. Also, the electronic processing

Electronic processing

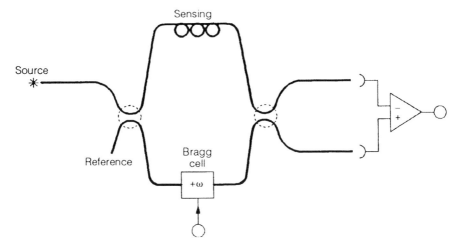

Fig. 11.15 Heterodyne signal processing using a Bragg cell frequency shifter.

involved is quite complex. An alternative approach which also yields a wide measurement range is heterodyne processing, and this is discussed below.

11.5.3 Heterodyne processing

Figure 11.15 shows one possible implementation of a true heterodyne processing scheme. The basic requirement is that a frequency offset is produced between the two beams, so that at one of the outputs

$$I_1 = \tfrac{1}{2} k I_0 \left[1 + V\cos\left(\varphi + \omega t\right) \right] \qquad (11.66)$$

where ω is the angular frequency offset. We see that equation (11.66) has the form of a phase modulated heterodyne carrier which can be demodulated using one of a number of established electronic techniques. Methods for producing the offset frequency are discussed in section 11.4.3.

11.5.4 Noise considerations

The resolution of a sensor is defined as the minimum detectable change in measurand. The resolution is set by the noise floor of the system, and for the monomode fibre sensor the principal sources of noise are:

1. Environmental effects on the sensing element and on other parts of the system;
2. Receiver noise;
3. Source noise.

446 Monomode fibre optic sensors

The most serious environmental perturbations arise from changes in ambient temperature and vibration, which acting on the sensing element cause phase changes indistinguishable from those produced by the measurand. Temperature changes are less serious than vibration, because they are usually of a much lower frequency than the measurand and hence their effect may be removed by filtering. Minimization of environmentally produced noise is an important aspect of the sensor design. However, the actual noise levels in practical systems are very much dependent on the specific application, and it is not possible to state a typical value.

Receiver noise arises from two different sources. A fundamental limit is set by the photodetector shot noise, and is imposed by the photon statistics of light. This matter is discussed in (a) below. Noise is also produced within the signal-processing electronics.

The laser source itself introduces two types of excess noise into the system; intensity noise and frequency noise (Dandridge et al., 1980a; Dandridge and Tveten, 1981). Intensity noise is simply fluctuation of the power of the laser, and is discussed in section 11.5.4(b). Frequency noise is a fluctuation of the optical frequency of the source, described in section 11.5.4(c). Diode laser sources are preferred for many applications, and our discussion will be restricted to these.

(a) Detector noise

The shot noise current produced by any photodetector is given by

$$i_{SN} = (2ei\,\Delta f)^{\frac{1}{2}} \tag{11.67}$$

where e is the electronic charge, i is the mean photocurrent and Δf is the detector bandwidth (Senior, 1985). We shall now determine the signal-to-noise ratio arising from shot noise. We shall consider the case of a two-beam interferometer held at quadrature in the presence of a small high-frequency measurand. The intensity at one output of the interferometer is thus

$$I_1 = \frac{1}{2} I_0 \left[1 - V \sin(\varphi_{om} \sin \omega_m t) \right]. \tag{11.68}$$

If the photodetector has an efficiency η, then in the small signal limit the photocurrent at the frequency ω_m is

$$i_s = \frac{1}{2} \eta I_0 V \varphi_m, \tag{11.69}$$

whereas the shot noise current is

$$i_{SN} = (2ei_1 \Delta f)^{\frac{1}{2}}, \tag{11.70}$$

where

$$i_1 \approx \frac{1}{2} \eta I_0 \tag{11.71}$$

so that from (11.67)

$$i_{SN} = (e\eta I_0 \Delta f)^{\frac{1}{2}}. \tag{11.72}$$

We can therefore calculate the signal-to-noise ratio (SNR) from equations (11.69) and (11.70), obtaining

$$\text{SNR} = \tfrac{1}{2}(\eta I_0/e\Delta f)^{\frac{1}{2}} V \varphi_m. \tag{11.73}$$

It is instructive to consider the minimum phase change which can be detected, which we shall assume to correspond to a SNR of unity. In the optimum case of unity visibility

$$\varphi_m \text{ (min)} = 2 \, (e\Delta f/\eta I_0)^{\frac{1}{2}}. \tag{11.74}$$

Some numerical examples are given in Table 11.3. It may be seen that the resolution depends only on the optical power, and improves as the power is increased.

Table 11.3 Minimum detectable phase, expressed in μrads and for a signal-to-noise ratio of unity, as limited by various noise sources. The results are based on the laser noise data of Fig. 11.16 and 11.17, and assume a detector efficiency of 0.5 A W^{-1}, with a (Hz)$^{1/2}$ detector bandwidth

Noise source	Measurand frequency	Optical path difference			
		.01 m		1.0 m	
		Optical power			
		1 μW	1 mW	1 μW	1 mW
Detector shot noise	10 Hz	1.1	0.04	1.1	0.04
	1 kHz	1.1	0.04	1.1	0.04
Intensity noise	10 Hz	8.4	8.4	8.4	8.4
	1 kHz	0.6	0.6	0.6	0.6
Phase noise	10 Hz	53	53	5300	5300
	1 Hz	3.0	3.0	300	300

(b) Intensity noise

Intensity noise is power fluctuation of the source, and is described by the excess noise parameter.

$$\xi(f, \Delta f) = P(f, \Delta f)/P_0, \tag{11.75}$$

where $P(f, \Delta f)$ is the r.m.s. optical power, present as fluctuations with frequencies between f and $f + \Delta f$, and P_0 is the mean source power. Some data for a typical diode laser are shown in Fig. 11.16 (Jackson and Jones, 1986a), plotted in decibels, where

$$\xi \text{ (dB)} = 20 \log \xi. \tag{11.76}$$

448   *Monomode fibre optic sensors*

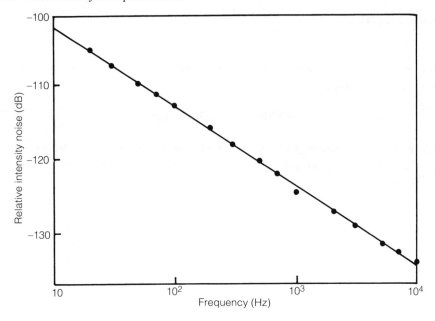

Fig. 11.16 Typical laser diode intensity noise spectrum.

It is now possible to calculate the SNR, which we shall do for the case of the two-beam interferometer held at quadrature for a small high-frequency measurand. The photocurrent for one output of the interferometer is, by analogy with equation (11.68) and using (11.75),

$$i_1 = \tfrac{1}{2}\,\eta\, I_0\,(1+\xi)\,[1 - V \sin(\varphi_{om} \sin\omega_m t)]. \tag{11.77}$$

The signal current at frequency ω_m is given by equation (11.69), and the corresponding intensity noise current is

$$i_{IN} = \tfrac{1}{2}\,\eta\, I_0\,\xi \tag{11.78}$$

giving

$$\text{SNR} = V\,\varphi_m/\xi \tag{11.79}$$

so that the minimum detectable phase (corresponding to SNR = 1) when the visibility is unity is

$$\varphi_m\,(\min) = 1/\xi. \tag{11.80}$$

Some typical values for the resolution when limited by intensity noise are given in Table 11.3.

Electronic processing 449

In practice, it is fairly straightforward to compensate for the effects of intensity noise (Dandridge and Tveten, 1981). It is only necessary to derive an intensity reference from the source (for example, by using an additional directional coupler or beamsplitter), and then to electronically divide the intensity output of the interferometer by this reference signal.

(c) Frequency noise

A serious source of noise is created by frequency jitter in the output of the laser. Some data for a typical diode laser are shown in Fig. 11.17, which shows that the amplitude of the optical frequency variations Δv diminishes as the observation frequency is increased. Therefore frequency noise worsens the resolution for low–frequency measurands. We see from equation (11.41) that the effect of frequency noise in an interferometer with an optical path imbalance of nl will cause phase changes indistinguishable from those produced by the measurand. The corresponding resolution is thus

$$\varphi_m \text{ (min)} = (2\pi nl/c)\Delta v \qquad (11.81)$$

It is evident that the effect of frequency noise is reduced by minimizing the path imbalance, and for a perfectly balanced interferometer frequency noise has no effect. However, many signal processing schemes demand a path imbalance in the

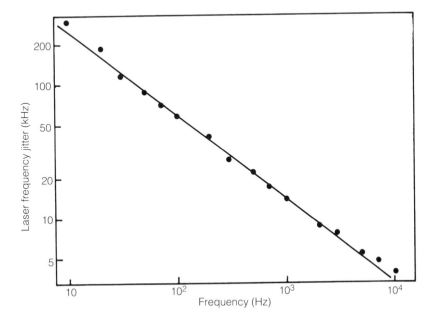

Fig. 11.17 Typical laser diode frequency noise spectrum.

interferometer, in order that phase modulation may be produced by controlling the frequency of the source, as explained in section 11.5.1. Furthermore, any processing scheme based on source modulation demands some minimum range of phase modulation amplitude, so that given the finite possible frequency modulation amplitude of the source, some minimum path imbalance requirement is imposed.

It is possible to compensate for frequency noise in a manner analogous to that used for the compensation of intensity noise. However, the compensation is practically more difficult, because it is necessary to derive a frequency reference from the source. A technique which has been successfully demonstrated (Favre and LeGuen, 1980) is to use a conventional high-finesse Fabry–Perot interferometer to monitor the source frequency and provide an error signal used in a feedback system to stabilize the source. In this way frequency noise reductions in the range 10 to 30 dB have been demonstrated.

11.6 APPLICATIONS

The previous sections have presented the physical principles of operation of optical fibre sensors in some detail; a range of practical applications to a variety of measurands will now be considered.

11.6.1 Intrinsic sensors

We have seen that intrinsic sensors are those in which the measurand acts either directly or indirectly on an optical fibre to modify its waveguiding properties in some predictable manner. The measurand thus induces a change in either the phase or state of polarization of the fibre-guided beam. Previous sections have discussed the basic transduction principles, and have shown how the phase or state of polarization may be determined in fibre systems. We now consider how the sensing element itself may be designed to enhance sensitivity to a specific measurand.

(a) Temperature sensors

The design of temperature–sensing elements is particularly straightforward because of the direct and sensitive thermal dependence of path length within the fibre. In fact, the fibre itself forms an adequate sensing element. Many researchers have chosen temperature sensing as an illustrative example to demonstrate different kinds of fibre interferometer and signal–processing schemes. The earliest reported work concerned the use of a fibre Mach-Zehnder as a temperature sensor (Hocker, 1979a), but since then almost all other types of interferometer, such as the Michelson (Corke *et al.*, 1983) and Fabry–Perot (Akhavan Leilabady *et al.*, 1986a) have also been used. Temperature resolutions in the millikelvin range have been achieved (Corke *et al.*, 1984). Sensors based on polarimetry have also received attention (Corke *et al.*,

1984). Polarimetric devices have inherently less sensitivity than interferometric ones, but with a concomitantly increased unambiguous measurement range corresponding to a phase change of 2π radians. Designs which exploit highly birefringent fibre, and in which both phase and polarization information are recovered simultaneously, combine the sensitivity of the interferometer with the enhanced measurement range of the polarimeter. Different optical arrangements which allow localization of the sensing region, and reduced thermal sensitivity in the fibre connecting lead, have been described for both interferometric (Yeh et al., 1990) and polarimetric devices (Corke et al., 1984).

The thermal response may be adjusted by using a coating on the sensing element. Metal coatings have a higher thermal expansivity than the fibre, and hence produce an axial strain when heated. The combined direct thermal and indirect strain effect enhances temperature sensitivity and also improves the linearity of the phase dependence on temperature (Corke et al., 1983).

Interferometric temperature sensors are receiving attention for some specialized applications. In chemical sensing, the temperature-sensitive fibre is coated with a catalyst so that, in the presence of a specific analyte, an exothermic reaction is produced, thus causing a detectable temperature rise. In this application the very high sensitivity of the fibre is essential. Also, fused quartz is chemically inert and an ideal substrate for coating. This concept has been demonstrated with a platinum coating for the detection of hydrocarbon gases (Farahi et al., 1987), and with an enzyme coating for biosensing (Tatam et al., 1987b).

(b) Pressure sensing and hydrophones

An uncoated quartz fibre shows a direct sensitivity to pressure (Giallorenzi et al., 1982). Hence the fibre itself is pressure responsive. However, direct measurement of quasistatic pressure is difficult in normal environments because of the strong cross sensitivity to temperature. The pressure sensitivity of the fibre may be enhanced with respect to the temperature sensitivity by bonding the fibre to a conventional pressure responsive element such as a diaphragm or tube, or by using a coating (see below). Interferometers may then be constructed with one arm having enhanced pressure sensitivity and the other in the same environment at nominally equal temperature, thus allowing some common mode rejection. However, progress in this area is relatively limited.

In contrast, sensing of acoustic pressure by interferometry is of great importance (Bucaro et al., 1977; Cole et al., 1977). Indeed, the hydrophone is one of the best-developed interferometric sensors. Optical fibre allows the fabrication of flexible sensing elements of great length, providing a distributed sensing capability. The fibre may be wound and configured in appropriate shapes for point sensing, for distributed sensing with tailored directional properties, or for acoustic field-gradient measurement (Giallorenzi et al., 1982). Most work on fibre hydrophones has exploited the

Mach–Zehnder interferometer, although other arrangements such as the Fabry–Perot have also been used (Dakin, 1987).

The sensitivity scales with length of the sensing element, and about 1 km is required to match the sensitivity of the best piezoelectric hydrophones. However, the sensitivity of the fibre can be enhanced by applying a compliant coating. Improvements in sensitivity by a factor of about 100 have been reported using silicone rubber coatings (Hocker, 1979b). Conversely, the sensitivity of the fibre can be reduced using stiff coatings. A low pressure sensitivity for the reference arm of the interferometer is desirable. The fibre pressure sensitivity arises from the change in physical dimensions and the strain optic effect; the two effects have opposite signs and the former dominates for uncoated fibres at low acoustic frequencies. Stiff metal coatings enhance the strain optic effect, and with the appropriate coating thickness a zero pressure sensitivity is achievable (Lagakos and Bucaro, 1981).

It was noted above that at frequencies up to about 10 kHz, the pressure-induced change in physical dimensions of the fibre is the dominant transduction mechanism. At intermediate frequencies in the range of 10 kHz–1 MHz, the isotropic refractive index change produced by the strain-optic effect dominates. At frequencies above about 1 MHz, the acoustic wavelength is smaller than the fibre dimensions so that the refractive index change is anisotropic, thus producing birefringence. It is hence possible to construct ultrasonic hydrophones based on polarimetry (Giallorenzi, 1987).

(c) Strain sensing

Table 11.1 shows that optical path length is strongly dependent on axial strain, and length changes below 10^{-12} m have been measured using fibre interferometers (Jackson et al., 1980). Somewhat less sensitive strain gauges have been demonstrated using highly birefringent optical fibres and polarimetric techniques (Rashleigh, 1983). However, it is difficult in practice to exploit the high strain sensitivity of fibre interferometers because of their strong cross sensitivity to temperature. This cross sensitivity may be alleviated by constructing sensors which measure simultaneously temperature and strain. Available techniques involve measuring two phases within the interferometer which are related to the two measurands by a pair of simultaneous equations. Inverting these equations thus yields temperature and strain. The two phases correspond either to illuminating the interferometer at two different wavelengths (Meltz et al., 1987), or, when a single wavelength is used, the two phases may correspond to the two eigenmodes of a birefringent fibre (Farahi et al., 1990).

Fibre strain sensing has received a new stimulus for the testing of advanced composite materials, used chiefly for aerospace applications (Culshaw, 1989). Conventional inspection techniques cannot reveal internal damage, such as delamination, in these materials. However, fibres embedded in the matrix of the material can be used to determine internal strains. Research has concentrated on embedding techniques which transfer strain accurately from the matrix to the fibre. Problems with thermal cross sensitivity remain to be solved fully. However, one test technique involves using pulsed laser radiation to induce ultrasound in the test specimen. In this technique very high frequency strains are generated in the material, which are easily measured by fibre interferometers and are well outside the bandwidth of thermal effects.

Another special example of the use of fibres in the measurement of oscillatory strain is the vortex-shedding flowmeter (Akhavan Leilabady *et al.*, 1985). When a bluff body is placed in a fluid stream, then in the appropriate Reynolds number range vortices are shed alternately from each side of the bluff at a frequency approximately proportional to the flow velocity. The shedding of a vortex creates a pressure change and produces a small strain in the bluff. Fibre interferometers have been used in vortex-shedding flowmeters either to detect the oscillatory strain in a bluff body, or with the sensing fibre itself acting as the bluff body.

(d) Magnetostrictive magnetometers

The magnetostrictive magnetometer is a special application of the fibre strain sensor (Yariv and Winsor, 1980). The primary sensing element is a magnetostrictive material, such as nickel, to which the sensing fibre is attached. A magnetic field produces a dimensional change in the sensing element and hence strains the fibre (Dandridge *et al.*, 1980b). Materials such as certain metallic glasses show very high coefficients of magnetostriction, and have enabled magnetic fields in the 10^{-5} A m^{-1} range to be measured, hence providing sensitivity comparable to high-quality flux-gate magnetometers (Koo and Sigel, 1982).

Many applications demand the measurement of weak magnetic fields varying at low frequencies of only a few hertz. Therefore, problems with temperature cross sensitivity arise. These may be resolved by exploiting the non-linearity of the magnetostrictive response (Kersey *et al.*, 1983c; Koo *et al.*, 1983). The magnetostrictive strain as a function of applied field has an approximately parabolic form for small fields. In operation, the sensing element is biased with a static field $\boldsymbol{H}_B$ and a low-amplitude dither field $\boldsymbol{H}_A$ is applied, thus producing an oscillatory strain of amplitude σ. The application of the measurand field $\boldsymbol{H}_X$ shifts the operating point on the strain-field characteristic and hence produces a change in σ from which $\boldsymbol{H}_X$ may be deduced. The dither field $\boldsymbol{H}_A$ effectively produces a carrier signal which is amplitude modulated by the measurand field $\boldsymbol{H}_X$. A dither frequency is chosen beyond the bandwidth of thermal effects. The measurand is recovered either by demodulating the AM carrier directly, or by using a feedback loop to control the bias field $\boldsymbol{H}_B$, hence locking the operating point. Closed loop operation effectively reduces hysteresis effects (Kersey *et al.*, 1985).

(e) Electromagnetic sensors

Apart from the magnetostrictive techniques noted above, a number of electromagnetically induced polarization phenomena have been exploited to form sensors.

The best developed of these is the magnetic field sensor based on the Faraday effect (Smith, 1978). When a polarized beam of light propagates through a medium in a magnetic field $\boldsymbol{H}$ the polarization azimuth is given by

$$\theta = \int V \boldsymbol{H} \cdot d\boldsymbol{l} \qquad (11.82)$$

where V is a dispersive constant specific to the medium which is called the Verdet constant and l is the distance travelled.

Faraday effect magnetometers have received considerable attention for the measurement of high electric currents, often at high potentials. One important application is in the electricity distribution system. The basic sensing element is a loop or coil of fibre enclosing the current-carrying conductor. Provided that the optical system is not birefringent in the absence of magnetic fields, then equation (11.82) together with Ampère's law shows that the polarization azimuth rotation is directly proportional to the enclosed current. The Verdet constant in fused silica fibre is low ($\sim 4 \times 10^{-6}$ rad m^{-1} at a wavelength of 633 nm), so that the polarization rotation is small. A substantial advantage of the fibre current sensor is that it is dielectric, so that electrical isolation is straightforward. Also, unlike the conventionally used current transformer, the sensor does not saturate at high magnetic fields and the rise time is very fast. The Faraday effect sensor is therefore suitable for fault location in power distribution systems, and as a diagnostic tool in large-scale plasma experiments.

Many signal-processing schemes have been employed. Early work used simple open-loop techniques, where a polarizer was used to measure the output azimuth from variations in intensity transmission, being manually adjusted to the position of maximum sensitivity (Rogers, 1983). Closed-loop (Berwick *et al.*, 1987) and heterodyne (Kersey and Jackson, 1986) techniques have also been used, involving the polarization modulation schemes described in section 11.4.

The presence of linear birefringence in the optical system produces changes in sensitivity (Donati and Annovazzi-Lodi, 1984). Changing linear birefringence is caused by ambient temperature changes and vibration, thus producing measurement errors. Efforts have been made to overcome these problems by using special spun fibres which have been twisted in manufacture to quench the linear birefringence, or have helical cores (Rashleigh and Ulrich, 1979; Birch, 1987).

An alternative solution to the birefringence problem is to use extrinsic sensors in which the sensing element is a block of some optical material with a relatively high Verdet constant (Rogers, 1973) such as flint glass or bismuth silicon oxide. However, complicated optical arrangements are required to allow the sensor to enclose the conductor.

Electric field, and hence voltage, can be measured using a variety of electro-optic effects. However, none of these are strong enough to be useful in fused silica fibres. All reported fibre based electric field sensors have therefore been extrinsic devices (see, for example, Mitsui (1986) and Kanoi (1986)).

(f) Gyroscopes

The fibre optic gyroscope is the best developed of all monomode fibre sensors (Culshaw and Giles, 1983). Almost all designs are based on the Sagnac interferometer described in section 11.3.1. The fibre gyroscope is a commercial reality and is competitive with the ring laser gyroscope. Although it does not offer such high performance, it is potentially much cheaper.

The basic arrangement for the Sagnac interferometer is shown in Figure 11.4. The dependence of phase shift on angular velocity may be written, from equation (11.19), as

$$\phi_s = 2\pi LD\Omega/\lambda c, \tag{11.83}$$

assuming that the fibre of length L is wound into a circular coil of diameter D.

The ideal Sagnac interferometer is reciprocal, being such that the clockwise and anticlockwise optical path lengths are identical in the absence of rotation. This condition of reciprocity must be met to a very good approximation, or the relatively small changes in phase produced by rotation will be submerged by other environmental effects such as temperature and strain. For example, consider an interferometer using 1 km of fibre wound with a diameter of 90 mm, illuminated at a wavelength of 840 nm. A rotation rate of 1.4 rad s^{-1} will produce a phase shift of π radians. Conversely, a similar phase shift is produced by a temperature change of 1 K in a fibre length of only 0.03 m; hence the clockwise and anticlockwise paths must be equivalent to 30 p.p.m. The fundamental resolution is set by shot noise in the photodetector (section 11.5.4) and for received powers in the 1–10 μW range, the rotational resolution is $\sim$1.5–5 x 10^{-7} rad s^{-1} Hz$^{-1/2}$. This demands a reciprocity within $\sim$10^{-13}, such that the phase shift produced by a 1 K temperature change is equal to the minimum detectable rotation rate as limited by shot noise.

Sources of non-reciprocity are discussed below. However, a primary requirement is the use of a mode filter to ensure that both beams propagate around the loop in the same spatial mode (Ulrich, 1980). A short length of monomode fibre is a suitable mode filter. Reciprocity also demands that the optical output from the interferometer cannot be taken from D$_2$ as shown in Fig. 11.4. Instead, a second directional coupler must be inserted into the input lead of the interferometer to give an output at D$_1$, where the beams have each experienced one coupling and one transmission at the coupler within the Sagnac loop. The use of a second coupler inevitably causes a loss of optical power.

The transfer function of the Sagnac interferometer is the same as for any two-beam interferometer (see equation (11.14)) and therefore requires phase biasing to achieve adequate sensitivity. Static phase modulation using the techniques of section 11.4.1 is inappropriate because their effects are reciprocal. However, the piezoelectric phase modulator can be used dynamically (Martin and Winkler, 1978). The modulator is inserted close to one end of the Sagnac loop and is driven harmonically, usually at a mechanical resonance. A wave packet entering the interferometer is split into clockwise and anticlockwise components. Because of the finite propagation time through the loop, the two components see the modulator at different times and hence experience different phase retardances. Dynamic phase biasing is thus achieved (Bergh et al., 1981).

Reciprocity can only be maintained for beams which are transmitted through the entire length of the Sagnac loop. Unwanted reflections within the loop seriously damage the reciprocity. Splices and connectors in the loop must be avoided, and reflections at interfaces between fibres and other components, such as integrated optic phase modulators whose use is described below, must be suppressed. One technique for reducing these reflections is to polish the end of the fibre at an angle, so that internally reflected power is outside the acceptance angle of the fibre core and is radiated into the cladding.

In all fibres an unavoidable source of backward propagating power is scattering. A component of the clockwise beam which is Rayleigh scattered within approximately the coherence length of the source will mix coherently with a similar component of the anticlockwise beam to give a non-reciprocal contribution to the interference at the output (Cutler et al. 1980). This problem is solved by using a source whose coherence length is as short as possible, but which has adequate spatial coherence to allow efficient power coupling into the fibre (Bohm et al., 1981). Superluminescent light-emitting diodes are suitable for this purpose. The reciprocity of the interferometer allows high-visibility Sagnac interference, even with short coherence length sources.

A serious cause of non-reciprocity arises from fibre birefringence effects, given that the polarization modes of the fibre have unequal optical paths (Kintner, 1981). Even when highly birefringent fibre of the highest quality is used, there is a cross talk between the modes with a maximum isolation of about 20 dB, which is insufficient to satisfy reciprocity requirements. However, when a short coherence source is used in conjunction with birefringent fibre, the path imbalance between the polarization modes is great enough to prevent the polarization cross terms from interfering (Frederiks and Ulrich, 1984).

The time-of-flight effects exploited in the phase-biasing scheme discussed above indicate that dynamic perturbations of the fibre loop, such as vibrations close to one end, will cause non-reciprocal phase shifts. This phenomenon has been exploited with advantage to produce new types of optical hydrophone (Udd, 1983), and for intrinsic (Dakin et al., 1987) and extrinsic vibration sensing (McBride et al., 1990a). Nevertheless, it is an unwanted effect in gyroscopes, and is minimized by winding the fibre coil symmetrically from its mid-point (Shupe, 1980).

A number of intrinsic non-reciprocal effects exist which must also be minimized to achieve high performance. For example, stray magnetic fields cause a non-reciprocal phase shift through the Faraday effect. Indeed, Sagnac interferometers have been used as current sensors (Akhavan Leilabady et al., 1986b). An ideal non-birefringent loop will show no Faraday-induced phase shift unless it encloses a current source, because if the source is outside the loop the integrated Faraday rotation around the loop is zero. However, linear birefringence distributed through the fibre causes different sections of the loop to exhibit different sensitivities to the Faraday effect, so that the Faraday rotation does not cancel, and even the earth's magnetic field can cause a measurable phase shift (Bergh et al., 1981). This effect can be minimized by using highly birefringent fibre to obtain as much linear birefringence in the loop as possible (Hotate and Tabe, 1986).

A fundamental source of non-reciprocity is the non-linear Kerr effect; that is, the dependence of path length on optical intensity (Ezekiel et al., 1981). This effect causes the path length of the beam propagating in one direction to be modulated by the intensity of the beam propagating in the opposite direction. The effect can be minimized by using sources pulsed with an appropriate duty cycle. A more convenient alternative is to use a source with thermal statistics. This may be approximated by a short coherence length source such as a superluminescent diode (Bergh et al., 1982).

In practical gyroscopes the limited bandwidth of piezoelectric dynamic phase modulators dictates the use of open-loop signal-recovery schemes. More recently, integrated optic wide-band phase modulators have been used to permit closed loop operation, thus providing high accuracy and linearity of scale factor over a wide dynamic range (Lefevre et al., 1986).

This section has concentrated on the use of the Sagnac interferometer. However, gyroscopes may also be formed using ring resonators as described in section 11.3.2 (Carroll et al., 1986). In a ring resonator, high finesse allows optical power to recirculate through the loop many times, so that shorter fibres are required than in the Sagnac interferometer. Although this is a promising area of current research, the Sagnac gyroscope is at present much better developed.

11.6.2 Extrinsic Sensors

An extrinsic sensor is a device in which a measurand may be recovered from the modulation which it imposes on a beam of light scattered or reflected from, or transmitted through, a remote measurement volume or sensing element. Fibres are used to guide light to the transmitting optical system, thus illuminating the measurement volume or sensing element, and then to guide the received light to the photodetector.

In extrinsic sensors the fibres can fulfil a number of different functions. At the simplest level they allow the source and detector, and other active components, to be separated from the passive sensor by flexible optical links. However, they may also be used to facilitate signal processing using the modulators described in section 11.4. When fibre components are used, then optical systems too complex to be implemented in bulk optics can be realized.

There are many examples of extrinsic sensors. An important class use a bulk-optic sensing element, such as those used for electromagnetic sensing discussed in section 11.6.1 above. Fibres have also been used to illuminate conventional interferometers (Gerges et al., 1988) or micro-optical assemblies (Venkatesh and Culshaw, 1985) or integrated optical devices (Hosokawa et al., 1988) which serve as sensing elements.

Other extrinsic sensors are best described as adaptations of classical optical techniques. For example, fibre systems have been developed for ellipsometry, polarimetry, refractometry and holography (Jackson and Jones, 1986b). Of special interest is the use of fibre techniques in laser velocimetry and vibrometry, which are described in more detail below.

All extrinsic sensors based on coherent optical techniques have similar general requirements. Monomode optical fibre provides a beam of high spatial coherence, with no need of further spatial filtering. Birefringent fibre may be used in applications where the state of polarization must be preserved. However, highly birefringent fibre components, such as directional couplers, are more expensive and less readily available than those made from normal fibres. It is therefore often more convenient to use normal fibres together with polarization controllers, as described in section 11.4.

A common requirement in extrinsic sensors is to maintain constant optical path lengths to avoid phase drifts, often caused by ambient thermal effects on the fibre. Suitable phase locking techniques have been described in section 11.5.1.

Laser velocimetry and the use of optical fibres for vibration sensing are described in detail in Chapter 6. Laser velocimetry is discussed briefly below as an illustration of the features of extrinsic sensors.

(a) Laser velocimetry

As seen in section 6.1, laser velocimetry is an important and well-established technique for the measurement of fluid and surface velocities (Durst *et al.*, 1976). The capabilities of the technique have been enhanced through the use of fibres and related components. Many optical arrangements have been used for different applications in laser velocimetry. One which has received considerable attention is the Doppler difference technique used chiefly for the measurement of transverse velocity components in fluids but also for in-plane vibration measurements on surfaces.

Fibre-based laser velocimeters can be classified into three basic groups which make progressively greater use of fibres: links, probes and all-fibre systems. Fibre-linked systems use fibres simply to separate the optical source and detector from a remote assembly containing the transmitting and receiving optics, and any active bulk-optic modulator used to aid signal processing (Buchave and Knuhtsen, 1982; Jackson *et al.*, 1984; Hironaga *et al.* 1985). For example, many Doppler difference laser velocimeters use a Bragg cell as a frequency shifter to allow heterodyne processing. In a fibre probe system, all active components, including the modulator, are separated from the passive probe to which they are linked by a cable containing several fibres (Hironaga *et al.*, 1985; Sasaki *et al.*, 1980; Knuhtsen *et al.*, 1982; Nguyen and Birch, 1984; Nakatani *et al.*, 1985). In an all-fibre system only passive bulk-optic components are used, with modulation achieved using fibre devices (Chan *et al.*, 1985). Fibre links and probes are well developed and systems are available commercially. All-fibre systems have the potential advantages of being exceptionally compact and rugged.

Reference-beam techniques for the measurement of out-of-plane vibration have paralleled the development of the Doppler difference technique. Apparatus using fibre links, probes and all-fibre systems have been realized. A number of different types of fibre interferometer have been used for different applications. The Michelson (McBride *et al.*, 1990b) and Mach–Zehnder (Lewin *et al.*, 1985) are most commonly employed, but the Fabry–Perot has also been used (Laming *et al.*, 1986); see section 6.4.3 for a more detailed description. Most recently, an adapted Sagnac interferometer has been demonstrated, yielding an optical output dependent on surface velocity with passive phase biasing (McBride *et al.*, 1990a).

(b) Further applications

Fibre optics have been exploited in a range of other techniques of engineering metrology. In particular, fibre optic systems for holographic interferometry (Chapter 7) have been developed (Jones *et al.*, 1984; Corke *et al.*, 1985). Fibres have also been used in the related technique of electronic speckle pattern interferometry (Chapter 8). New applications continue to be reported including techniques as diverse as flow visualization (Philbert *et al.*, 1986) and photon correlation spectroscopy (Brown, 1985).

The introduction of monomode fibre components into optical systems has led to a new class of sensor. In their simplest implementation, the fibre serves basically as a flexible link that conveys optical signals to and from a conventional optical sensor. At a higher level the fibre serves not only as the transmission medium but also as the sensor itself.

Although fibre optic sensors are relatively new, considerable progress has been made in the transfer from laboratory experiments into fully engineered prototypes and a number of products are now available. Some impetus for this transfer comes from defence applications, and the fibre gyroscope and hydrophone are well developed. A number of manufacturers of laser velocimetry equipment now offer fibre-based products, which are taking an increasing market share. Many other products which exploit fibres to make existing optical instruments more versatile are also being commercialized; for example, in remote vibration measurement. Much of the pioneering work on monomode fibre sensors is now entering a more mature phase of engineering development. However, there remains considerable scope for basic research, chiefly concerned with the problem of transferring more of the necessary signal processing from the electronic to the optical domain.

REFERENCES

Adams, M. J. (1981) *An Introduction to Optical Waveguides*, Wiley.
Akhavan Leilabady, P., Jones, J. D. C. and Jackson, D. A. (1985) Monomode fibre optic interferometric techniques in flow velocity measurement. *Optica Acta*, **32**, 233.

Akhavan Leilabady, P., Jones, J. D. C. and Jackson, D. A. (1986a) Combined interferometric-polarimetric fibre optic sensor capable of remote operation. *Opt. Comm.* **57**, 77.

Akhavan Leilabady, P., Wayte, A. P., Berwick, M., Jones, J. D. C. and Jackson, D. A. (1986b) A pseudo-reciprocal fibre optic Faraday rotation sensor. *Opt. Comm.* **59**, 173.

Al-Chalabi, S. A., Culshaw, B. and Davies D. E. N. (1983) Partially incoherent sources in interferometric sensors. *IEE Conf. Publ.*, **221**, 132.

Beasley, J. D., Moore, D. R. and Stowe, D. W. (1983) Evanescent wave fibre optic couplers: three methods. *Proc. OFC83*, ML5.

Bergh, R. A., Lefevre, H.C. and Shaw, H. J. (1980a) Single mode fibre optic polariser. *Opt. Letts.*, **5**, 479.

Bergh, R. A., Koder, G. and Shaw H. J. (1980b) Single mode fibre optic directional coupler. *Electron. Letts.*, **16**, 260.

Bergh, R. A., Lefevre, H. C. and Shaw, H. J. (1981) All single mode fibre optic gyroscope with long term stability. *Opt. Letts.*, **6**, 502.

Bergh, R. A., Lefevre, H. C. and Shaw, H. J. (1982) Compensation of the optical Kerr effect in fibre optic gyroscope. *Opt. Letts.*, **7**, 282.

Berwick, M., Jones, J. D. C. and Jackson, D. A. (1987) Alternating current measurement utilising the Faraday effect. *Opt. Letts.*, **12**, 293.

Bilodeau, F., Hill, K. O., Johnson, D. C. and Faucher, S. (1987) Compact low loss fused biconical taper couplers: overcoupled operation and asymmetric supermode cut-off. *Opt. Letts.*, **12**, 634.

Birch, R. D. (1987) Fabrication and characterisation of circularly birefringent helical fibres. *Electron. Letts.*, **23**, 50.

Bohm, K., Russer, P., Weidel, E. and Ulrich, R. (1981) Low drift fibre gyro using a superluminescent diode. *Electron. Letts.*, **17**, 352.

Bricheno, T. and Fielding, A. (1984) Stable low loss single mode couplers. *Electron. Letts.*, **20**, 230.

Brown, R. G. W. (1985) Laser fibre optics in biotechnology, *Trends in Biotechnology*, **8**, 200.

Bucaro, J. A., Dardy, H. D. and Carome, E. F. (1977) Fibre optic hydrophone. *J. Acoust. Soc. Am*, **62**, 1302.

Buchave, P. and Knuhtsen, J. (1982) Fibre optic laser anemometry measurements. *First International Symposium on Applications of Laser Anemometry to Fluid Mechanics*, Ladoan, Lisbon, p. 13.

Carroll, R., Coccoli, C. D., Cardarelli, D. and Coate, G. T. (1986) The passive resonator fibre gyro and comparison to the interferometer fibre gyro. *Proc. SPIE*, **719**, 169.

Chan, R. K. Y., Jones, J. D. C. and Jackson, D. A. (1985) A compact all-optical-fibre Doppler difference velocimeter. *Optica Acta*, **32**, 241.

Cole, J. H., Danver, B. A. and Bucaro, J. A. (1982) Synthetic heterodyne interferometric demodulation. *IEEE J. Quantum Electron.*, **QE18**, 694.

Cole, J. H., Johnson, R. L. and Bhuta, P. G. (1977) Fibre optic detection of sound. *J. Acoust. Soc. Am.*, **62**, 1136.

Corke, M., Jones, J. D. C., Kersey, A. D. and Jackson, D. A. (1985) All single-mode fibre optic holographic system with active fringe stabilisation. *J. Phys. E*, **18**, 185.

Corke, M., Kersey, A. D., Jackson, D. A. and Jones, J. D. C. (1983) All fibre Michelson thermometer. *Electron. Letts.*, **19**, 471.

Corke, M., Kersey, A. D., Liu, K. and Jackson, D. A. (1984) Remote temperature sensing using polarisation preserving optical fibre. *Electron. Letts.*, **20**, 67.

Culshaw, B. (1989) Applications of fibre optic sensors in the aerospace and marine industries, in *Optical Fibre Sensors: Systems and Applications* (eds. B. Culshaw and J. Dakin) Vol. 2, Ch. 18, Artech House.

Culshaw, B. and Giles, I.P. (1983) Fibre optic gyroscopes. *J. Phys E.*, **16**, 5.

Cutler, C. C., Newton, S. A. and Shaw, H. J. (1980) Limitation of rotation sensing by scattering. *Opt. Letts.*, **5**, 488.

Dakin, J. P. (1987) Multiplexed and distributed optical fibre systems. *J. Phys. E*, **20**, 954.

Dakin, J. P., Pearce, D. A., Wade, C. A. and Strong, A. (1987) A novel distributed optical fibre sensing system enabling location of disturbances in a Sagnac loop interferometer. *Proc. SPIE*, **838**, 18.

Dandridge, A. and Tveten, A. B. (1981) Noise reduction in fibre optic interferometric systems. *Appl. Opt.*, **20**, 2237.

Dandridge, A., Tveten, A. B., Miles, R. O. and Giallorenzi, T. G. (1980a) Laser noise in fibre optic interferometric systems. *Appl. Phys. Letts.*, **37**, 526.

Dandridge, A., Tveten, A. B., Sigel, G. H., West, E. J. and Giallorenzi, T. G. (1980b) Optical fibre magnetic field sensors. *Electron. Letts.*, **16**, 408.

Digonnet, M. J. F. and Shaw, H. J. (1983) Wavelength multiplexing in single mode fibre couplers, *Appl. Opt.*, **22**, 484.

Digonnet, M. J. F. and Shaw, H. J. (1982) Analysis of a tunable single mode optical fibre coupler. *IEEE J. Quantum Electron.*, **QE18**, 746.

Donati, S. and Annovazzi-Lodi, V. (1984) A fibre sensor for current measurements in power lines. *Alta Frequenza*, **III**, 310.

Durst, F., Melling, A. and Whitelaw, J. H. (1976) *Principles and Practice of Laser Doppler Anemometry*, Academic Press.

Dyott, R. B. and Bello, J. (1983) Polarisation-holding directional coupler made from elliptically cored fibre having a D section. *Electron. Letts.*, **19**, 601.

Eyges, L. and Wintersteiner, P. (1981) Modes of an array of dielectric waveguides. *J. Opt. Soc. Am.*, **71**, 1351.

Ezekiel, S., Davis, J. L. and Hellwarth, R. W. (1981) Intensity dependent non-reciprocal phase shift in a fibre optic gyroscope. *Springer-Verlag Series in Optical Sciences*, **32**, 332.

Farahi, F., Akhavan Leilabady, P., Jones, J. D. C. and Jackson, D. A. (1987) Optical fibre flammable gas sensor. *J. Phys. E*, **20**, 453.

Farahi, F., Webb, D. J., Jones, J. D. C. and Jackson, D. A. (1990) Simultaneous measurement of temperature and strain: cross-sensitivity considerations. *J. Lightwave Tech.*, **LT8**, 138.

Favre, F. and LeGuen, D. (1980) High frequency stability of laser diodes for heterodyne communication systems. *Electron. Letts.*, **16**, 709.

Feit, M. D. and Fleck, J. A. (1981) Propagating beam theory of optical fibre cross coupling. *J. Opt. Soc. Am.*, **71**, 1361.

Fowles, G. R. (1975) *Introduction to Modern Optics*, Holt, Rinehart and Winston.

Frederiks, R. J. and Ulrich, R. (1984) Phase error bounds of fibre gyro with imperfect polariser-depolariser. *Electron. Letts.*, **20**, 332.

Gerges, A. S., Newson, T. P., Farahi, F., Jones, J. D. C. and Jackson, D. A. (1988) A hemispherical air cavity fibre Fabry-Perot sensor. *Opt. Comm.*, **68**, 157.

Giallorenzi, T. G. (1987) Optical fibre interferometer technology and hydrophones, in *Optical Fibre Sensors, NATO ASI Series E*, **132**, 35, Martinus Nijhoff Dordrecht.

Giallorenzi, T. G., Bucaro, J. A., Dandridge, A., Sigel, G.H., Cole, J. H., Rashleigh, S. C. and Priest, R.G. (1982) Optical fibre sensor technology. *IEEE J. Quantum. Electron.*, **QE18**, 626.

Goldberg, L., Taylor, H. F., Dandridge, A., Weller, J. F. and Miles, R. O. (1982) Spectral characteristics of semiconductor lasers with optical feedback. *IEEE J. Quantum Electron.* **QE18**, 555.

Gruchmann, D., Petermann, K., Staudigel, L. and Weidel, E. (1983) Fibre optic polarisers with high extinction ratio. *Proc. ECOC9*, Elsevier, Amsterdam, 305.

Hironaga, K., Maromoto, H., Hishida, K. and Maeda, M. (1985) LDV system using single-mode fibres and applications. *Proc. Int. Conf. Laser Anemometry – Advances and Applications*, BHRA, 387.

Hocker, G. B. (1979a) Fibre optic sensing of temperature and pressure. *Appl. Opt.*, **18**, 1445.

Hocker, G. B. (1979b) Fibre optic acoustic sensors with composite structure. *Appl. Opt.*, **18**, 3679.

Hosokawa, H., Takagi, J. and Yamashita, T. (1988) Integrated optic microdisplacement sensor. *International Conference on Optical Fibre Sensors*, New Orleans, p. 16.2.

Hotate, K. and Tabe, K. (1986) Drift of an optical fibre gyroscope caused by the Faraday effect. *Appl. Opt.*, **25**, 1086.

Jackson, D. A. and Jones, J. D. C. (1986a) Fibre optic sensors. *Optica Acta*, **33**, 1469.

Jackson, D. A. and Jones, J. D. C. (1986b) Extrinsic fibre optic sensors for remote measurement, *Opt. & Laser Tech.*, **18**, 243.

Jackson, D. A., Jones, J. D. C. and Chan, R. K. Y. (1984) A high power fibre optic laser Doppler velocimeter. *J. Phys. E*, **17**, 977.

Jackson, D. A., Kersey, A. D., Akhavan Leilabady, P. and Jones, J. D. C. (1986) High frequency non-mechanical optical polarisation scale rotator. *J. Phys. E*, **19**, 146.

Jackson, D. A., Kersey, A. D., Corke, M. and Jones, J. D. C. (1982) Pseudo-heterodyne detection scheme for optical interferometers. *Electron. Letts.*, **18**, 1081.

Jackson, D. A., Priest, R., Dandridge, A. and Tveten, A. B. (1980) Elimination of drift in a single mode optical fibre interferometer using a piezo-electrically stretched coiled fibre. *Appl. Opt.*, **19**, 2926.

Jones, J. D. C., Akhavan Leilabady, P. and Jackson, D. A. (1986) Monomode fibre optic sensors: optical processing schemes for recovery of phase and polarisation state information. *Int J. Opt. Sensors*, **1**, 123.

Jones, B. E., Medlock, R. S. and Spooncer, R. C. (1989) Intensity and wavelength based sensors and actuators, in *Optical Fibre Sensors: Systems and Applications* Vol 2, (eds. B. Culshaw and J. Dakin), Artech House, Norwood, Mass, p. 431.

Jones, J. D. C., Corke, M., Kersey, A. D. and Jackson, D. A. (1984) Single-mode fibre optic holography. *J. Phys. E*, **17**, 271.

Kaminow, I. P. (1981) Polarisation in optical fibres. *IEEE J. Quantum Electron.* **QE17**, 15.

Kanoi, M. (1986) Optical voltage and current measurement system for electrical power systems. *IEEE Trans. Power Delivery* **PWRD 1**, 91.

Kawachi, M., Kawasaki, B. S., Hill, K. O. and Edahiro, T. (1982) Fabrication of single polarisation single-mode fibre couplers. *Electron. Letts.*, **18**, 962.

Kawasaki, B.S., Hill, K. O. and Lamont, R. G. (1981) Biconical taper single mode fibre coupler. *Opt. Letts.*, **6**, 327.

Kersey, A. D. and Jackson, D. A. (1986) Current sensing utilising heterodyne detection of the Faraday effect in single mode optical fibre. *IEEE J. Lightwave Tech.*, **LT4**, 640.

Kersey, A. D., Jackson, D. A. and Corke, M. (1983a) A simple fibre Fabry–Perot sensor. *Opt. Comm.* **45**, 71.

Kersey, A. D., Jackson, D. A. and Corke, M. (1983b) Demodulation scheme for interferometric sensors employing laser frequency switching. *Electron. Letts.* **19**, 102.

Kersey, A. D., Jackson, D. A. and Corke, M. (1985) Single mode fibre optic magnetometer with DC bias field stabilisation. *IEEE J. Lightwave Tech.*, **LT3**, 836.

Kersey, A. D., Jackson, D. A., Corke, M. and Jones, J. D. C. (1983c) Detection of DC and low frequency AC magnetic fields using an all-single-mode-fibre interferometer. *Electron. Letts.*, **19**, 469.

Kim, B. Y., Blake, J. N., Engan, H. E. and Shaw, H. J. (1986) All fibre-optic acousto-optic frequency shifter. *Opt. Letts.*, **11**, 389.

Kintner, E. C. (1981) Polarization control in optical fibre gyroscope. *Opt. Letts.*, **6**, 154.

Knuhtsen, J., Olldag, E. and Buchave, P. (1982) Fibre optic laser Doppler anemometer with Bragg frequency shift utilising polarisation preserving single mode fibre. *J. Phys. E*, **15**, 1188.

Koo, K. P. and Sigel, G. H. (1982) Characteristics of fibre optic magnetic field sensors employing magnetic glasses. *Opt. Letts.*, **7**, 334.

Koo, K. P., Tveten, A. B. and Dandridge, A. (1982) Passive stabilisation scheme for fibre interferometers using 3 × 3 directional couplers. *Appl. Phys. Letts.*, **41**, 616.

Koo, K. P., Dandridge, A., Tveten, A. B. and Sigel, G.H. (1983) A fibre optic d.c. magnetometer. *IEEE J. Lightwave Tech.*, **LT1**, 524.

Lagakos, N. and Bucaro, J. A. (1981) Pressure desensitisation of optical fibres. *Appl. Opt.*, **20**, 2716.

Laming, R. I., Gold, M.P., Payne, D. N. and Halliwell, N. A. (1986) Fibre optic vibration probe. *Proc. SPIE*, **586**, 38.

Langeac, D. (1982) Temperature sensing in twisted single mode fibres. *Electron. Letts.*, **18**, 1022.

Lefevre, H. C. (1980) Single mode fibre fractional wave devices and polarisation controllers. *Electron. Letts.*, **16**, 778.

Lefevre, H. C., Vatoux, S., Papuchon, M. and Puech, C. (1986) Integrated optics: a practical solution for the fibre-optic gyroscope. *Proc. SPIE*, **719**, 101.

Lewin, A. C., Kersey, A. D. and Jackson, D. A. (1985) Non-contact surface vibration analysis using a monomode fibre optic interferometer incorporating an open air path. *J. Phys. E*, **18**, 604.

Liu., K., Sorin, W. V. and Shaw, H. J. (1986) Single mode fibre evanescent polariser/amplitude modulator using liquid crystals. *Opt. Letts.*, **11**, 180.

McBride, R., Harvey, D., Barton, J. S. and Jones, J. D. C. (1990a) Velocity measurement using fibre optic Sagnac interferometers. *Fifth International Symposium on Applications of Laser Techniques to Fluid Mechanics*, Lisbon, p. 36.1.

McBride, R., Barton, J. S., Jones, J. D. C. and Borthwick, W. K. D. (1990b) Fibre optic interferometry for acoustic emission sensing in machine tool wear monitoring. *Proc. Conf. Frontiers in Electro Optics*, Cambridge University Press, p. 89.

McIntyre, P. O. and Snyder, A. W. (1973) Power transfer between optical fibres *J. Opt. Soc. Am.*, **653**, 1518.

Martin, J.M. and Winkler, J. T. (1978) Fibre optic laser gyro signal detection and processing technique. *Proc. SPIE*, **139**, 98.

Meltz, G., Dunphy, J. R., Morey, W. W. and Snitzer, E. (1983) Cross-talk fibre optic temperature sensor. *Appl. Opt.*, **22**, 464.

Meltz, G., Dunphy, J. R., Glenn, W. H., Farina, J. D. and Loenberger, F. J. (1987) Fibre optic temperature and strain sensors. *Proc. SPIE*, **798**, 104.

Millar, C. A., Miller I. D., Mortimore, D. B., Ainslie, B. J. and Urquhart, P. (1988) Fibre laser with adjustable fibre reflector for wavelength tuning and variable output coupling. *IEE Proc. J. (Optoelectron.)*, **135**, 303.

Miller, I. D., Mortimore, D. B., Urquhart, P., Ainslie, B. J., Craig, S. P., Millar, C. A. and Payne, D. B. (1987) A Nd^{3+}–doped cw fibre laser using all-fibre reflectors. *Appl. Opt.*, **26**, 2197.

Mitsui, T. (1986) Development of fibre optic voltage sensors and magnetic field sensors. *IEEE Trans. Power Delivery*, **PWRD2**, 87.

Nakatani, N., Tokita, M., Izumi, T. and Yamada, T. (1985) LDV using polarisation preserving optical fibre for simultaneous measurement of multidimensional velocity components. *Rev. Sci. Inst.*, **56**, 2025.

Nguyen, T. T. and Birch, L. N. (1984) Fibre optic laser Doppler anemometer. *Appl. Phys. Letts.*, **45**, 1163.

Pannell, C. N., Tatam, R. P., Jones, J. D. C. and Jackson, D. A. (1988) A fibre optic frequency shifter utilising travelling flexure waves in birefringent fibres. *J. IERE*, **58**, 592.

Parriaux, O., Gidon, S. and Kuznetsov, A. A. (1981) Distributed coupling on polished single mode optical fibres. *Appl. Opt.*, **20**, 2420.

Payne, D. N., Barlow, A. J. and Ramskov-Hansen, J. (1982) Development of low and high birefringence optical fibres, *IEEE J. Quantum Electron.*, **18**, 477.

Philbert, M., Devon, R., Faleri, J-P, Bion, J-R and Salun, H. (1986) Optical fibres application to visualisation of flow separation inside an aircraft air intake in wind tunnel. *Proc. SPIE*, **586**, 21.

Pleibel, W., Stolen, R. H. and Rashleigh, S.C. (1983) Polarisation preserving couplers with self-aligning birefringent fibres. *Electron. Letts.*, **19**, 825.

Rashleigh, S. C. (1983) Polarimetric sensors: exploiting the axial stress in high birefringence fibres, *IEE Conf. Publ.*, **221**, 210.

Rashleigh, S. C. and Ulrich R. (1979) Magneto-optic current sensing with birefringent fibres. *Appl. Phys. Letts.*, **34**, 768.

Risk, W. P., Youngquist, R. C., Kino, G. S. and Shaw, H. J. (1984) Acousto-optic frequency shifting in birefringent fibre. *Opt. Letts.*, **9**, 309.

Rogers, A. J. (1973) Optical technique for measurement of current at high voltages. *Proc. IEE*, **120**, 261.

Rogers, A. J. (1983) Optical fibre current measurements. *Proc. SPIE*, **374**, 196.

Sasaki, D., Sato, T., Abe, T., Mizuguchi, T. and Niwayama, M. (1980) Follow-up type laser Doppler velocimeter using single mode optical fibres. *Appl. Opt.*, **19**, 1306.

Senior, J. (1985) *Optical Fiber Communications: Principles and Practice*, Prentice-Hall.

Shupe, D. M. (1980) Thermally induced non-reciprocity in the fibre optic interferometer. *Appl. Opt.*, **19**, 654.

Smith, A. M. (1978) Polarisation and magneto-optic properties of single-mode optical fibre. *Appl. Opt.*, **17**, 52.

Steel, W. H. (1983) *Interferometry*, Cambridge University Press.
Stolen, R.H. (1986) Polishing induced birefringence in single mode fibres. *Appl. Opt.*, **25**,344.
Stolen, R. H. and Turner, E.H. (1980) Faraday rotation in highly birefringent optical fibres. *Appl. Opt.* **19**, 842–5.
Stone, J. (1985) Optical fibre Fabry-Perot interferometer with a finesse of 300. *Electron. Letts.*, **21**, 504.
Tai, S., Kyuma, K., Hamanaka, K. and Nakayama, T. (1986) Applications of fibre optic ring resonators using laser diodes. *Optica Acta*, **33**, 1539.
Tatam, R. P., Pannell, C. N., Jones, J. D. C. and Jackson, D. A. (1987a) Full polarisation state control utilising linearly birefringent monomode optical fibre. *J. Lightwave Tech.*, **LT5**, 980.
Tatam, R. P., Rollinson, G., Jones, J. D. C. and Jackson, D. A. (1987b) High resolution optical fibre thermometer: applications to biotechnology. *Biotechnology Techniques*, **1**, 11.
Tatam, R. P., Hill, D.C., Jones, J. D. C. and Jackson, D. A. (1988) All-fibre-optic polarisation state azimuth control: application to Faraday rotation *J. Lightwave Tech.*, **6**, 1171.
Udd, E. (1983) Fibre optic acoustic sensor based on the Sagnac interferometer. *Proc. SPIE*, **425**, 90.
Ulrich, R. (1980) Fibre optic rotation sensing with low drift. *Opt. Letts.*, **5**, 173.
Varnham, M. P., Payne, D. N., Barlow, A. J. and Tarbox, E. J. (1984) Coiled birefringent fibre polarisers. *Opt. Letts.* **9**, 306.
Varnham, M. P., Payne, D. N., Birch, R. D. and Tarbox, E. J. (1983) Single polarisation operation of highly birefringent fibres. *Electron. Letts.*, **19**, 246.
Venkatesh, S. and Culshaw, B. (1985) Optically activated vibrations in a micromachined silica structure. *Electron. Letts.*, **21**, 315.
Villaruel, C. A. and Moeller, R.P. (1981) Fused single mode fibre access coupler *Electron. Letts.*, **17**, 243.
Villaruel, C. A., Abebe, M. and Burns, W. K. (1983) Polarisation preserving single mode fibre coupler. *Electron. Letts.*, **19**, 17.
White, B. J., Davis, J. P., Bobb, L. C., Krumboltz, H. D. and Larson, D. C. (1987) Optical fibre thermal modulator. *J. Lightwave Tech.*, **LT5**, 1169.
Wilson, J. and Hawkes, J.F.B. (1983) *Optoelectronics: An Introduction*, Prentice Hall.
Yariv, A. and Winsor, H.V. (1980) Proposal for detection of magnetostrictive perturbation of optical fibres. *Opt. Letts.* **5**, 87.
Yeh, Y., Lee, C. E., Atkins, R. A., Gibler, W.N. and Taylor, H. F. (1990) Fibre optic sensor for substrate temperature monitoring. *J.Vac Sci. Technol.*, **A8**, 3247.
Yokohama, I., Kawachi, M., Okamoto, K. and Noda, J. (1986) Polarisation-maintaining fibre couplers with low excess loss. *Electron. Letts.*, **22**, 929.
Yu, M. H. and Hall, D. B. (1984) Low loss fibre ring resonator. *SPIE Proc. Fibre Optic and Laser Sensors II*, p. 104.

Index

Abbe
  error 167
  prism 251–3
Abel transform 259
Accelerometer
  fibre optic 204–91
  limitations of 187, 276, 322
  use with holography 244
Accidents, recording of 131
Acousto-optic transducer, *see* Bragg cell
Acoustic scanning 367
Adhesion, testing of 236, 380
Aerofoil surface 240
Aerosols, holograms of 228
Air, *see* Atmosphere
Aircraft
  engine 332
  fan
    assembly 247–53, 256–8, 261–2
    blades 233–4, 238–40, 245–9, 267
  tyres 240–2
  wing
    flap 329–30
    tips 236
Airy disc 281
Alignment
  laser 25–43, 93–4
  telescope 88–103
  three-point system 25–40, 43
Aluminium alloy, strain in 380
Analyser, polarization 427
Anemometry, laser Doppler 179–83
Angle
  comparison 107–8
  gauges 108
  measurement 164–5
Antenna, radar 145
Aperture
  effect on speckle 281–2
  position detection using 49–56
  slotted, in moiré photography 349–52
Array, CCD, *see* Photodiode array
Aspheric surface, measurement of 402
Astigmatic lens in focus detection 58
Asymmetry, elimination of 26, 74, 88, 110
Atmosphere
  dispersion 47–8
  refractive index 43, 109, 156–7, 169
    gradient 44–8, 109, 156
    turbulence 43, 109, 156
Autocollimation
  prism 81–2
  laser 74–9
Autocollimator
  electronic 105–6
  visual 79–80, 103–8
Automobile
  acoustic noise 320
  body shell, vibration of 276–7, 324–5
  brake, vibration of
    disc 322–4
    drum 324
  combustion chamber, deformation of 319–20
  cooling fan 252
  cylinder bore, deformation of 316–8
  door, vibration of 370–1

Automobile *(contd)*
  engine block
    deformation of   319
    vibration of   187, 322
  floorpan   325
  tyres 330–1
Auto reflection   92–3, 99
Axicon
  camera   132–4
  lens   43

Ball, golf, shape of   398–400
Barrel, rifle, straightness of   52–6
Beam, laser
  Bessel   40–43
  curvature   13–15
  displacement due to air   44–7
  elliptical   172
  Gaussian   12–15, 22, 156
  position detection   49–64
  propagation of   11–25
  reference, *see* Reference beam
Beam, wooden, deformation of   358–61
Beam expanders   23–4
Bearings,
  crankshaft, alignment of   94–8
  squareness of   84–5
  wear in   237–8
Bed, straightness of   105–6
Bedplate, flatness of   100–1
Bessel
  beam   40–3
  function in vibrometry
    holographic   244–5, 285–6
    laser Doppler   208
    TV holographic   300, 404–5
Benton, Stephen   217
Billets, gauging of   65
Biosensing   451
Biostereometrics   342
Birefringence
  circular   420
  in fibres   419, 434
Blade
  aero engine fan   233–4, 238–40, 245–53, 267
  circular saw   193
  helicopter rotor   236

Bleaching
  of grating image   353
  of hologram   221–2
Block, engine
  deformation of   319
  vibration of   322
Block gauge   168–71
Board, circuit   236
Body shell, car   276–7, 324–5
Bones, deformation of   236
Books on metrology   2
Bore
  alignment   94–8
  deformation of   316
  fixture   94–5
  straightness of   52–6
Bottle, glass, strain in   332
Bragg
  cell
    in interferometry   395, 434, 445
    in vibrometry   183, 206
    see also Modulator, acousto-optic
  planes   218
Brake
  disc   322
  drum   324
Bridges
  deformation of   148–9, 363
  movement of   27–30
Buckets, integrated   305
Buckling, measurement of   342–4
Building, subsidence of   363–4
Bulb, light
  air flow in   261
  contouring of   265
Bundle solution   123–8

Cables, gauging of   65
Caesium
  as diode laser stabilizer   174
  time standard   159
Camera
  axicon   132–4
  calibration of   141–4
  Hasselblad   223
  holographic   223–3, 247
  image dissector   396

Camera *(contd)*
  metric 130–3
  Nikon 222
  photographic 342–4
  photogrammetric 125–6, 130–3
  pinhole 119–22, 125
  Rollei 131
  thermoplastic 225–7, 242
  video, in holography 242, 260
  see also Photodiode array
Car, *see* Automobile
Carbon dioxide, in air 156–7
Carbon fibre composites 367, 380
  see also Composite materials
Cascade, turbine blade 256, 258, 393–4
Cat's eye reflector 69
Cavity, resonant
  as coherence improver 219, 269
  confocal 25, 174
  fibre optic 425–6
  in wavelength comparison 162
CCD array, *see* Photodiode array
Cellular automata method 311
Centring of lens 83–4
Centroid of laser beam 24, 61
Chamber, combustion
  shape of 269
  deformation of 319–20
Channel Tunnel 64, 68
Chirp, frequency 238
Chopper disc 50, 62
Church walls, deterioration of 325–8
Clinometer 108
Coherence 154–5
  diode laser 174, 422
  length 219, 269, 422
  spatial, in fibres 415
Coin, contouring of 268, 270
Compact disc player 58, 171
Comparator, shape 386
Composite materials, testing of
  by embedded fibres 452
  by holography 236
  by moiré 358, 367, 381
  by TV holography 295, 328
Compressor
  aero engine 261
  reciprocating 88, 201

Computer, micro, in TV holography 301, 404
Concrete, crack propagation in 366
Confocal etalon 25, 174
Conical lens 43, 66
Conjugate image 22
Construction
  lasers in 71
  monitoring of 117
Contact, optical 168
Contour fringes
  of density change 254
  of displacement 235–42
  holographic 230–1
  of surface shape 262–70, 400–2
  two-angled illumination 264–6
  two refractive index 266–7, 402
  two wavelength 268–70, 402
  of vibration 242–53
speckle
  of displacement 316–20, 325–30
  map analogy 298
  of surface shape 332–6
  of vibration 320–5, 335–6
moiré
  of strain 357, 372–3
  of surface height 341, 398–400
Convection, visualization of 315
Convolution filtering 308
Coordinate measuring machine, *see* Measuring machine
Cosine error 167
Count, bidirectional fringe 157
Coupler, fibre optic 421, 436–8
Courses in metrology 2
Cow, meat content of 342
Crack
  detection of 356
  fatigue 373
  propagation of 237, 366
Crankshaft 99, 194, 196–7
Crates, beer 236
Cube corner reflector 69, 157
Current, electric, measurement of 454
Curvature
  of beam wavefront 13–15, 156
  of laser beam 44–7
Cylinder bore, car 316–8

468 Index

Cylinder, piezoelectric, *see* Piezoelectric transducer
Cylindrical
  lens 58, 389
  mirrors 332

Dam, movement of 27–33, 68
Deadpath, interferometer 167–8
Debrazing, testing for 238, 386–7
Defects, testing for 238–42
Deformation, measurement of
  in diaphragm 239–40, 404
  explosively-induced 367–9
  high-speed 365
  by holography 228–32
  plastic 386
  radial 342–4
  by TV holography 316–20
  see also Displacement; Photogrammetry
Deformity, spinal 342
Delamination, testing for 236, 240–2, 328, 330–1
  see also Honeycomb panel
Denisyuk, Yuri 218
Density
  of fluid 254–6
  of holograms 221
Dental records, holographic 228
Depth of field, improvement of 348–51
Derotator 250–3, 324, 332
Detection
  heterodyne, *see* Heterodyne detection
  synchronous, *see* Phase-sensitive detection
Diameter, monitoring 82–3
Diaphragm
  deformation of 239–40
  as pressure sensor 451
  vibration of 238, 404–5
Differentiation, optical
Diffuser, nylon 55
Diffraction
  efficiency of holograms 221
  holographic 215
  optical 14–15, 281
Dimensional gauging
  in holography 219
  in metrology 4–7

Diode, laser
  in holography 219
  in interferometry 171–7
  noise in 447–9
  in TV holography 325–8, 333–5
Disbond, *see* Delamination
Disc
  drive 236
  oscillating 199–201
  rotating, strain in 332
  scattering, as frequency shifter 184–6
  slotted 50, 62, 194, 196–7
Dish, telescope 145–8
Dispersion
  by a hologram 217
  due to air 47–9
Displacement
  components of 232
  in-plane 294–5, 325–8, 344–8
  out-of-plane 294, 325–8, 340–4
  of beam by air 46–7, 109
  small-scale 381
  see also Alignment; Contour fringes
Distance measurement, electronic 174–5
Distortion, *see* Deformation
Dividing table, calibration of 108
Door, car, vibration of 370–1
Doppler
  anemometry 179–83
  frequency shift 163, 180, 207–8
  signal processing 186–7
  spectral broadening 161
  velocimetry 179–83, 458–9
  vibrometry 183–210, 244, 336, 458–9
Dove prism 74
Drift of laser beam 24, 78
DSPI, *see* Holography, television

Edlén equation 157
Eigenmodes of fibre 419, 423, 435, 438
Ellipsoid, error 126–7
Ellipsometry, fibre optic 458
Emulsions, photographic
  for holography 220–2
  stripping 364–5
Engine
  aircraft 332
  automobile

Index 469

Engine *(contd)*
    cooling fan 25
    deformation of block 319
    vibration of block 187, 322
  diesel
    bed plate 101
    crankshaft 94–9, 194, 196–7, 201
  gas turbine 289
ESPI, *see* Holography, television
Etalon, *see* Cavity, resonant
Evanescent field 433–4
Exciter, vibration
  electrodynamic 208, 249, 320, 324
  piezoelectric 322
Expansion
  radial, measurement of 332
  thermal, effect of 109
Eulerian strain 346
Eyepiece
  laser 67
  parallax errors 110–11

Fabry–Perot interferometer, *see* Cavity, resonant
Face mask, contouring of 408–9
Faceplate, camera 287
Fan blade, *see* Aircraft
Far field of laser beam 13–15, 388
Faraday effect 420, 424, 454
Fatigue, detection of 367
Feedback, optical 157, 174, 424
Fibre, optical
  birefringence of 419, 434
  monomode 415
  multimode 169, 416
Fibre optics in TV holography 291–2
Field
  electric, measurement of 454
  evanescent 433–4
Film
  high-resolution 352
  for holography 222
    thermoplastic 223–6
  stability of 352
  stripping 365
Filter
  optical 61
  spatial 24, 291

Filtering in fringe analysis 391
Finesse of etalon 426
Finite element model 240, 249, 410
Fizeau interferometer 74, 336
Flatness
  optical 74–9, 170
  mechanical 99–101, 106–7, 164
Flaw, detection of 329–30, 386–7
  see also Delamination
Flood fill algorithm 310
Floorpan, car, vibration of 325
Flow
  measurement 179, 186
  meter, fibre optic 453
  visualization 253–62, 393–4, 459
Fluid mechanics 179
Focus, detection of 58–60
Fourier transform
  digital 235, 302–4, 307, 408–9
  optical 15, 22
Fracture characterization 295
Frame store, timing of 137, 293
Frequency
  beat between lasers 162
  spatial 345, 350
Fresnel
  biprism 376
  lens 63–4
  reflection 423
  zone plate 27–30
Fringes
  analysis of 231–5, 302–16, 381–2, 385–410
  carrier 235, 302, 395, 408–9
  contour, *see* Contour fringes
  contrast 156, 422
  counting 157, 391–5
  formation of 154
  holographic
    frozen 228–30
    live 230, 237, 260
    stabilization of 401
    vibration 233–4, 238, 242–53
  moiré
    difference 348
    interferometric 375–81
    strain 372–3
    vibration 370–1

Fringes *(contd)*
  speckle
    addition   286
    deformation   316–20, 325–30
    frozen   284
    sawtooth   404
    subtraction (correlation)   284–5, 307, 311
    vibration   285–6, 290–1
  skeletonizing   390–3
  tracking   233, 301, 390–4
  wedge   156, 168–71, 389

Gabor, Dennis   213, 279
Galactic metrology   8–9
Galilean beam expander   23
Gas
  density   254–62
  refractive index   254–5
Gate, liquid, in holography   231
Gauges
  angle   108
  length   168–71, 389–90
  strain   194, 244, 339, 366
Gauging, dimensional   65
Gaussian intensity distribution   12–15
Gear wheel   199
Gearbox   99, 230, 237–8
Gladstone–Dale constant   255
Gradient, air density   45
Grating
  circular   43
  in moiré   340
    fill-in of defects   374–5
    harmonics of   374–6
    high temperature   381
    photo-etched   348
    replica   376
    self-imaged   347
    virtual   375
    zero thickness   381
  phase reversal   29
  radial
    as frequency shifter   183
    as modulator   62
Gyroscope, fibre optic   425, 455–7

Hazard, ocular, *see* Safety, laser
Helicopter rotor blade   236
Helium   47
Heterodyne method
  detection   181, 188, 199
  holography   233–5
  interferometry   162–4, 395–7, 434–6, 445
  TV holography   300
Hole, position detection with   49, 55
Holocamera   225
  see also Camera, holographic
Holo-diagram   9
Hologram
  amplitude   221
  double-exposure   228–9
  dual reference beam   233, 400–1
  flow   228, 253–62
  image plane   217, 256, 258
  particle   228
  phase   221–6
  processing   221–2
  rainbow   217
  reflection   218
  security   217
  time-average   233, 242–6
  transmission   216–7
  white light   217–8
Hologrammetry   228, 262
Holographic interferometry   228–70, 275–8, 386–7, 392–5, 400–3
  see also Fringes
Holography
  photographic   213–70, 275–8
  television   275–336, 403–6
Honeycomb panel, testing of   238, 329, 386–7, 392–5, 400–3
Hubble Space Telescope   5–6
Humidity   46, 156–7
  sensor   166–7, 169
Hydrocarbon gases, detection of   451
Hydrophone, fibre optic   444, 451–2, 456

Illumination of target   96
Image
  centre, location of   49–64, 134–5, 139–41

Image
  distortion, analysis of 125–6
  holographic 217–8, 228
  processing 301, 325, 385
    see also Fringes, analysis of
  processor, synchronization of 137, 293
Infra-red telescope 146–8
Integrated optics 4
Interference, optical 153–6, 214
Interferogram 230, 281, 385
Interferometer, laser
  common path 165
  errors in 15–16, 166–7
  Fabry–Perot, see Cavity, resonant
  fibre optic 204–6, 415–39, 450–7
  Fizeau 74, 336
  gauge block 169–71
  heterodyne 162–4, 395–7, 434–6, 445
  holographic 228–70, 275–8, 386–7, 392–5, 400–3
  homodyne 440–5
  Mach–Zehnder 204–6, 254, 421–3
  Michelson 153–4, 423–4
  Michelson stellar 348
  moiré 375–82
  multiple beam, see Cavity, resonant
  multiple wavelength 162, 168–70, 175–6, 430
  polarization 163–4, 432–4
  remote sensing 430
  speckle pattern 275, 278–95
  Sagnac 424–5, 455–7
  scanning 69
  shearing 295, 330
  two beam, for distance measurement 153–69
  two wavelength 429–30, 452
  Twyman–Green 74, 168, 395–8
  white light 170–1
  see also Fringes
Inversion, principle of 26, 74, 88
Iodine wavelength stabilization method 161

Joints
  adhesive 380
  dovetail 380

Joints (contd)
  human limb 237
  pipe 236
  welded 358
Jones matrix

Kerr
  cell 186
    see also Bragg cell
  effect 457
Knock sensors, location of 321–2

Labyrinth seal 258, 260
Lagrangian strain 346
Lamp, air flow in 261
Laser
  argon 219, 291
  diode, see Diode, laser
  dye 333
  helium-neon 153–71, 219
  pulsed 219, 286
  ruby 219, 288, 330
    see also Coherence
Lamb dip 161
Latent image regression 353
Least squares in photogrammetry 123–9, 140–1
Length
  coherence 154–5, 219, 269, 422
  standard of 158
Lens
  astigmatic 58
  conical 43, 66
  cylindrical 58, 389
  field 59, 63
  Fresnel 63–4
  Goerz 374
  Olympus 365
  ring 43
Lens system
  afocal 38–41
  variable focus 36–8
Limb joints, artificial 237
Linewidth, laser 155, 269
Lloyd's mirror 377–8
Look-up table, arctangent 305, 310, 312

472 Index

Loop, phase locked 62, 187
Loudspeaker
  as exciter 248
  vibration of 193, 314

Machines, rotating, vibration of 194
Machine tool calibration 164
  see also Measuring machine
Mach number 255
Mach–Zehnder interferometer 204–6, 254, 421–3
Magnetometer, magnetostrictive 453
Map, phase 303, 312
Mapping, stereo 115, 118
Mask
  face, shape of 408–9
  in phase unwrapping 310, 312
  slotted, in moiré 349–52
Masonry, deformation of 358, 363
Material, woven, moiré in 371
Maxwell, James Clerk, telescope 146–8
Measurand 415
Measurement, optical 3–7
Measuring machine
  three-axis 6, 65, 263
    calibration of 40, 69
  two-axis 134
  see also Machine tool
Mechanics, experimental 340
Mekometer 175
Mesh
  graphical display 313, 320, 400
  as moiré stencil 366–8
Metre, definition of 158
Metrologist, optical 1–3
Metrology
  dimensional 4–7
  galactic 8–9
Michelson interferometer 153–4, 423–4
Micrometer, optical
  see Plate, tilting glass
Milling machine 102
Mine, as stable environment 277
Mirror, Lloyd's 377–8
Missile fin, examination of 328–9
Modal retardance in fibre 419, 428
Mode hopping 172–4, 431

Modes
  laser, longitudinal 269
  optical fibre, see Eigenmodes
  vibration 242, 246–53, 277–8, 405–6
Modulation
  phase
    in homodyne processing 443
    sawtooth 300–1, 436
    sinusoidal 298–300, 436
  polarization 416, 432
  stroboscopic 297, 300, 312
Modulator
  acousto–optic 61, 164, 169
  electro–optic 61, 402
  see also Bragg cell
Moiré
  imaged 342–4, 347–8
  interferometry 375–82
  meaning of 340
  multiplicative 407–8
  photographic 348
  photography, high resolution 348–75
  projection 342, 386, 398–400, 409
  reflection 342
  shadow 340–2, 400
  transmission 347
  zone plate 33–6
Monomode optical fibre 415
Motor, speed fluctuation of 201–3
Movement
  components of 232, 249
  freezing of 219
  see also Displacement
Mounting of optical components 227
Multimode optical fibre 416
Multiple reflections 72, 425–7
Mural painting, deterioration of 325–8

Nanotechnology 8
Near field of laser beam 13
Networks, neural 316
Newton–Raphson method 124
Nodal line 233, 244–6, 248
Noise
  acoustic, in car 320
  coherent optical 290–1, 306–8
  in fibre sensors 445–50

Noise (contd)
   in fringes   391
Non-destructive testing   326, 386
Notch, strain around   332
Notched specimen   295
Nuclear reactor   88
Numbering of fringes   391–5

Offshore platforms   100–3
Oscillations, torsional   194–203, 320
Oscillator, voltage controlled   62, 187

Painting, mural, deterioration of   325–8
Panel, honeycomb, testing of   238, 329, 386–7, 392–5, 400–3
Paper processor, alignment of   79–80, 98–9
Paraboloid, antenna   145–8
Parallax   110–11, 217–8
Particles, holograms of   228
Pattern
   moiré dot   354
      paper   362–3
      printed   366
      random   369–70
      scratched   370
      stencilled   366–7
   wax, measurement of   263
Pentagonal prism   70–2, 75–9, 88–91, 98–103
Periscope   68, 76, 79–80
Perspective, analysis of   120–2
Phase
   meter   396
   modulation, see Modulation, phase
   stepping, see Stepping, phase
Phase-sensitive detection   51–2, 62, 135, 161
Photodetector, quadrant   57–9, 75
Photodiode array
   linear, as position detector   64, 82
   two dimensional
      faceplate   287
      in interferometry   168–71
      in moiré   342–4, 375
Photodiode array   (contd)
   in photogrammetry   136–44
   in position detection   62, 67
   in TV holography
Photographic materials for holography   220–2
Photography
   cine, of live fringes   260
   in measurement   113–35, 144–51
   moiré   348–75
   speckle   369, 388–9
   stereoscopic   113–6, 214
Photogrammetry, close range
   applications   144–51
   mathematical formulation   117–29
   real time   135
   sequence of operations   118–9
   stereo   113–16
   video   135–44
Photomultiplier, position detection with   57
Photopotentiometer   60–1
Photoresist, in moiré interferometry   376
Phototheodolite   117
Phototransistor, position detection with   55
Piezoelectric transducer
   as cavity stabilizer   161
   cylinder as fibre stretcher
      in phase stepping   292, 298–9, 308
      in sensing   204, 431
   as frequency shifter   186
   as phase stepper   308–9, 398, 400–2
   as vibration exciter   322
Pigment, white, for moiré   366–8
Pipe joints   236
Plastic
   artefacts, measurement of   263, 358
   deformation of   366
Plate
   aluminium, strain in   373
   bending, measurement of   311
   half-wave, as polarization rotator   427–8
   half-wave rotating, as phase shifter   308
   holographic   215–8, 220–6
   photogrammetric   131
   steel, deformation of   368
   tilting glass   52–5, 75, 88, 382
   vibration, measurement of   312, 402–3
Plumb lines   97–8
Plummet, optical   149

474 Index

Pockels cell   186, 297
  see also Bragg cell
Poisson's ratio   359
Polarimetry, fibre optic   458
Polarization
  in optical fibres   419–20, 427–30, 432–4
  in position detection   63
  in phase stepping   308, 382
  in speckle interferometry   290
  modulation   416, 432
Polygon
  angle standard   107–8
  scanning   65, 69
Porro prism   68, 82, 251, 308, 376
Position, detection of image   49–64
Pressure, air   45–6, 156–7
Pressure sensor   166–7, 169
  acoustic   451–2
Principal point of camera   121, 142
Prism
  Abbe   251–3
  anamorphic pair   172
  autocollimation   81–2
  derotator   251–3
  Dove   74
  pentagonal   70–2, 75–9, 88–91, 98–103
  periscope   68, 76
  Porro   68, 82, 251, 308, 376
  wedge, see Wedge, glass
  Wollaston   165
Probe, optical   65
Processing
  of holograms   221
  of moiré photographs   353
Profile, surface   400–2
  see also Shape
Profilometer   74–9
Projection moiré   342, 386, 398–400, 409
Proximity gauge   59
Pump, mud   101–3

Quadrant photodetector   57–9, 75
Quadrature, fringe phase   157, 440–4
Quartz
  in frequency shifter   434
  in transducer   183
Quasi-heterodyne method   235, 397–405

Radar antenna   145
Ray tracing   17–22
Rayleigh, Lord   340
Reconstruction of wave fields   215–8
Rectification, phase-sensitive, see Phase-
    sensitive detection
Recording materials for holography   220
Reference beam
  in holography   216
  in TV holography   279–84
    adjustment of   292–3
    combining methods   286
    geometry of   290
    polarization of   290
  in velocimetry   181
Refraction
  air   44–7, 108–9, 112
  compensation   47–9
Refractive index
  air   45, 156–7, 169
  fluid   253–5
Refractivity, air   45
Refractometer
  air   157
  fibre optic   458
Reseau, in camera   131
Resin, epoxy, as grating replica   379
Resolution, camera, improvement of
    349
Resonance, torsional   320
Resonator
  ring   426–7, 457
  see also Cavity, resonant
Retardance, modal   419
Retroreflector
  cube corner   69, 157
  paint   69, 253, 370
  tape   184, 192, 201
Rifle barrel, straightness of   52–6
Robot, monitoring movement of   69
Rodolite   43
Rods, nuclear fuel   228, 262
Rollers, alignment of   79–82, 98–9
Rotating object   190–203, 250–3, 261–4
Rubber, silicone, for grating replica   376,
    379
Rubidium, as diode laser stabilizer   174

Sagnac interferometer   424–5, 455–7
Safety, laser   192, 219–20, 291, 416
Saw, circular, vibration of   193
Sawtooth fringes   404
Scanning
  acoustic   367
  laser beam   65–79, 82, 213–4
  max–min, in TV holography   306–7, 329
Scale, for alignment   92, 97–8, 101–2
Scheimpflug condition   343
Schlieren method   254
Scoliosis   342
Scratched patterns in moiré   370
Screen, for viewing image   26, 33, 49
Seal, labyrinth for gas   258, 260
Second of time, definition of   159
Segmentation of image   39
Sensors, fibre optic
  applications   450–9
  extrinsic   203, 206–10
  interferometric   420–7
  intrinsic   203–6
  polarimetric   419–20, 427–9
  strain   419
  temperature   419
Sensors, knock   321–2
Shadow moiré   340–2, 400
Shadowgraph   254
Shaker, electrodynamic   208, 249, 320, 324
Shape, measurement of   262–70
  see also Moiré, projection; Photogrammetry
Shearing interferometer, speckle   295, 330
Shearography   330
Ship
  alignment on   94–8, 100–1
  bending by sun   109, 112
Shock
  tube   256
  wave   228, 262, 264
Shrinkage, measurement of   327
Shutter
  acousto-optic   169
  CCD camera   297
  see also Bragg cell

Skeletonizing, fringe   390–3
Slabs, pavement, movement of   363
Slideway   5, 104–6
  see also Straightness, measurement of
Slit
  fringes from pair   42
  image position detection with   50, 52
  vibrating   50, 59
Slotted aperture   349–52
Source
  white light   155, 218, 256–7, 339
  see also Laser
Spatial filter   24, 291
  in moiré   254–6
Spatial frequency   345, 350
Speckle
  characteristics of   179, 188–9
  interferometry   275–336, 403–6
  photography   369, 388–9
  size of   281–2
  white light pseudo   369–71
Speed of light   159
Sphere, capped   69
Spherical mirror, measurement of   5
Spindle, alignment of   99, 102
Spot size of laser beam   12–13
Square, optical, see Pentagonal prism
Squareness, machine   106
Stability
  of instrument   93, 112
  of laser beam   24, 78, 93
  of TV holography system   287–9
Stabilization
  fringe pattern   401
  laser wavelength   159–62, 166, 174
Steel, strain in   348
Stepping, phase   397–405
  in holography   235
  in moiré   357, 381–2
  spatial   405–9
  in TV holography   298, 304–16
Stereoscopic photography   113–6, 214
Stinger   320
Stonework, degradation of   325–8
Straightness, measurement of   38–40, 165
  see also Slideway

476 Index

Strain
  components of   378
  fibre sensing of   419–20, 452–3
Strain *(contd)*
  linear   346
  shear   346–7
Strain gauges
  conventional   194, 244, 339, 366
  fibre optic   452
Strain-optic effect   419, 452
Stress intensity factor   295
Stroboscopic
  holography   402–3
  TV holography   297, 300, 312
Submarine, bending by sun   112
Subsidence, investigation of   363–4
Surface
  rotating   190–203
  velocity of   179–83
  see also Flatness; Shape; Vibration
Survey
  air   113, 116, 127–8
  photogrammetric   113–51
  theodolite   117

Table, arctangent look-up   305, 310, 312
Table
  dividing, calibration of   108
  surface, flatness of   106–7
Talbot imaging   347, 399
Tank, fuel storage   149–51, 342–4
Target
  ball   150
  illumination of   96
  photogrammetric   134, 147–50
  retroreflecting   147–9
  sighting   91–2, 94–9
Telescope
  alignment   88–103
  infra-red   146–8
Television holography   275–336, 403–6
Temperature, air   45–7, 108–9, 112, 156–7
Temperature sensor   166, 169, 419
Template, for moiré analysis   157
Testing, non-destructive   236, 386
Textile, moiré in   371
Theodolite   67, 80–2, 117, 138–9

Thermo-optic effect   419, 431
Thermoplastic   220, 223–6, 277
Three-axis measuring machine, *see*
    Measuring machine
Three-dimensional image   214–8, 228
Three-point alignment system   25–40, 43
Threshold, speckle modulation   310
Tile, in phase unwrapping   311
Tilt fringes, *see* Fringes, wedge
Tilt, monitoring of   40, 72
Timber structure, deformation of   358–61
Time
  coherence   154–5
  standard of   159
Tomography in flow measurement   259
Torsiograph   194
Torsional oscillations   194–7, 320
Total survey station, laser   70, 72
Traceability, laser wavelength   158–9, 162, 167
Tracker, frequency   187
Tracking, fringe   233, 301, 390–4
Transducer
  acousto-optic, *see* Bragg cell
  fibre optic   417–20
  see also Piezoelectric transducer
Transfer function, fibre sensor   417, 439
Transient events, study of   314
Transonic cascade   256, 258, 393–4
Triangulation   65–7, 113
Tube
  gauging of   65
  piezoelectric, *see* Piezoelectric transducer
  as pressure sensor   451
Tuning
  of camera lens   349–50
  of diode laser   172
Tunnel
  Channel   64, 68
  piston   256
  profile   65
  wind   256
Tunnelling machine   64
Turbine blade, *see* Aircraft
Turbulence
  atmospheric   43, 109, 156
  fluid, holograms of   256